THINGS TO MAKE *and Do* in the *Fourth Dimension*

Matt Parker

PENGUIN BOOKS

PENGUIN BOOKS

UK | USA | Canada | Ireland | Australia
India | New Zealand | South Africa

Penguin Books is part of the Penguin Random House group of companies whose addresses can be found at global.penguinrandomhouse.com.

First published by Particular Books 2014
Published in Penguin Books 2015
001

Artwork by Richard Green

Set in 11.34/13.68 pt Garamond MT Std
Typeset by Jouve (UK), Milton Keynes
Printed in Great Britain by Clays Ltd, St Ives plc

A CIP catalogue record for this book is available from the British Library

ISBN: 978–0–141–97586–3

$$9 + (7 \times 3) + 8 + (0 \times 3) + 1 + (4 \times 3) + 1 + (9 \times 3) + 7 + (5 \times 3) + 8 + (6 \times 3) + 3 = 130 = 0 \pmod{10}$$

www.greenpenguin.co.uk

Penguin Random House is committed to a sustainable future for our business, our readers and our planet. This book is made from Forest Stewardship Council® certified paper. All numbers are made from recycled prime factors where possible.

Dedicated to Keith and Nona Pallot,
my maternal grandparents, who inspired
me to make and teach.

Contents

Zero

THE ZEROTH CHAPTER

Have a look around you and find a drinking vessel, like a pint glass or a coffee mug. Despite appearances, almost certainly the distance around the glass will be greater than its height. Something like a pint glass may look like it is definitely taller than it is round, but a standard UK pint glass is actually around 1.8 times greater in circumference than in height. A standard 'tall' takeaway cup from omnipresent high-street coffee shop Starbucks is actually 2.3 times further around, but yet they refuse my requests to rename it the 'squat'.

Using this to your advantage is easy enough: when you are next drinking in a pub, café or whichever drinking establishment serves the sort of beverages you enjoy getting for free, bet someone that their drinking vessel is further around than it is high. If there is a 'pot' beer glass (the ones with handles) in the pub, or an obscenely large mug in the café, then you're sorted: they are typically three times as far around as they are tall, so you can dramatically stack three of them and claim it is still further around than up. Producing a tape measure at this point may cause your victims to question the spontaneity of the whole exercise, so use a nearby straw, or its protective paper sleeve, as a makeshift ruler.

This works for all glasses, except for only the skinniest of champagne flutes. If you'd like to subtly check your drinking glass without arousing suspicion, try to wrap your hand all the way around it. Your fingers and thumb will not meet on the other side. Now, with your thumb and index finger, try to span the height of the glass. You will most likely succeed (or, at worst, come very close). This is a dramatic demonstration of how much shorter glasses are than they are around.

This is exactly the sort of maths I wish more people knew about: the surprising, the unexpected and, most importantly, the type that wins you free drinks. My goal in this book is to show people all the fun bits of mathematics. It's a shame that most people think maths is just what they were subjected to at secondary school: it is so much more than that.

In the wrong situation, maths can indeed border on the tedious. Walk into a school at random in search of a maths class, and you'll most likely find a room where the majority of the students are *not* excited. In the least. You will shortly be asked to leave the premises, and the police may even be called. You're probably on some kind of list now. The point is, those students in the classroom are following in a long line of generation after generation of uninspired maths students. But there will be a few exceptions. Some of them will be loving maths and will go on to be mathematicians for the rest of their lives. What is it that they're enjoying which everyone else is missing?

I was one of those students: I could see through the tedious exercises to the heart of maths, the logic behind it all. But I could sympathize with my fellow students, and specifically, the 'sporty ones'. At school, I dreaded football drills in the same way that other people dreaded maths class. But I could see the purpose of all that messing around dribbling a football between traffic cones: you're building up a basic repertoire of skills so that you're better when it comes to an actual game of football. By the same token, I had an insight into why my sporty classmates hated maths: it's counterproductive to make pupils practise the basic skills needed for maths but then not let them loose into the field of mathematics to have a play around.

That is what the maths kids knew. This is why people can make a career out of being a mathematician. If someone works in maths research, they're not simply doing harder and harder sums, or longer divisions, as people imagine. That would be like a professional footballer merely getting faster at dribbling up the field. A professional mathematician is using the skills they've learned and the techniques they've honed to explore the field of mathematics and discover new things. They might be hunting for shapes in higher dimensions, trying to find new types of numbers, or exploring a world beyond infinity. They are not just practising arithmetic.

Herein lies the secret of mathematics: it's one big game. Professional mathematicians are playing. This is the goal of this book: to open up this world and give you the freedom to play with maths. You too can feel like a premier-league mathematician, and if you were already one of those kids who embraced maths, there are still plenty of new things to discover. Everything here starts with things you can really make and do. You can build a 4D object, you can dissect counterintuitive shapes and you can tie unbelievable knots. A book is also an amazing piece of technology with a state-of-the-art pause function. If you do want to stop and play around with a bit of maths for a while, you can. The book will sit here, the words static on the page, waiting for you to return.

All the most exciting bits of cutting-edge technology are mathematical at heart, from the number-crunching behind modern medicine to the equations that help carry text messages between mobile phones. But even technology which relies on bespoke mathematical techniques still ultimately rests on mathematics that was originally developed because a mathematician thought it would be fun to try to solve a puzzle.

This is the essence of mathematics. It is the pursuit of pattern and logic for their own sake; it is sating our playful curiosity. New mathematical discoveries may have countless practical applications – and we may owe our lives to them – but that's rarely why they were discovered in the first place. As the Nobel Prize-winning physicist Richard Feynman allegedly said of his own subject: 'Physics is a lot like sex; sure it has a practical use, but that's not why we do it.'

I also hope to bring the maths you did learn at school into focus. Without it, all the other interesting bits of maths would not be within reach. Every student has vague memories of learning about the mathematical constant pi (roughly equal to 3.14), and some may recall that it is the ratio of the diameter to the circumference of a circle. It is because of pi that we know the distance around a glass is over three times greater than the distance across it. And it is the distance across which most people use when judging how big a glass is, forgetting to multiply by pi. This is more than memorizing a ratio, this is taking it for a test-run in the real world. Sadly, very little school maths focuses on how to win free drinks in a pub.

The reason that school maths cannot be completely dismissed is that the more exciting bits of mathematics rest on the less exciting bits. This is partly why some people think maths is so hard: they've missed a few vital steps along the way, and without them the higher ideas seem impossibly out of reach. But if they had tackled the subject one step at a time, in the correct order, it would have been fine.

No one bit of mathematics is that hard to master, but sometimes it's important that you do things in the optimal order. Sure, getting to the top of a very high ladder may take a lot of effort from start to finish, but each individual rung is no more work to reach than the last one was. It's the same with mathematics. Step by step, it's great fun. If you understand prime numbers, then exploring prime knots is much easier. If you get to grips with 3D shapes first, then 4D shapes are not that intimidating. You can imagine all the chapters in this book as a structure, where each one rests upon several of the previous chapters.

You can even choose your path through the chapters, as long as, before tackling a later chapter, you've read all the ones that support it. As the book goes on, the chapters cover more advanced mathematics, the sorts of things you generally won't hear about in a classroom. This can be daunting at first glance. But as long as you pass through them in the right order, by the time you reach the far-flung corners of mathematics, you'll be fully equipped to enjoy all the delights and surprises there are on offer.

Above all else, remember that the motivation for climbing this structure should be merely to enjoy the view as you go. For too long, maths has been synonymous with education; it should be about fun and exploration. One puzzle at a time, one maths game after another, and soon we'll be at the top, enjoying all the maths most people never know even exists. We'll be able to play with things beyond normal human intuition. Mathematics allows access to the world of imaginary numbers, to shapes that exist only in 196,883 dimensions, and objects beyond infinity. From the fourth dimension to transcendental numbers, we'll see it all.

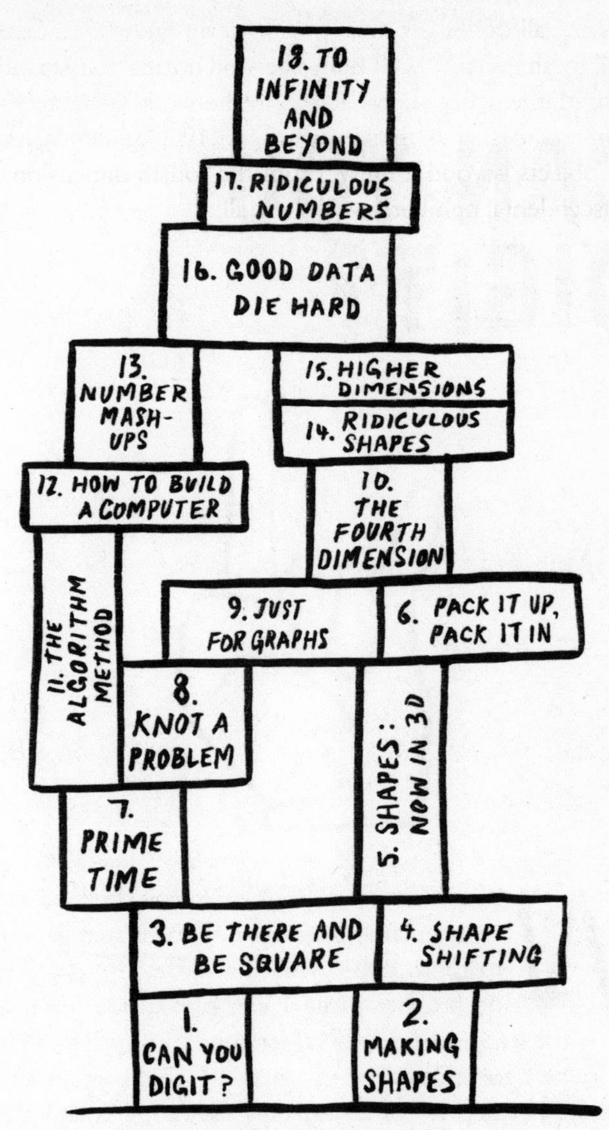

This book as a tower of chapters resting on each other.
Choose your path wisely.

One

CAN YOU DIGIT?

When I have to go to the dentist's, I like to have some mental distraction while a stranger tries to climb into my mouth, normally some kind of number game I can play inside my head. So, on the way to the dental surgery one day, I put a call out on Twitter for a good maths puzzle I could try to solve without needing to write anything down. A friend challenged me to order all nine digits so that the first two are a multiple of 2, the first three a multiple of 3, and so on, up to all nine digits making a multiple of 9. There is only one solution.

Before getting too comfortable in the dentist's chair, I'd figured out that the traditional, boring order 1 2 3 4 5 6 7 8 9 doesn't work. Even though 12 is divisible by 2, and 123 divisible by 3, it stops there. 1,234 is not evenly divisible by 4. By the end of my dental procedure, I had some but not all of the digits worked out, but, apparently, you're not allowed to stay in the dentist's chair after they're finished. At home, I confirmed that the only arrangement that works is 381,654,729.

(If you don't care about using all nine digits, and you use zero as well, there are loads more options, for example, 480,006. Because so many contiguous combinations of their digits are divisible, these numbers are called polydivisible numbers. There are 20,456 polydivisible numbers, the largest of which is 3,608,528,850,368,400,786,036,725.)

Interestingly, this puzzle works only because of the digits we happen to use these days. If you'd given this brainteaser to an ancient Roman, it would not have helped distract them during a dental procedure. Not only do they use different digits, such as V and X, but their digits have the same value regardless of where they appear in a number. V always represents 5; X always represents 10. Not so with our numbers: the 2 in 12 represents 2, whereas the 2 in 123 represents 20. Thankfully, Roman dentistry was crude and swift.

There is a dirty secret when it comes to number puzzles and, indeed, a lot of the maths you learn at school: much of it works only because of how we happen to write numbers down. In our current number system, if you multiply 111,111,111 by itself, you get the rather pleasing 12,345,678,987,654,321 (all the digits count up from 1 to 9 and then back down again). It works for small runs of 1s as well: 11,111 × 11,111 = 123,454,321; and 111 × 111 = 12,321. Try writing down numbers in a different way, though, and the pattern evaporates. 111 is CXI in Roman numerals and CXI × CXI is the rather *dis*pleasing $\overline{\text{X}}$MMCCCXXI.

What this all goes to show is that there is a difference between the word 'number' and the word 'digit'. For example, here is the number three: 3; and here is the digit three: 3. They may look identical (primarily because they do), but they are subtly different things. A number is just what you think it is: it represents a number of things: 3 is a number; 3,435 is also a number. Numbers are abstract concepts, and to write them down we use digits. So a digit is just a symbol to communicate a number in writing, in the same way that letters are the symbols we use to write words. The number 3,435 uses the digits 3, 4 and 5. All the maths you learn and see around you can be put into two categories: actual maths, based on intrinsic properties; and results which are merely a byproduct of the way we happen to put pen to paper.

Getting Tricky

A great place to start (and a great way to make this less like a school maths lesson) is with the 37 Trick.

Pick any digit and write it down three times. You'll now be looking at something like 333 or 888. Add those three digits together: 3 + 3 + 3 = 9; or 8 + 8 + 8 = 24. Nothing exciting yet. So far, you're just adding numbers. Now divide your original three-digit number (333, or 888) by the sum of its digits (9, or 24). You can do this either by using a calculator or through the sheer power of thought. (The calculator is quicker.) Whatever your method, and regardless of what digit you started with, you will get an answer of 37. Which is why this is often called the 37 Trick.

As I say, this works for any digit you choose to start with. However, this free choice is quickly cancelled out. It is absolutely determined that, at the end of the calculation, you will come up with 37. There is a sly bit of algebra going on behind the scenes. Typing the same digit three times is the same as multiplying it by 111. If you picked the digit 8, then 888 is the result of 8×111. Adding the three digits means you are inadvertently multiplying it by 3: $8 + 8 + 8 = 3 \times 8 = 24$. So, dividing 888 by 24 is the same as dividing 111 by 3 because the 8s 'cancel out'. And so with any digit . . .

. . . and yet not quite. If a Roman had picked the digit V, the 37 Trick would no longer yield an answer of 37 and therefore no longer be known as the 37 Trick, or indeed as any trick at all. Thankfully, in this case at least, our current system of ten digits is used almost exclusively now, but, should you have wanted to impress an ancient Babylonian, this trick would not be a good option, because they wrote numbers down completely differently to how we do today. Should we ever be visited by hypothetical aliens, who might write numbers down in all sorts of strange ways, it would almost certainly not work for them either. This trick is a combination of what we would consider the 'fundamental' properties of numbers (those which don't change when they are written in different ways), and a quirk of our current system of expressing numbers.

So, why is that? Well, 111 is divisible by 3 regardless of how you write the figures down. CXI is divisible by III, 一百十一 is divisible by 三, and any aliens anywhere in the universe will know that one hundred and eleven is divisible by three. The answer is always 37 (or XXXVII, or 三十七, or some alien squiggle for 'thirty-seven'). If you have a pile of one hundred and eleven rocks you will always be able to split them into three piles of thirty-seven rocks. And because this property is detached from any one expression of the numbers, mathematicians consider it to be one of the more important, abstract properties.

On the other side of the coin, the fact that writing the same digit three times is the same as multiplying it by 111 is nothing more than an unintended side-effect of how we happen to write things down. In Roman numerals, writing the same digit three times is the same as multiplying it by 3, not by 111. (VVV = III × V.)

Part of the strength of mathematics is that it expresses universal truths but is able to express them in different ways. Ancient Mayans and Romans were learning the same maths, but they were writing their numbers down using very different systems to our modern one. To explore the world of mathematics, we need to know what language everyone is speaking. Let's start with the number system we use today – which is not necessarily the best one.

What is a Number?

What is the largest number you can count to on your fingers? Well, most people stop counting on their fingers when they hit ten, mainly because they've run out of fingers. But not everyone uses the rather limited system of raising their fingers to count on and not putting them back down. If you do allow fingers to go back down, then you can count to 3 using only your first two fingers. Lift your first finger for 1, your second finger for 2 and both for 3. Now your third finger is available to be raised up, alone, for 4; then your third and first fingers for 5, and so on. You can get all the way to 16 before needing to use even the fifth finger on your first hand.

Using this system, you can count all the way from 0 to 1,023 just with your fingers. But our personal digital abacus can do better. If you use each finger in the down, halfway up and fully up positions, you can count from 0 to 59,048. Take it one step further and use four positions (down touching your palm, down not touching your palm, halfway up and fully up), and you get a range from 0 to 1,048,575. That's over a million, counted on your fingers. That's a greater than 100,000 times improvement in performance, with only a slight increase in risk of arthritis.

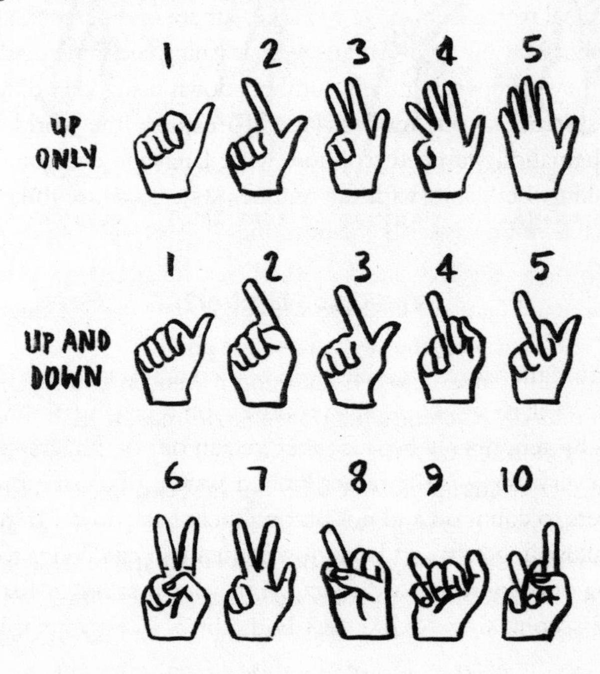

And why stop there? Using eight positions for each finger means not only have you acquired previously unknown levels of digital dexterity, but that you can count from 0 to 1,073,741,823: over a billion! Of course, on the down side, you might end up accidentally joining a street gang.

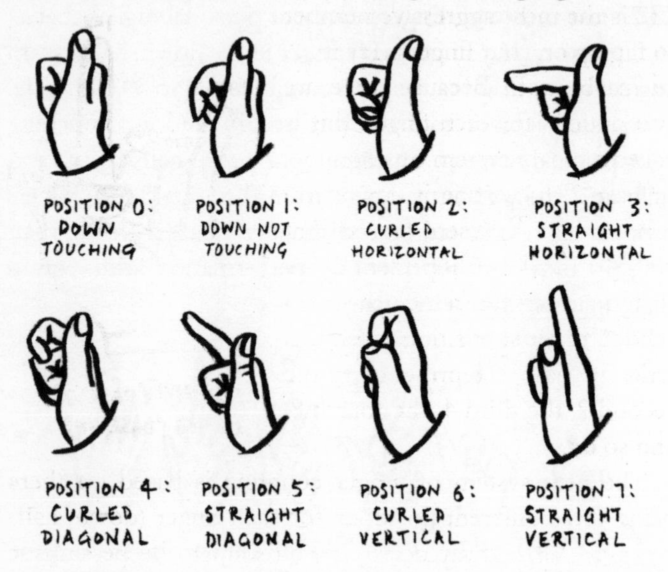

This was the last one I worked out, but just how much higher could it go? For those with extremely flexible fingers and nimble minds, the sky may be the limit.

The difference between counting to ten on your fingers and suddenly reaching a billion is that, instead of each finger being a glorified tally, the *position* of each finger now matters. When we count up to three with our first two fingers, rather than 'normal' counting (where all fingers are equal, i.e. they each represent 1, or 1 more), the first finger still represents 1 when it's raised, but the second finger represents 2 when it's raised. If we carry on, following the 'up and down' directions in the diagram above, the third finger represents 4, the fourth finger 8 and the fifth finger 16, and we can see a sequence emerging: each finger in its upright position is worth twice the previous finger's upright value. With a bit of experimenting, you can find a way to represent every possible number using your fingers in these two positions. (Tip: 132 is the most aggressive number to flip up on your fingers. Try it . . . or maybe not.) Because there are two options for each finger, this is called a binary system of counting, or base-2. In writing it down, you can use 0 to represent fingers that are down and 1 to represent those that are up. If you remember from school, the first position when you write in base-2 represents 1, the second 2, the third 4, the fourth 8, and so on.

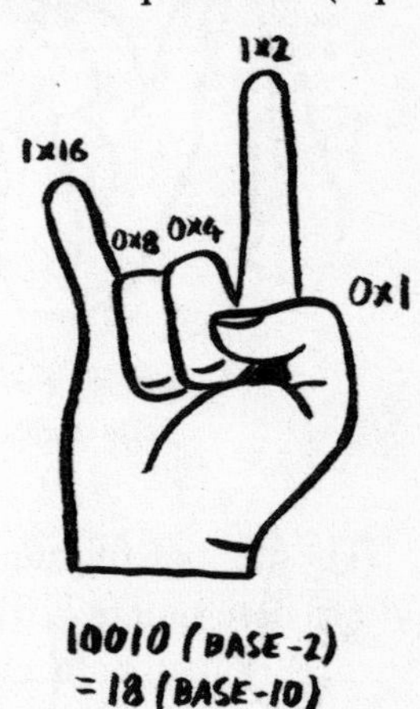

The next system of finger counting is based on there being three different positions for each finger (down; halfway up – or halfway down, if you want to be pessimistic about it; and up), so it is generally known as base-3. And you can follow the system through: four finger positions give you base-4; eight positions, base-8. And just to consolidate (make sure we are all paying attention): the value of each position, be it finger or written, is the previous position multiplied by the base, so, in base-3, the sequence is 1, 3, 9, 27 . . . and we can use the three digits 0, 1 and 2 to represent successive stages of finger position; and in, say, base-8, the sequence runs 1, 8, 64, 512 . . . and the eight digits 0, 1, 2, 3, 4, 5, 6 and 7 are represented by the different finger positions. In base-8, the representation of a billion can thus be written as 7,346,545,000.

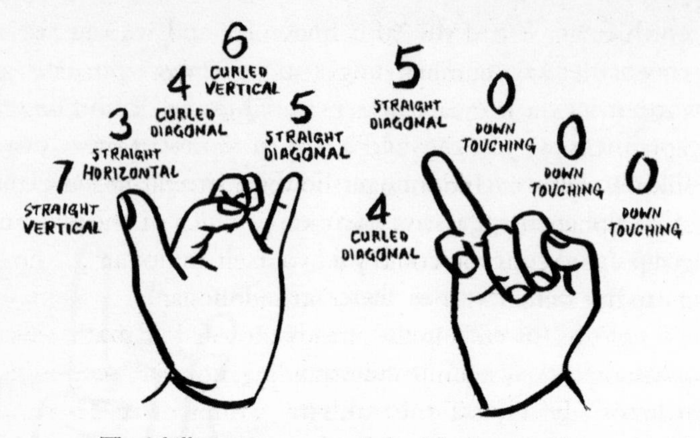

The 1 billion arrangement of the eight finger positions (also doubles as a maths-gang hand sign).

These types of numbers form a whole family of 'positional base systems', which are completely different to things such as Roman numerals, where the position of a digit does not make a jot of difference to what it is representing. The digit V represents 5 wherever it appears in a number, whereas, in the base-10 number 3,435 the digit 3 represents both 3,000 and 30, depending on where it is positioned. Roman numerals are an over-engineered tally system, now used far beyond their original capacity. Positional number systems are much more powerful, as they can express a number of any size with ease. Of course, it's the base-10 system of numbers that we now use almost exclusively in the modern world, but – I'll say it again – it is only one of many different options.

When you're using different bases, there's more than enough room for any amount of confusion. I can translate numbers into a system which uses completely different digits, such as going from base-10 to Roman numerals (e.g.

3,435 becomes MMMCDXXXV), and it's easy to spot what system the latter number is written in: it's akin to translating a word into a language which uses a completely different alphabet, say, English to Japanese. If you translate a word from English to Indonesian, however, it will still use the same alphabet and, if you don't know which of the two languages you're in, you could find yourself in hot air . . . no, I mean hot water ('a-i-r' is 'water' in Indonesian).

I can't resist retelling the already too-oft-told maths 'joke' based on just such a misunderstanding, normally seen in the form of a hilarious T-shirt with this written on it: 'There are 10 types of people in the world, those who understand binary and those who don't.' To explain the hilarity: '10' means 2 in a binary number system, so only those who understand binary would know it means 2 and not 10. I'll give you a second to finish laughing and compose yourself.

I am slightly scathing of this joke solely because, being a mathematician who also works as a stand-up comedian, I am told it *all the time*. Normally, people start with, 'OK, have you heard this joke? Well, it kinda doesn't work when you *say* it, but . . .' and then proceed to try to tell a joke which only works when it's written down. That's the problem with binary jokes: they either work, or they don't. However, whatever its comedy value, it's a fantastic example of how different numbers can be written using the same digits in the same order, depending on what system you're using.

Anyhow, here we are, dead set on using base-10. People say this is a consequence of our having ten fingers: if you use your fingers as a tally, then you need to 're-set' each time you reach ten, and keep separate track of how many sets of ten you've gone through. If a friend was keeping track for you, they would also run out of fingers after counting ten tens, so they would need someone to count how many hundreds have rolled over. Keeping track of numbers in multiples of tens is therefore, it is said, natural for humans (or at least for humans with friends). So it seems the Mayans also counted on their toes, because they used a base-20 system.

And here is the reason we use the same word – 'digits' – for both our fingers and the squiggles we write numbers with. Intelligent creatures evolving elsewhere in the universe may equally well not have ten fingers; hypothetically, they could have evolved three arm-like limbs each ending in four finger-like protrusions for grasping. These hypothetical life forms may well write numbers using a base-12 system.

Even here on Earth, there are a few protesting humans who are adamant that we should switch over from base-10 to base-12. 'Dozenalists' extol the benefits of using twelve as a base for our counting system (for example, more numbers divide into twelve than ten, and so it is easy to write fractions) while overlooking the major upheaval such a change would require. Were we to switch, base-12 requires twelve digits, so we would have to add, for example, 'A' for 10 and 'B' for 11, making 3,435 in base-10 the same as 1BA3 in base-12.

It's extremely unlikely we will ever swap. Other base-number systems remain the playthings of mathematicians, while almost all humans use base-10 pretty much exclusively. The only time different number systems cross over from the realm of mathematical curiosity to the real world is when it comes to computers. Binary (or base-2) is useful for computers because of its limited set of digits: you can hardly get more minimalist than a system that uses only 0 and 1. Modern computers have developed in such a way that they are founded on situations in which there are only two options. Current is either flowing down a wire in a circuit; or it isn't.

A magnet on a hard drive has either a north magnetic pole facing one way; or a south. Everything is thus either 0 or 1. Thankfully, all numbers can be converted into a series of 1s and 0s in base-2.

There's a balance to be struck, however, between having a limited enough set of digits so a number system is usable but there still being enough to make writing numbers down efficient. For intelligent creatures, be they humans or hypothetical aliens, around ten or twelve digits works nicely. Computers need the limited-digit set of binary to be able to run: all smartphones, digital televisions and (even) microwave ovens are doing their counting and calculating in binary numbers behind the scenes. However, when they need to interact with humans, they kindly convert the numbers back to base-10, for our convenience.

The very first, crude computers were not so considerate. I once had the honour of meeting an elderly gentleman who, back in his youth, was the last mathematics student Alan Turing taught before he died in 1954. Turing is rightfully known as the 'Father of Computing' and he wrote one of the earliest operating systems for one of the first ever computers while based at the University of Manchester. Apparently, Turing's first operating system required anyone using the computer to be fluent in binary, which Turing was. When a new version of the operating system was introduced which converted binary number inputs into base-10, Turing, until his dying day, insisted on the computer being re-set to binary when he was using it.

Even though the binary numbers used in computers have been pushed deep below layers of user interfaces, you can still see some evidence of them. You will have come across the fingerprint of binary numbers when dealing with computers, with everything from 16- and 32-gig memory cards to

1,024-pixel computer screen resolutions. Much as humans like numbers to be round – 1,000 and 1,000,000 are just somehow so appealing – so do computers: only, they like round binary numbers. Because all the position values in binary are powers of 2, you will see them start to pop up all over the place around computers: $2^5 = 32$ and $2^{10} = 1,024$.

Sometimes, computers will accidentally slip out a base-16 number, for example in Wifi passcodes, but most people do not notice. 'Hexadecimal' – fancy for 'base-16' – uses the symbols 0 to 9 and then the letters from A to F. These numbers are less obvious compared to spotting powers of 2, but they are still there. If you look on the back of a WiFi router, the original passcode is normally a series of digits (0 to 9) and letters (A to F). Also, if you look at the numerical values for colours in painting or photo-editing software you will see hexadecimal values. And, of course, now you know about them, you will see the letters A to F appearing occasionally in the middle of numbers used by computers. And if you're like me, also in your dreams . . .

Hexadecimal is used when numbers need to be stored a bit more efficiently than they can be in binary but are going to be seen only by computer programmers and other highly technical users of computers: 16 is the go-to base in these circumstances. This may seem a bizarre choice – why not simply base-10? – but it was selected because 16 is itself a power of 2 and it is very easy to switch between two bases when one is a power of another. Normally, when you go from one base to another, the new number position values are completely different. When the new base is a power of the previous one, however, some of the old position values are skipped, but there are no new ones. In the case of base-16, every group of four binary digits is always swapped for the same, single, new symbol.

0000 → 0	1000 → 8
0001 → 1	1001 → 9
0010 → 2	1010 → A
0011 → 3	1011 → B
0100 → 4	1100 → C
0101 → 5	1101 → D
0110 → 6	1110 → E
0111 → 7	1111 → F

Binary to hexadecimal, so 1011110000100001 becomes BC21.

Once you understand different number systems, it is easy to drill down to the actual maths beneath them. In fact, it is easier to decipher foreign numbers than foreign languages. When the Mayan cities were rediscovered in the 1800s and huge amounts of unintelligible writing unearthed, the numbers were deciphered and translated well before the writing was understood – even though they were in that strange base-20 system. If we did meet galactic-travelling hypothetical aliens, once we worked out what symbols they use for digits, we could happily communicate numerically – but, if we wanted to share a maths puzzle with them, we'd need one which works regardless of how the numbers are written down.

Totally Addicted to Base

Can you find the only number between ten and twenty that is not the sum of consecutive numbers?

It can't be 13, because 13 = 6 + 7, and 6 and 7 are consecutive numbers; similarly, 18 is ruled out, because

18 = 5 + 6 + 7. If you've already come up with the answer, find one between thirty and forty. Work out the next one, just past sixty, and you'll begin to spot the pattern in these numbers. What's interesting about this 'sum of consecutive numbers' puzzle is that it works however you write the numbers down. An ancient Roman could solve it, using their numbers, and so could an ancient Mayan and our extraterrestrial 'Hypotheticals'.

The first answer you should have found is 16: there is no way to add any consecutive numbers and come up with a total of 16. And 16 joins an elite group of other such numbers, for example, 8 and 32 (there is a guide to why these numbers have this property in 'The Answers at the Back of the Book').

If it's another puzzle you're after, and one that can be translated from a base-specific conundrum into a quandary in another number system, let's go back to polydivisible numbers which use all the non-zero digits once each, but in a base other than 10. (Don't feel bad if you have to go back a few pages and remind yourself what a polydivisible number is. I just have.) If humans had evolved differently and we used base-4, there would have been two solutions for me to find while sitting in that dentist's chair: 123 and 321. There are no solutions in base-5; but two again in base-6 (14,325 and 54,321); none in base-7; a staggering three in base-8 (3,254,167; 5,234,761; and 5,674,321); none in base-9; and the single one we have already met in base-10 (381,654,729). And that is the last solution until base-14, which has 9C3A5476B812D. Phew.

I was surprised to find that there were no solutions in base-12. If we assume that our hypothetical visitors from outer space are indeed twelve-digited (I don't know about you, but I feel like we're getting to know them quite well), then this puzzle would be completely lost on them. (One more reason *not* to follow the Dozenalists.) So, when I wrote my computer program to look for these numbers for me (I had a spare weekend), the sudden appearance of a solution in base-14 was even more surprising; I'd thought that, if base-12 didn't work, then there might be no further solutions in higher bases. I tweaked my program as much as I could and made it all the way up to checking that there are no more solutions in base-15 or base-16. I don't know if there are more solutions in bases beyond sixteen. *Please*, if anyone has more free time than me or better programming skills, let me know.

This quest to take a problem and see what happens in different situations is called generalizing, and it is this force that drives mathematics forward. Mathematicians constantly want to find solutions and patterns which apply to as many situations as possible, i.e. are as general as possible. A maths puzzle is not complete when you merely find an answer, a maths puzzle is complete when you've then tried to generalize it to other situations as well – and minds including Leonhard Euler and Lord Kelvin have excelled in mathematics by displaying just this kind of curiosity. Because mathematicians like the puzzles which work on the pure number rather than the symbolic digit and the system we happen to be writing our numbers down in, there is a sense that, when a puzzle works only in one given base, there is something rather, well, 'second class' about it. Mathematicians do not like things which work only in base-10; it is only because we have ten fingers that we find that system interesting at all. Mathematics is the search for universal, not base-specific, truth.

Having said that, there are some *great* number puzzles that do work only in base-10. Just don't tell the great British mathematician G. H. Hardy. In 1940, he pointed out the interesting fact that 'There are just four numbers, after unity, which are the sums of the cubes of their digits,' before sweeping this statement aside with 'but there is nothing in them which appeals to the mathematician'. He did admit that such number facts are 'very suitable for puzzle columns and likely to amuse amateurs' but absolutely drew the line at them being mathematics. Which is a shame, because I, personally, like such pointless number curiosities.

The four numbers are:

$$1^3 + 5^3 + 3^3 = 153$$
$$3^3 + 7^3 + 0^3 = 370$$
$$3^3 + 7^3 + 1^3 = 371$$
$$4^3 + 0^3 + 7^3 = 407$$

Aside from $1^3 = 1$, there are no other such numbers, but people soon started generalizing them, albeit only in base-10. (So Hardy failed in his bid to have them dismissed.) It was noticed that the only four numbers this works for happen to be three-digit numbers, and so the hunt was on for four-digit numbers which are the sum of the fourth power of their digits. (There are three, it turns out, and 8,208 is one of them.) Because these numbers seem to be self-obsessed and always reflecting on how many digits they have, they are called narcissistic numbers. If you're interested, 54,748 is one of three five-digit narcissistic numbers, and there are many more to find beyond that. With a lot of computational power, you can show that the biggest narcissistic number is the 39-digit monster 115,132,219,018,763,992,565, 095,597,973,971,522,401.

Don't bother going the other way; it's far less exciting. Oh, all right then. There are no two-digit narcissistic numbers: no two digits when squared add up to give you themselves back as a two-digit number. I know, I've checked. And all one-digit numbers are, technically, narcissistic numbers because any number raised to the power of 1 equals itself. These are called trivial narcissistic numbers. 'Trivial' translates as 'boring' in maths language. It means you have technically fulfilled the requirements, but the result is dull. There is nothing interesting to be learned from a trivial solution.

Of course, Hardy would consider *all* narcissistic numbers to be trivial. Sure, they are not the greatest of maths discoveries, but they are a great diversion along the way. However, Hardy was right that there needs to be a line between interesting maths patterns and things which are purely coincidence. Playing with the properties of numbers in this way can be fun, as long as you keep the number-base limitations in mind. There is a greased gradient slipping inexorably down from base-specific properties to the badlands of numerology – and we most definitely do not want to go there, to be suckered into reading meaning into digits which is not there. Forget not that maths is primarily about *numbers*, not digits.

This is one reason that mathematicians have gone one step further and stripped numbers not only of the digits we use to write them down but of any link to physical objects. If you really want to know what a number is, this is the bottom line. I could talk about numbers as piles of rocks as a nice concrete example, but I could also pick something else. Five ducks are as good an example of 5, and so would be a collection of five cups of tea. They all involve some Earth-centric object, though. The process of separating a mathematical pattern or concept from physical reality is known in maths as abstraction. But the abstract concept of what '5' means when it's not directly associated with a group of five things is hard to pin down. Thankfully, though, there is a solution.

Mathematicians have agreed that, effectively, '5' is the name we give any collection of all possible sets of five things. The theoretical set of all groups of five things is what we mean when we say '5'. When we write $5 + 3 = 8$, we mean that if you take any group of five things and combine them with any group of three things (from the set we call '3'), the resulting group of stuff belongs in the set called '8'. What it gains in universality, this definition loses from long-windedness . . .

It is this universality of mathematics which means that if any space-faring civilization of Hypotheticals does reach our small planet, numbers may be the only thing we have in common. When we cannot bond over being carbon-based, or even both see things in the same visual spectrum of light, we can still swap number puzzles. Should you be the first on the scene when the first Hypothetical lands, here are two options.

The first is one that Hardy would have hated: the number 3,435. It is a Münchhausen number, which is a bit like a narcissistic number, which is why it is named after Baron Münchhausen, an eighteenth-century German who used to tell long, rambling stories about how great he was. To get a Münchhausen number, you raise each digit to the power of itself and then add them all together. Sure enough, $3^3 + 4^4 + 3^3 + 5^5 = 3{,}435$. And this is the only such base-10 number this works for, apart from the trivial 1. It does, however, work for other bases and, fortunately for our twelve-fingered Hypotheticals, there is one base-12 Münchhausen number: 3A67A54832. The real winners, however, would be any creatures with thirteen fingers, as they get to have four different solutions (33661, 2AA834668A, 4CA92A233518 and 4CA92A233538).

Secondly, you can go for the 37 Trick, if you give it a slight tweak. Even though it does involve the digits (rather than just the numbers themselves), it can be made to work for any base. Entering the same digit a given number of times and then dividing by the sum of those digits will always return the same answer in any position-value base system – it just might not be a whole number.* So, not all digit tricks are limited to just the one base. Well, I think that's all the bases covered . . .

* You get a whole-number answer when the amount of times a digit is entered divides evenly into the number of the base minus 1. So, for base-10, the base minus one is nine, and three divides into nine. To get a whole-number answer in base-12, the digit would need to be entered eleven times.

Two

MAKING SHAPES

First things first: there is a serious flaw in the way most people cut up pizzas. When faced with a pizza, the orthodox method is to make a series of straight cuts that cross in the centre, resulting in equally sized pieces. This is considered fair because everyone gets a piece that is not only the same size as everyone else's but also exactly the same shape: a triangle with one side rounded off (and made of crust). The problem with this system is that, while all the pieces are the same shape, they all touch the very middle of the pizza. This means that if there's a topping in the centre of the pizza that you don't like, there's no way to choose a piece that doesn't contain it. A better way of cutting a pizza would be one that produces identically shaped and sized pieces but not all of them including the very middle.

To find a way to do this, you're going to need a pizza. The pizza can be real or imaginary, and you may find it easier to use a circle drawn on a piece of paper. Your challenge is to find a way to cut up the pizza, or dissect the circle drawn on the page, into same-shaped pieces which do not all touch the centre. This isn't a trick question: the solution isn't to start with a square pizza or get less fussy friends, it can be done with your run-of-the-mill circular pizza.

A pizza divided the boring way, and a blank pizza for you to slice a better way.

Well . . . there *are* a few qualifications. We're going to consider a perfectly circular pizza which has a uniform coverage of the same topping and is without a crust; also, its base is extremely thin (so you can't cheat by cutting horizontally through it). To sum up, then, the pizza can be described as perfectly circular, homogenous, infinitely thin and (because it's two-dimensional) terrible value for money. It can also be frictionless and in a vacuum if you like, but that would make it substantially harder to eat. Although possibly much easier to digest.

This problem has all the ingredients to make it a perfect puzzle. Initially, it seems impossible, but as soon as you start playing around with it there's a hint of the solution. Then, suddenly, you'll find it – you may even need to eat an actual pizza to bring you down from the maths high. Also, it involves one of the best shapes in mathematics: the circle. Arguably, with no angles and one side, the circle is the simplest shape. It's certainly one of the most ancient. From the human eye to the sun glaring down on us, there are perfect circles in nature in a way that there aren't regular triangles, squares and pentagons.

If you're attempting the pizza problem, you may need to draw a number of circles. The easy option is to sketch them freehand and, as long as the two ends meet and your line forms a loop, it'll do. Otherwise, use a compass.* A compass is one of my favourite pieces of mathematical equipment, not only because the pointy end has entertained generations of school students (as a handy way to etch words into desks, or to act as a shank), but because it embodies what a circle is: a line that is always the same distance from a central point. You set the spacing of the compass to what will be the radius of the circle, stab the tip into the page and draw a line through all the points that are exactly one radius away from it. *Et voilà!*: a perfect circle. Or not . . .

* Officially, a 'compass' is only half of the device; the full kit is a 'pair of compasses' (after all, we don't talk about a 'scissor'). I assume there is some word-history reason why scissors got to keep their plural and their 'pair' but the compass did not. (Except in the minds of extreme pedants, that is, of which I am happy to include many among my acquaintance. One of them was once asked by airport security what he had in his bag, and after he replied, 'A pair of compasses,' had a long and awkward conversation trying to explain why the other one couldn't be found.)

Technically, though, a compass will not draw an exact circle on a piece of paper. If you were to zoom in close enough, you would see slight irregularities in the line because the surface of the paper is never perfectly smooth; also, any slight looseness in the pivot or screw of the compass will introduce variation into the radius. You'll get used to this: in mathematics, we like to distinguish between the perfect, ideal situation and what actually happens in imperfect, messy reality. Conceptually, there *is* such a thing as a 'perfect circle', consisting of an exact line which curves perfectly around a precise centre. In reality, we *can* draw an approximation on paper in which the imperfections are too small to be of concern.

You can use the compass not only to draw the circumference of the pizza, but also to draw the solution to the pizza problem. Instead of cutting up a pizza with straight cuts, make cuts of the same curvature as the circumference of the pizza, as in the diagram below. Unlike many mathematics puzzles, this one has a practical solution: you can physically cut a pizza up this way. I have. Next time you go to a pizzeria, make a copy of the diagram and tell them to cut your pizza up properly! (Of course, reactions to this request will vary.)

How to cut a pizza fairly, in two stages.

Making a Pentagon, Knot

Not all shapes are as well-behaved as circles. Using a compass to bang out a circle or two is easy, but drawing a five-sided pentagon is hard. Using a ruler to draw a five-sided shape is simple enough, but if you want all the sides to be exactly the same length, making a 'regular' pentagon, it all goes pear-shaped. Geometry as we know it started with the ancient Greeks, and they loved regular pentagons. Being able to draw one even allowed you access to a secret maths club. I can't offer you that, but I can show you an extremely easy cheat.

Take a long strip of paper and tie it in a simple over-and-under knot. Slowly pull the paper tight, flattening it at the same time, and the knot will settle into a regular pentagon. You can measure the edges if you don't believe me; and, if

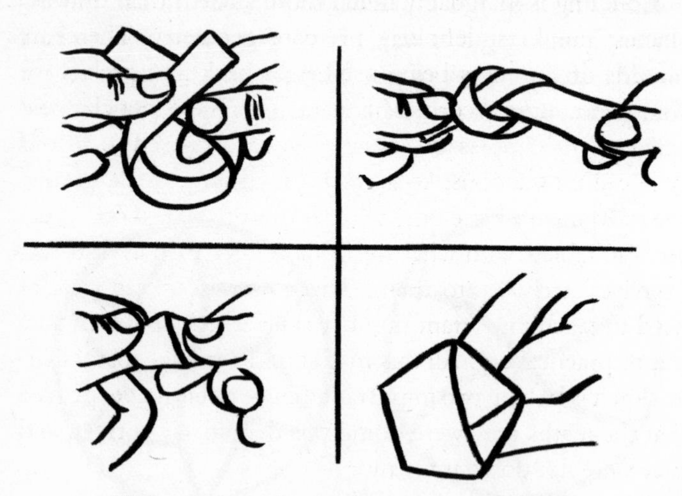

Tie a knot in a strip of paper and get a free pentagon.

you still don't believe me, I demonstrate that the edges are, theoretically, of identical length in 'The Answers at the Back of the Book'. The fact that you can whip up this pentagon pretty much anywhere in such a short amount of time has led to it being given the exciting sobriquet the 'emergency pentagon'.

Unfortunately, the emergency pentagon would not gain you access to the ancient Greeks' secret maths club (or be much use in the vast majority of emergencies). By around 300BCE, the Greeks had become obsessed with drawing shapes using only a compass and a ruler. And we are still obsessed with the ancient Greeks and how they constructed shapes because they were among the first mathematicians and geometry was the first area of mathematics. *And* there's a very good reason why the compass and ruler can be considered to be the genesis of mathematics, *and* it relates to why numbers alone never quite cut it as 'real maths'.

Counting is inarguably much more ancient than drawing shapes; numbers definitely pre-date geometry. There are records of numbers being used right back into prehistory. They were often recorded by making marks in wet clay, and there exist examples of ancient clay tablets used to record financial transactions, keep track of farming stock, predict the movement of the tides with the moon, and so on. There are also tablets with what we would understand as 'learning exercises' etched into them. These exercises were puzzles used to teach important number skills which could then be put to practical use. All this may sound a lot like mathematics, but it is lacking two important things: nobody ever proved that the maths they were doing was definitively correct, and they were not doing it for fun.

When people first started using geometry, it was for

practical reasons, such as dividing fields or building structures. Shapes joined numbers as merely another tool which humans employed to advance civilization. This all changed with the ancient Greeks. They decided to do mathematics purely for the sake of it. For them, it was a game. Not only that, but the focus was not only on finding the answer but also on proving that it really really without a doubt was the correct answer. This new approach to mathematics was embodied in one man: Euclid.

Euclid was born around the year 300BCE, as far as we know (one thing the ancient Greeks didn't do on a regular basis was to record birthdates). In fact, for all we know, 'Euclid' could be the pseudonym for a whole team of people. In any case, he (or they) did write thirteen books, and we do still have them today. Known collectively as 'Euclid's Elements of Geometry' (or simply as 'The Elements'), they were Euclid's attempt to capture all current human mathematical knowledge *and* prove that it was all correct.

Euclid didn't want anyone to have to trust him or take anything on faith: every step had to be rigorously proved. Unfortunately, you cannot prove *everything* from scratch: you have to start with a few things which are assumed to be true and see where you can go from there. So Euclid picked the most obvious things to assume, things so clearly true that they require no justification or evidence. The first was that it is possible to use a ruler to draw straight lines; the second that it is possible to use a compass to draw circles. You wouldn't believe how far these two propositions can take you.

Kit yourself up, and see if you can use your ruler and compass to draw a triangle with three sides of exactly the same length (an equilateral or regular triangle). If that's too easy, go for a square (obviously, all four sides must be the same length), or a regular hexagon. The real challenge begins when you try to draw a regular pentagon . . . What all this experimentation goes to show is that you do not have to assume on faith that triangles exist. The first proof in Euclid's *Elements* is how to use a compass and ruler to draw a triangle; once you accept that lines and circles exist, then the existence of regular triangles, squares, pentagons and hexagons flows as a natural consequence.

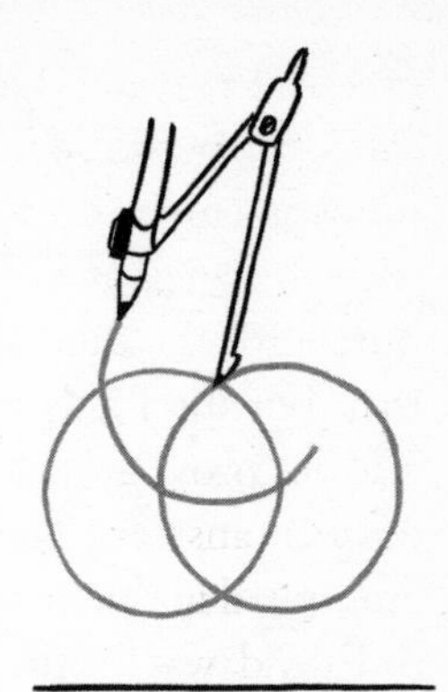

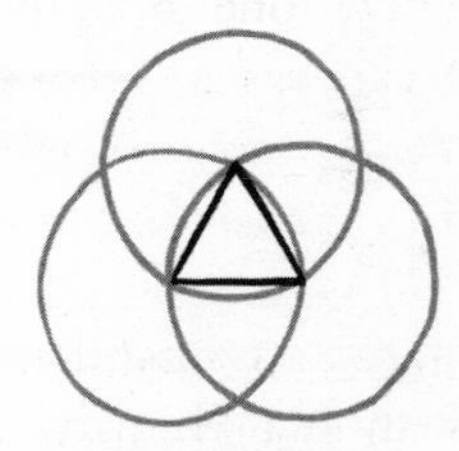

Three circles later and you have an equilateral triangle!

This is another instance of abstraction, just as we saw with numbers. Maths as we understand it began when people detached themselves from physical reality and tried to understand shapes in their abstract form. A right angle went from being something that existed only whenever two fences physically intersected each other at the corner of a field to a generalized concept. Unlike numbers, which are expressed in a given base system, shapes appear to be free of this restriction. Circles are circles, straight lines are straight lines, however you express them. Those hypothetical aliens could argue with us about how to write down a number, but they would have to agree that a pentagon is a pentagon (in whatever language; let's stick to the language of geometry, the ruler and the compass. Know where we are then).

Interestingly, though, the Hypotheticals may have started with different assumptions to the Greeks. Tying a knot in a strip of paper to make a regular pentagon would not count as a valid method to Euclid because it goes beyond his

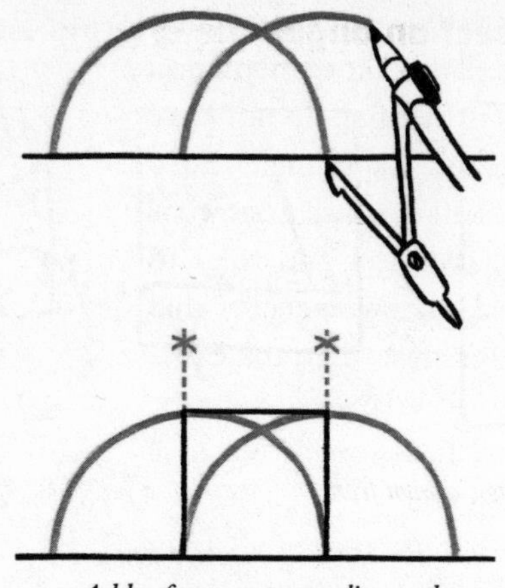

Add a few arcs to any line and a square is all yours.

starting assumptions, but the Hypotheticals might be fine with that. Humans evolved a love of drawing on things, so our understanding of shapes relies heavily on how we can draw them, whereas the Hypotheticals may like to fold things and have more of an origami understanding of geometry. In fact, there is a lot more to shapes than Euclid was prepared to admit.

DON'T GREEK TOO SOON

There were some things the Greeks were unable to do with a compass and ruler, such as dividing an angle into thirds, or drawing a circle and a square with the same area, and it drove them *mad*. The problem was, they couldn't prove that these things were impossible (as we now know they are), so there was always the dangling possibility that the solution was there to be found, if only they were clever enough. The act of splitting an angle into three smaller, identically sized

Trisect an angle, the origami way

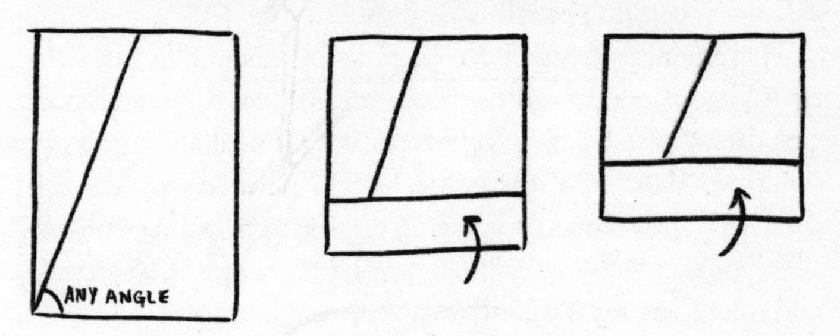

Start with any angle drawn from the corner of a piece of paper.

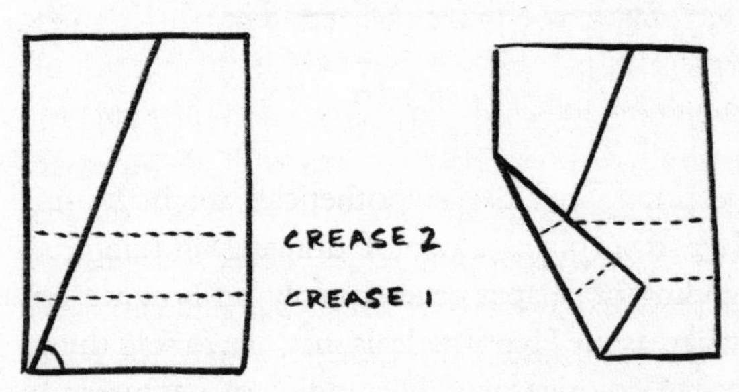

Fold it once and then over again so you have two parallel lines (crease 1 and crease 2).

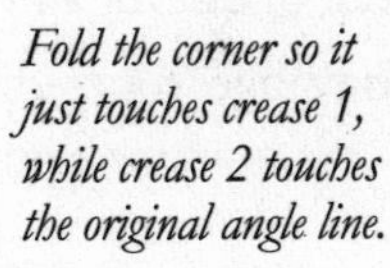

Fold the corner so it just touches crease 1, while crease 2 touches the original angle line.

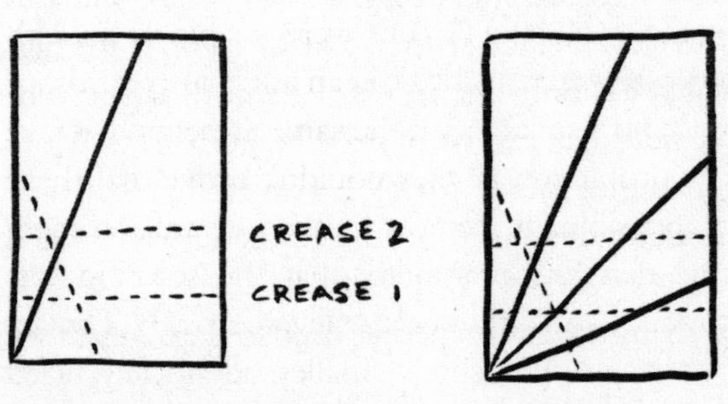

You will now have two points on crease 1: where the corner touched it and where the new fold crosses it. Join these to the corner and the angle is divided exactly in thirds!

angles certainly feels so straightforward, it'd be natural to assume it couldn't possibly be impossible.

Hypotheticals would have no problem doing this, however, by folding paper. On the previous page is just such a solution, first presented by Hisashi Abe from Hokkaido University, Japan, in 1980. The reason the Greeks never discovered this method is that it requires lining up two different points with two different lines at the same time. Beyond the scope of a compass and ruler then, but easily achievable through origami.

Another thing that drove the ancient Greeks *mad* was shapes in which the edges go through other edges. If you take the knot pentagon you made earlier and hold it up to the light, you'll see the pentagonal star within it, commonly called the pentagram, to avoid confusion. Both the pentagon and the pentagram are five-sided shapes with all the edges exactly the same length – except the pentagram is often dismissed because of its overlapping edges. There is a sense of it not being a 'real' shape. Which is a real shame. I'd argue it's a question of personal taste. I love shapes where the edges intersect.

If you don't like your edges crossing, you only get one heptagon, but if you don't mind a bit of casual intersection – and let's face it, who doesn't these days? – there are two more perfectly good regular heptagons to be found. All in, there are five different regular hendecagons (I'll help you out this time: it's an eleven-sided shape). The hexagon is the biggest shape which has only one regular form: any shape with more than six sides has two or more different forms. More sides do not guarantee lots of forms, however; there are only two regular dodecagons (just once more then: twelve-sided shapes).

I think it's a shame that when people need to draw a star they

The five different regular hendecagons.

always go with the five-sided-pentagram option. Why not draw seven- or even nine-sided stars? One thing: if you do want to practise drawing different stellated shapes, the hardest bit is getting the points equally spaced. Below are some pre-spaced dots for you to try drawing some stars on, and there are

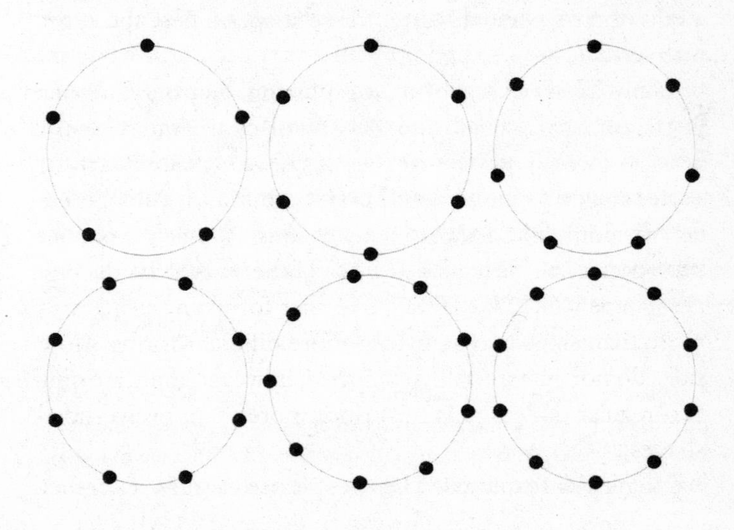

plenty more such cheat-sheets you can print out on the website makeanddo4D.com.

The last thing which (rightfully) stumped the ancient Greeks, with their stubborn rules and sticking points, was drawing a regular shape with seven sides. This got them *really* mad; after a lot of work, they cracked how to draw a pentagon, but the heptagon proved elusive. So they decided to leave it and try some other shapes first. Of the next few regular-sided shapes, those with eight and ten sides could be drawn with a compass and ruler, but nine-sided and eleven-sided shapes seemed impossible. After this, twelve-sided shapes were easy, but thirteen and fourteen were both impossible again.* Frustratingly, the Greeks did not know enough maths to prove that these shapes were definitely impossible to construct with a ruler and compass, so they continued to mount futile efforts.

I should point out that it wasn't as if the Greeks couldn't draw these shapes, they just couldn't draw them using the methods they wanted to use. After a bit of trial and error with a ruler, you can end up with a very accurate heptagon, certainly close enough for any practical heptagon needs. What annoyed Euclid and his friends was that it would never be a truly perfect heptagon. They were prepared to accept only straight lines and perfect circles on faith, and if they couldn't find a way to arrange them to make a regular heptagon, they were not happy. They refused to assume

* If you're looking for a pattern in which regular shapes cannot be drawn, the most common almost correct guesses involve prime numbers. Which is close. Bear in mind that not only do three and five work, but it is possible to construct a regular seventeen-sided shape.

heptagons existed without having proved it from the ground up.

To my mind, the Greeks became too obsessed. It is possible to trisect an angle using only lines and circles, but it requires drawing a circle in one place and then sliding it over to a different location. Because you cannot do this on a piece of paper with a ruler and compass, the Greeks didn't count it as a valid method. But that was their prerogative. Maths is, ultimately, a giant game, and the Greeks were obsessively playing within the rules they had made up. Maths is also a game where you cannot cheat;* you only ever add more rules.

That may be my favourite definition of maths: a game where you choose the starting rules – the things you assume to be true or which are allowed – and then try to prove as many new facts as possible from there. Numbers are easy: you assume they exist then set about proving that, for example, no base-10 narcissistic number can have more than thirty-nine digits. The world of numbers is very good value, as it has only one starting assumption: that numbers are a thing. Geometry, on the other hand, has much more scope for starting assumptions. The great thing about maths, though, is that once they've been picked, you'll get the same results as anyone else using the same assumptions anywhere in the universe, or possibly beyond. If we knew what assumptions the Hypotheticals used for geometry, we could derive identical results.

* Actually, there is one very big way to cheat fundamentally at maths which has caused a lot of controversy in the maths world: getting a computer to do the thinking for you.

PIECE OF CUBE

I once set a friend the pizza problem and they, on realizing that I wasn't going to give them the answer, posed me a similar puzzle, just to annoy me. Here's how it goes. There are five people waiting to eat a cube-shaped cake, which is a delicious homogenous cake on the inside and is covered in equal-depth delicious icing on all five of the exposed sides. As all five people love both cake and icing (after all, it is delicious by definition), they want to cut the cake into five pieces such that all the pieces have the same volume of cake and the same surface area of icing. Apparently, not only is this possible, but the puzzle can be adapted for any number of people sharing a cube cake (although I am still waiting for an opportunity to put this knowledge into practice at an actual cube-cake party).

It's not only the party-food/carbohydrate/calorific elements that this puzzle has in common with the pizza problem: they both have more than one solution. The traditional cube-cake solution involves cutting the pieces of cake into a shape in which the volume of the slice can change while the surface area of the icing remains the same by way of stacked pyramidal cuts. This solution didn't occur to me, however, until I had found a non-traditional way, which involved cutting the top of the cake up into triangles. Luckily, I had more than one virtual cube cake available.

My method involves marking the top perimeter of the cake into five equal lengths and then finding the exact centre of the top face. Cut from each mark on the perimeter to the centre point and slice directly down. The icing on the sides of the cake is thus cut into five rectangles of exactly the same surface area. The icing on the top, and the volume of each

The traditional way to cut a cube-cake

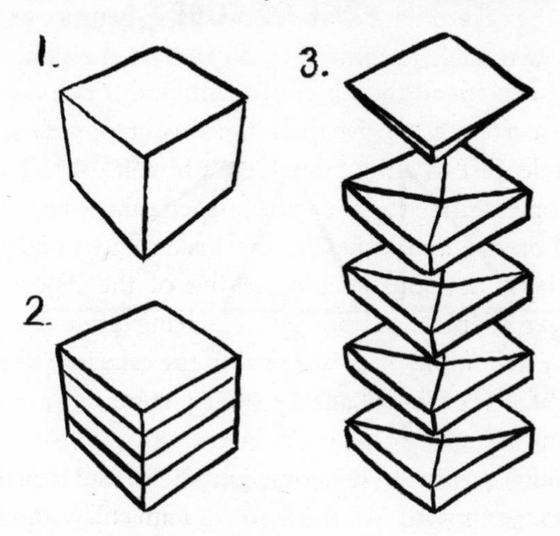

Step 1. Bake and ice a cube-shaped cake.
Step 2. Mark three equally spaced horizontal cuts around the sides (do not actually cut through the cake).
Step 3. Cut off the top surface of the cake, with a pyramid of one fifth of the cake below. Then cut off each new slice with whatever pyramid cut below gives you the next fifth of the cake.

A non-traditional way to cut a cube-cake

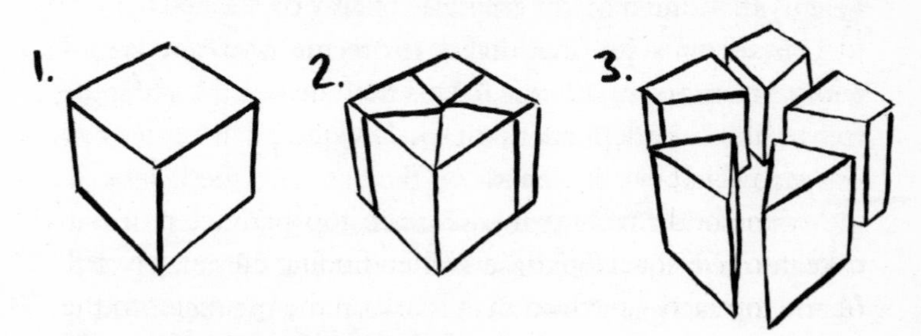

Step 1. Bake and ice a cube-shaped cake.
Step 2. Mark five equally spaced distances around the top edge and join them to the centre.
Step 3. Cut each line directly down to get five normal-shaped pieces of cake.

slice, will also be identical. Even though the triangles on the top of the cake are not all the same shape, because of a quirk with how triangles behave they do all have the same surface area.

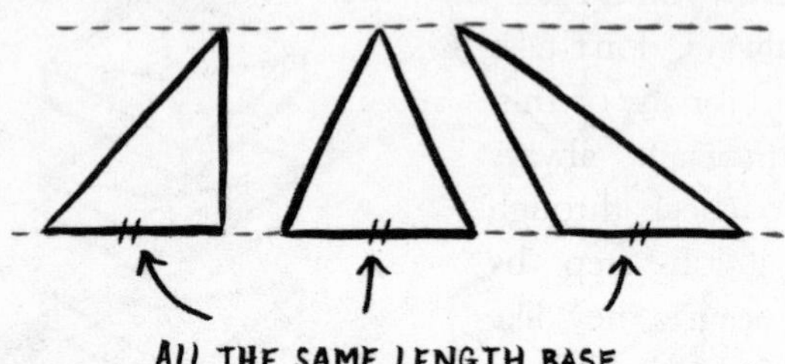

Students across the millennia have been told that the area of a triangle is half of the base multiplied by the height, often written as ½ × *base* × *height*, or ½*bh*. This is normally presented as a fact to memorize, which is a shame, because maths is all about proving why things work. How to calculate the area of a triangle is fairly obvious: you can see that the area taken up by a triangle is equal to half the total area of the rectangle which (conceptually) surrounds it. The area of the rectangle is the base of the triangle multiplied by its height, so the area of the triangle is half of that.

This formula for area makes no mention of any angles within the triangle, and this means that the area of a triangle is completely independent of the degree of its angles: it depends solely on the length of the base and the height of the other corner from that base. So, if there were a series of triangles with the same base and height but completely different angles, they would all still have the same area. This is exactly the case for our cube cake: all the triangles on the top have the same base and height.

The pieces that go around a corner of the cake appear to

be trickier but they can be considered to be two adjacent triangles and shown to have the same area when combined. This does mean that the corner pieces are not actually triangular like I promised, unless you're prepared to accept a triangle with an extra corner. Yeah, I didn't think you would.

As always, don't take my word for any of this. Mathematicians always want to check through the solution step by step, because they like rigorous proof. Bake a cuboid* cake – why not? There is a complete guide to cutting cuboid cakes in 'The Answers at the Back of the Book', so the next time you're dividing such a cake, make sure you have a ruler handy as well as a knife. Personally, I had a ruler engraved on the back of my cake knife to ensure ultimate precision.

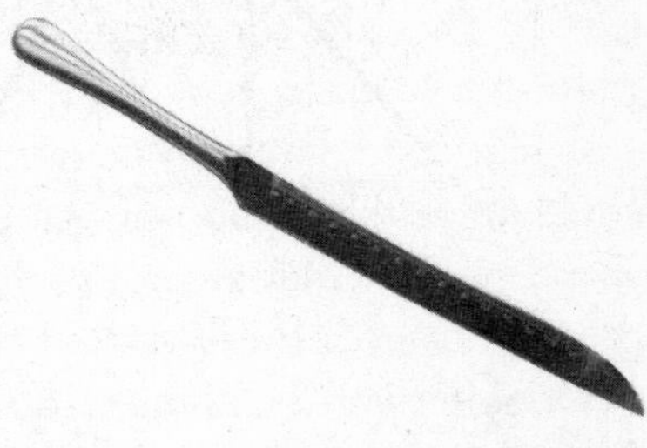

No more arguments over cake-cutting precision!

MORE THAN ONE WAY TO SLICE A PIZZA

Not only is there more than one way to cut a cake, but there is more than one way to slice a pizza. In fact, it's ridiculous how many different ways there are to solve the pizza problem. If you didn't find the first solution, of course, it may not help to know you didn't miss only two possible answers but

* Mathematicians use '-oid' to indicate when something is a similar, more general form to the norm of whatever that something happens to be (it's stolen from the Latin suffix *-oides*). In terms of cakes, as long as we keep the right angles at all of the corners at exactly 90°, the lengths of the edges can change and the overall shape still looks enough like a cube for mathematicians to call it a cuboid.

you missed infinitely many. Some of them are even better than the first solution I showed you.

The down side to the first solution is that some of the pieces are mirror images of each other. In maths, we would still consider these to be the same shape, because if you pick two mirror-image pieces of pizza up you can line them up exactly. They are 'congruent', a word based on the Latin *congruere* for 'coming together'. Being congruent is a more stringent condition than merely being the same shape. You could argue that a snooker ball and the moon are the same shape, because they are both spheres. For shapes to be congruent, however, they also have to be the same size.

One alternative solution starts the same way – dividing the circle into the same six curved triangles – but then, instead of joining the three curved lines with straight lines, you can use another curved line. Now, none of the pieces is a mirror image of another, so there can be no arguments: this method gives twelve pieces which are all absolutely identical, except that half touch the centre and half do not.

The first solution uses two mirror-image shapes; the second needs only one shape.

From here on, the solutions start to get much more complicated, and they all have more than twelve pieces.* There are two different ways to cut a pizza into forty-two absolutely identical pieces, as it happens, such that not all of them touch the centre. There are ways to cut pizzas up into twenty, thirty, forty, fifty . . . and every multiple-of-ten number of identical pieces. It is a puzzle with a rich and never-ending number of different ways it can be solved. All the solutions will *not* be found in 'The Answers at the Back of the Book'.

Mathematics can give the impression of being a very strict and rigid discipline; as if the rules were decided by the ancient Greeks and now everyone must follow suit. In reality, mathematics is all about adding new rules and then seeing what happens if you break them. You don't have to do maths on a flat piece of paper just because Euclid liked it that way. And even within the same set of rules there is often more than one way to do things, sometimes leading to many different solutions to exactly the same problem. Mathematicians can have their cake and solve it.

* There are more twelve-piece solutions to find, but they are all variations on a theme.

Three

BE THERE AND BE SQUARE

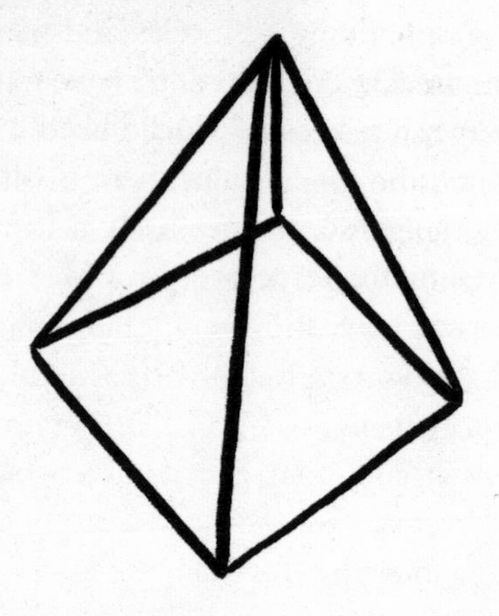

In 1994 NASA calculated a strange numerical code and concealed it in the backwaters of their website, where it remains to this day. If you go to the 'Astronomical Picture of the Day' section of NASA's public site, there is a hidden folder called 'htmltest/gifcity', within which is the cryptically named file sqrtz.10mil. If you find and open this file, your computer screen will be completely filled with digits. Here are the first few of the 10 million digits in that file:

1.4142135623730950488016887242096 9
807856967187537694807317667973799
0732478462107038850387534327641572...

From: http://apod.nasa.gov/htmltest/gifcity/sqrt2.1mil

Enough? I don't *know* why NASA produced all these digits, but I can make a pretty good guess. It is in fact a number reasonably well known to mathematicians, the square root of 2, or 'root 2' to its friends. When a number is multiplied by itself, we call it squaring that number. If you multiply root 2 by itself, you get an exact answer of 2. As for why NASA calculated the first 10 million digits of this number, my pretty good guess would be: purely because the NASA engineers thought it would be fun.

Mathematicians seem to have a strange fascination with square numbers. If you start with normal numbers like 1, 2, 3, etc., and you square them, you get the square numbers 1 (1×1), 4 (2×2), 9 (3×3), and so on. Instead of writing each number out twice whenever you square it, the convention is to place a small superscript ^{2} next to it, for example, $1^2 = 1, 2^2 = 4, 3^2 = 9$. These square roots and superscripts appear all the time in various maths puzzles and games. For instance: arrange the numbers 1 to 16 in such a way that if you add each adjacent pair of numbers you always get a square number. (Wait for it . . .)

I'm not immune to all this: I love squares. Once, in a pub with some friends, our table number for ordering food was 36, so I commented on it at the bar. Then it struck me: there are other-shaped numbers, and 36, as well as being a square,

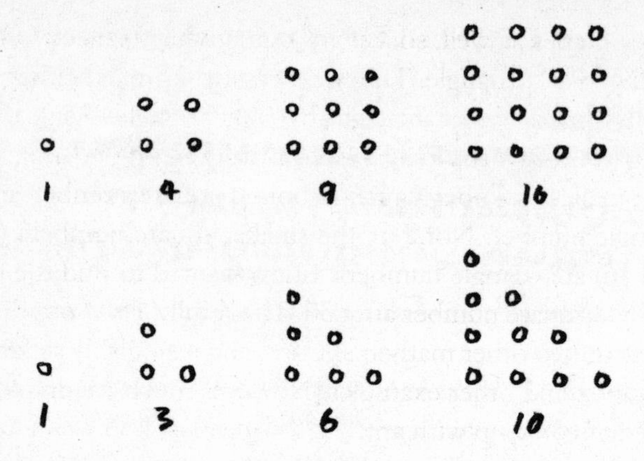

The first four square numbers and triangle numbers.

is also a 'triangle' number. I had never thought about numbers that were a triangle and a square at the same time, and it left me temporarily speechless. I can only assume the bar staff were struck with the same epiphany because they just stood there staring at me.

The link between square numbers and square shapes is fairly straightforward. If you have a square number of objects, you can arrange them in a square. This works for other shapes as well, so a triangle number of objects can be arranged as a triangle. Thirty-six can be arranged either as a six-by-six square or an eight-by-eight triangle. Aside from the trivial 1 (which I'm ignoring; it really is *so* mundane), 36 is the smallest number which is both a square number and a triangle number. None of the smaller square numbers (4, 9 and 16) are triangle numbers. I now wanted to find the next triangle-square number after 36. Thankfully, I was out drinking with two other mathematicians, and we quickly set about trying to find other examples. However much we drank, we couldn't come up with any.

None of the easy methods we could think of to find triangle-square numbers – short of getting our laptops out in the pub – worked. So we got our laptops out. Using a spreadsheet, we listed a few thousand square numbers in one column (see Chapter 1, footnote on 'Cheating'). First bit of algebra: if we represent any number as the letter n, then we know the square of that number is n^2. Triangle numbers are slightly more complicated, the nth triangle number is $n \times (n + 1) \div 2$ (because two triangles make one n by $(n + 1)$ rectangle). Spreadsheets, however, are great at calculating these things, and soon we had a second column of several thousand triangle numbers. What a night! (Not that I encourage drinking and deriving.) All we had to do now was look for the numbers that were in both lists. First after 36 was 1,225 then 41,616; 1,413,721 and 48,024,900.

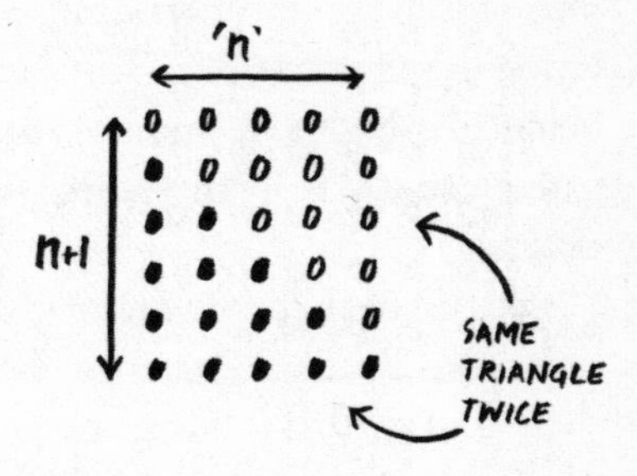

However, three mathematicians down the pub, even with their laptops out, still aren't quite at the mathematical cutting edge. Others will always insist on taking things just that little bit further and, sure enough, mathematicians have found numbers that match all types of shapes. Multi-sided shapes such as pentagons and hexagons are collectively called polygons, so there are all sorts of polygon numbers. Let's throw some shapes. (And there are more in 'The Answers at the Back of the Book'.) The first few pentagon numbers are 1, 5, 12, 22 and 35, and the first few hexagon numbers are 1, 6, 15, 28 and 45.(And there are more in The Answers at the Back of the Book'). The smallest square-pentagon number after 0 and 1 is 9,801 (followed by the impressively bulky 94,109,401); below 1 million (after 0 + 1) there are only two triangle-pentagon numbers: 210 and 40,755. No one has yet found a number which is triangle, square and pentagon simultaneously. Mathematicians have checked all the numbers with 22,166 digits or fewer for triangle-square-pentagon numbers without success . . . but until someone proves they don't exist, mathematicians keep on looking.

An obvious question now is 'Why?' Actually, several good questions now are all 'Why?' And it's a very good point: why *do* mathematicians spend their valuable time looking for these strange numbers? It's certainly not because they are useful: there is no practical reason why you would want these numbers. They are sought as a kind of game; finding them is its own reward. You can think of much of mathematics as being like train-spotting or stamp-collecting. Hmm, those may not be the two best examples I could have picked. I'm trying to sell this stuff to you. OK, maybe think of maths as being like big-game hunting or playing an amazing video game: the joy is in the search and in achieving something difficult.

Let's go back to our earlier conundrum.

If you did try to arrange the numbers from 1 to 16 in such a way that adding two adjacent numbers yields a square number, you're one step towards mathematicianhood, and this is what you will have come up with:

8-1-15-10-6-3-13-12-4-5-11-14-2-7-9-16

As you inched closer to the solution I hope you felt the excitement of catching sight of big game on the open savannah, or spotting a rare LNER Pacific steam engine in the wilds of Hertfordshire. Similar levels of excitement, to be honest. Either way, that's what impels humans to do maths. In fact, it was the fascination with square numbers that led, we believe, to the birth of mathematics. Since, mathematics has yielded all sorts of benefits for humankind, but that was not its original intent. Nor is it now. Sure, some maths is developed for purely practical reasons, but even then it will be rooted in some previous mathematics originally done for fun. As it should be.

Open a Can of Pythagoras

Euclid may have been the first mathematician to try to put all of mathematics together into one neat, rigorous bundle, but before him came the first superstar mathematician: the man – the legend – that was Pythagoras. His name is recognized by generations of schoolchildren as 'something to do with triangles', but he was in actual fact possibly the first person to do real, actual maths. He not only discovered patterns in shapes and numbers, he also proved that those patterns would always hold.

Despite his fame, we know surprisingly little about his life and work. Pre-dating Euclid, Pythagoras was born on the Greek island of Samos in the sixth century BCE, and by around 535BCE he was old enough to pop over to Egypt for a while. After that he went home for a bit to set up a club, The Semicircle of Pythagoras, before eventually ending up in Crotone, southern Italy, where he started a philosophical and religious society. The Pythagorean Society certainly did a lot of mathematics, but they were more of a religious group than anything else. They believed that 'at its deepest level, reality is mathematical in nature', but they were even more focused on using philosophy to achieve spiritual purification. There is no shortage of tales about Pythagoras's unusual beliefs (and major problems with beans), should you wish to look into them.

Mathematically, the members of this society did a lot of work we would now find familiar, including investigating triangle numbers. Pythagoras is also credited with developing the idea of ratios between musical harmonics. We believe much of what Euclid wrote about was first discovered by the Pythagorean Society, including almost all of Book 4. However, the society is most famous for Pythagorean Theorem, which states the eternal qualities possessed by all right-angled triangles. The proof of Pythagorean Theorem is the climax to Book 1 of *The Elements*, what everything else in it was building up to.

In short, Pythagorean Theorem states that if you take any right-angled triangle and square the lengths of all of the sides, the smaller two added together give the larger (the largest side of a right-angled triangle is commonly called the hypotenuse, so named by the ancient Greeks). To give it the classical phrasing: the square of the hypotenuse is equal to the sum of the squares of the other two sides. To quote Euclid's phrasing: 'in right-angled triangles, the square on the side subtending the right angle is equal to the (sum of the) squares on the sides surrounding the right angle'. To quote generations of schoolkids: 'Why do we need to know this?'

At the very start, maths was only done on a need-to-know basis; it was a practical tool used for getting things done. The first right-angled triangles would have been investigated by humans because they needed them, maybe to calculate how to build an upright wall or evenly divide up a field. Some people using triangles for practical purposes would have noticed, and utilized, the fact that two of the sides squared add up to give you the other one. Pythagorean theorem about right-angled triangles was known by the Babylonians well over a millennium before Pythagoras discovered it,* and such triangles were in use in Egypt when Pythagoras visited there.

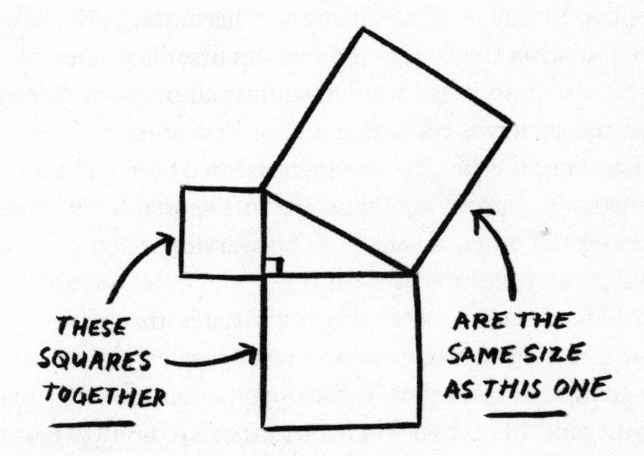

Pythagoras's gift to mathematics was not then the theorem itself, but rather that he proved it was true for all

* And no doubt in India and China as well as other ancient civilizations, as seems always to be the case with these things.

right-angled triangles. Prior to Pythagoras, all of maths was applied and directly linked to objects in the real world. He was the first person not to worry about practical applications and to approach maths as an abstract subject. He proved that the sum-of-squares pattern was not an accidental pattern which happened to work for the triangles humans had come across and checked but something true of any right-angled triangle anywhere in the (Euclidean geometry) universe.

Sadly, Pythagoras was not as open-minded about the possibilities of mathematics as he could have been. His theorem shows that if you have a right-angled triangle in which the two sides are both exactly one unit long (any unit works: centimetres, inches, miles, etc.), then the third side must be root two long (1.4142 centimetres, inches, miles, etc.). If you take the square you drew before, the diagonal line will be exactly root two long. But Pythagoras refused to accept that root two even existed.

The Pythagorean Society believed that, for any number whatsoever, you could find two factors which, when divided one into the other, would give that number. That is, for any number n, you can always find two other whole numbers, a and b, for example, so that $a \div b = n$. However, they were wrong. And root two is the counter-example: it can never be the answer to a number divided by another, different number. Legend has it that when a member of the Pythagorean Society realized they were wrong (supposedly a chap named Hippasus, but the ancient texts are not clear), he was drowned as a punishment (or it might have been for something else – time tends to blur the details). Whatever the case, the Pythagoreans may have invented what we know as 'mathematical proof', but they still chose to ignore it when they did not like the result.

Ship Shape

In the late 1500s the globe-trotting adventurer Sir Walter Raleigh (who, among other things, was instrumental in the English colonization of North America and supposedly introduced both the potato and tobacco to England) had a maths problem he needed solving. Thankfully, he kept a mathematician on his ship, so he asked him if there was a quick way to take a square pyramid of cannonballs and calculate the total number in it, instead of using the old-fashioned 'just count them all' method. His mathematical assistant was Thomas Harriot, also no historical lightweight, being the first person to observe both sunspots and the moon through a telescope.

You can try this yourself if you have sufficient cannonballs. If not, oranges make a good spherical substitute. The problem with oranges, though, is that they will try to form a triangle pyramid, not the square pyramid that Raleigh was interested in. Balls stack quite nicely if you start with a triangle number of them, then place smaller triangles of balls on top until you have a single ball at the apex. The formal name for a triangle pyramid is a tetrahedron, so the number of oranges in such a stack is called a 'tetrahedron number'. If you start with a square of balls, then stack

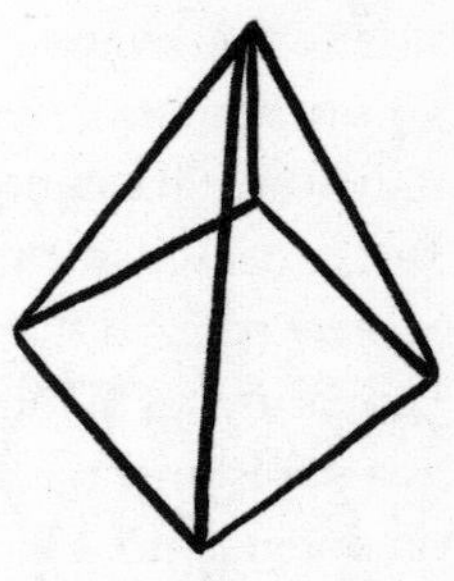

smaller squares of balls on top, you get a square pyramid, and it was these square-pyramid numbers that Raleigh wanted to calculate.

Harriot was able to calculate that a square-pyramid stack that is any number n cannonballs high would contain $n \times (n + 1) \times (2n + 1) \div 6$ cannonballs in total. You can check this if you want. Should you have made a square pyramid three oranges high, it will have required fourteen oranges and, sure enough, $3 \times (3 + 1) \times (2 \times 3 + 1) \div 6 = 14$ (or $3 \times 4 \times 7 \div 6 = 14$). I love the mental image of a ship's crew stopping in the heat of battle to do this calculation to make sure they had enough cannonballs in their reserve pile.

If you do have a square pyramid of cannonballs, you could now try to dismantle it and rearrange the cannonballs in a square shape. This was the puzzle into which Raleigh's question later mutated: which square-pyramid numbers of cannonballs can be rearranged in a flat square? The answers to this conundrum are called 'cannonball numbers', and I see them as the great-ancestor of all such multi-shape numbers. And my favourite cannonball number is . . . 4,900 – and it's not just my favourite cannonball number; I can guarantee that it is your favourite as well. 4,900 is the *only* possible solution to this puzzle: no other square-pyramid number is also a square number.

The 4,900 solution was found (or, at least, popularized) by the French mathematician Édouard Lucas in 1875. He suggested that it *may be* the only solution, but he couldn't prove it for sure. When a mathematician has a guess at something they believe to be true but cannot prove for certain, it's called a conjecture. It was thus Lucas's

conjecture that 4,900 is the only cannonball number. The problem with conjectures is that they may later be shown to be wrong. In this case, though, Lucas's hunch was correct: in 1918 it was *proved* that there are no cannonball numbers other than 4,900.

Lucas is one of my favourite mathematicians because, not only did he do some fantastic maths, he was also an early 'recreational mathematician'. A recreational mathematician is a kind of hobby-ist mathematician who does maths purely as a fun way to pass the time. Lucas's book *Récréations mathématiques* was published between 1882 and 1894 (sadly, still only available in French), and within its four volumes are all sorts of great maths games and diversions. Lucas also produced the first-ever description of the Dots and Boxes game which continues to waste endless student-hours in classrooms around the world.

As a recreational mathematician myself, I was inspired by Lucas to find other numbers similar to cannonball numbers, and I went for numbers that are both a polygon number and can form the pyramid based on that polygon. My first breakthrough was that 946 cannonballs can be arranged as

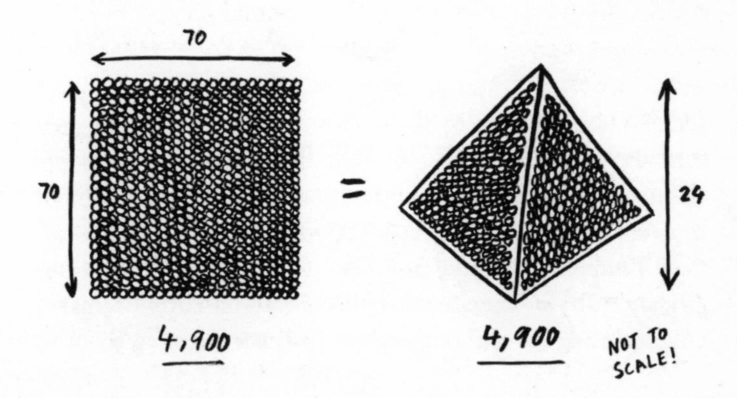

a hexagon with sides twenty-two cannonballs long, or turned into a hexagon pyramid eleven high. Then I discovered that both 1,045 and 5,985 are octagon numbers and octagon-pyramid numbers. This was becoming addictive. I needed more.

Instead of using a spreadsheet, I had written a computer program (I also enjoy programming as a hobby). Once I had it up and running, I thought I'd go wild and leave it overnight to see how far it could get. In the morning, the number 90,525,801,730 was waiting for me. It turns out that if you have over 90 billion cannonballs you can arrange them as a 31,265-agon with sides 2,407 cannonballs long – or rearrange them as a 31,265-agon pyramid 259 levels high.

I'm sure there are many big numbers that function a bit like 90,525,801,730, but this one is mine. As far as I know, I found it first. Since the dawn of time, 90,525,801,730 has existed as a number and been a 31,265-agon number as well as a 31,265-agon-pyramid number, but I was the first human to spot that. Should we ever meet our Hypotheticals, maybe even they haven't had enough free time to find this number, and it would be something new I could show them: my personal big-game win.

Taxi!

One of the greatest mathematicians of the twentieth century was almost entirely self-taught. He is also the star of my all-time-favourite story about cube numbers (numbers that can be arranged in a 3D cube). Growing up in India, Srinivasa Ramanujan could not help but do maths. Without guidance, he independently rediscovered all sorts of mathematical theories without anyone else ever showing them to him. He also found some new things in maths that no one else had ever seen.

Unfortunately, because he was self-taught he developed his own system of mathematics in isolation. His notation was so different to that of normal maths that when he wrote to other mathematicians they dismissed him as insane. Given his unusual notations contained cryptic notations such as '$1 + 2 + 3 + \ldots = {}^{-1}\!/_{12}$', you can see why. One of his letters was sent to G. H. Hardy on 16 January 1913 and Hardy was able to decipher what Ramanujan had written and recognized the impressive maths he was doing – and that he was doing some maths no one else had ever done. Hardy quickly organized for Ramanujan to come to Cambridge and share his maths with the rest of the world. Years later, at the end of a long, distinguished career, apparently Hardy was asked by the Hungarian mathematician Paul Erdős what his greatest contribution to maths was, and he replied, 'The discovery of Ramanujan.'

From 1914 until 1919 Ramanujan lived and worked in Cambridge with Hardy until, sadly, he fell ill. This, however, did nothing to stop him from doing maths. One time, Hardy took a taxi to visit him in a south London hospital and, upon arriving, remarked off-hand that the taxi's number had been the rather boring 1729 and he hoped that wasn't a bad omen. Ramanujan instantly replied that 1,729 was actually quite interesting as it was the smallest number which could be written as the sum of two cube numbers in two different ways. And, indeed, 1,729 does equal both $9^3 + 10^3$ and $1^3 + 12^3$. It is now known as a taxi-cab number, or the Ramanujan–Hardy number. The thought of Hardy enjoying a piece of recreational maths pleases me greatly. I've since checked the number of every taxi I've been in, but I'm yet to find the infamous 1729.

Of NASA and Numbers

Let's take it back to the beginning. This is why I think NASA calculated all those digits of root 2 for fun. Doing things for fun is what motivates most mathematicians and everyone who enjoys maths recreationally. The other big hint is that the webpage says so. The digits were in fact calculated by Robert Nemiroff at NASA 'during spare time on a VAX alpha class machine over the course of several weeks'. So it turns out the engineers at NASA had a spare cutting-edge computer lying around and thought they'd run a program on it over the weekends. For fun. They also calculated the square roots of several other numbers.

I can't blame them, and calculating square roots is a good challenge for a computer, partly because it's a task which is never complete. For square numbers, the roots are easy to find, because they are nice, neat whole numbers. Using the short-hand symbol $\sqrt{}$ for square root, we have $\sqrt{1} = 1$, $\sqrt{4} = 2$, $\sqrt{9} = 3$, and so on. But if you try and find the square root of a whole number which is not a square number, for example $\sqrt{2}$, then the digits carry on for ever. And this is why NASA could calculate over 10 million of them: the digits never end; the solution is never-ending. Even if you square the first hundred digits of root 2, the answer is still not exactly 2.

1.4142135623730950488016887242
0969807856967187537694807317
6679737990732478462107038850
387534327641572
×
1.4142135623730950488016887242
0969807856967187537694807317
6679737990732478462107038850
387534327641572

=
1.9999999999999999999999999
9999999999999999999999999
9999999999999999999999999
9999999999999999999999999
99792106690025646243817...

I was bored one day and decided to see if I could find numbers where the square root of those numbers started with the same digits. Sure enough, there were some! I then found a whole family of numbers this worked for. I called them grafting numbers, because the root grows out from the number itself. But merely finding them wasn't enough for me. Like a true maths junkie, I wanted an ever bigger high.

Finding and documenting numbers such as this is very much like stamp-collecting. All I was doing was writing these

$$\sqrt{764} = \underline{27.64}054992\ldots$$

$$\sqrt{76394} = \underline{276.394}6454\ldots$$

$$\sqrt{7639321} = \underline{2763.9321}63\ldots$$

numbers down and giving them a name. While this is perfectly valid maths fun, for the real maths rush you need to find a pattern behind something. This is the step from describing something to explaining why it works. With a bit of playing around, I realized where some of these grafting numbers were coming from. They all result from the expression $3 - \sqrt{5}$, each one is an odd number of digits, rounded up to the next whole number. I proudly named it the grafting constant.

GRAFTING CONSTANT:

$$3-\sqrt{5} = 0.763932022500210303590$$
$$8263312687237645593816403\ldots$$

GRAFTING NUMBERS:

$$\lceil (3-\sqrt{5}) \times 10^{2n+1} \rceil$$

764

76394

7639321

763932023

76393202251

7639320225003

763932022500211

76393202250021031

7639320225002103036

Other mathematicians have since done further research into my grafting numbers. For me, they capture those aspects of mathematics that make it an art, when it's not considered purely for its practical applications. The first aspect is hunting down and describing patterns; the second is finding out why those patterns are there and proving they will always work. Finally, all this is done simply for the fun, just for the sheer hell of it.

Four

SHAPE SHIFTING

Now it's time for some fun: get your scissors out. Nothing says it's time to make things with a small chance of slight injury like a pair of scissors. For this chapter, you will need your pair of compasses as well as a pair of scissors (so, one of each). Using some fairly thick but light corrugated cardboard from a cardboard box, cut out two circles 10cm in diameter (aka 5cm in radius). You can either use your compass for this, or cut the templates from this book, place it on the cardboard and cut around it. Once you've got those circles and have established the central point (a non-step if you used the compass, as you can use the hole made by the point) cut a slit in each 29 per cent of the way in towards the central point (for a 5cm radius, that's about 1.5cm). Make the slits just wide enough so the two circles slot together. Place the slotted-together circles on a table and watch as your contraption miraculously rolls away with a strange wobbling action.

It's called the Wobbler, and it only works if the two discs intersect each other by the perfect amount. If they intersect too much, the Wobbler will come to a stop in Position A, as in the diagram, with both discs making a 45° angle to the ground. If they do not intersect enough, it will come to a halt in Position B, with one disc standing straight up, at right angles to the ground. Only with the slits at exactly 29.2893 per cent will the intersection be perfect and the Wobbler keep rolling for ever. Well, obviously, it will slow down

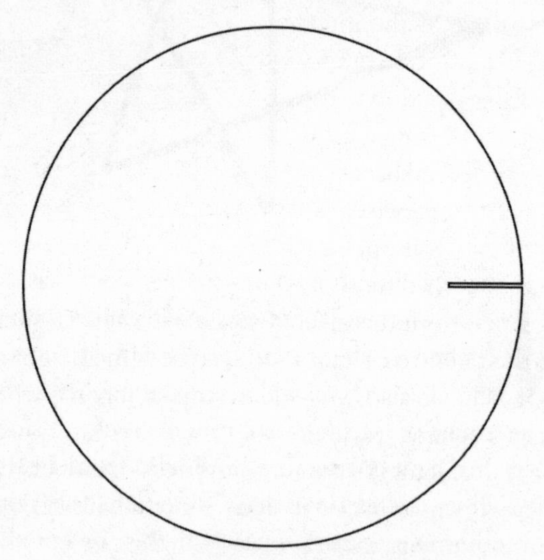

Cut this Wobbler Disc template out twice

eventually because of friction and air resistance (and it's probably nigh on impossible to cut a cardboard slit accurately to several decimal places) but, if it's good enough, blowing on it gently will at least propel your Wobbler across a table or even the room.

Position A

It works because, as the Wobbler moves forward, its centre of gravity (the point in the exact middle of its mass) moves neither up nor down. (It stays as level as a level thing which just won first prize at the international being-level competition.) Objects don't roll uphill because lifting a mass against gravity takes effort and energy. Likewise, an object will not roll on a flat surface if you need to lift its centre of mass as it turns. This is why circles make such great wheels: their centre of mass stays at the same height as they roll along. This is also why squares make terrible wheels. If you bump a square, it will not roll; it will take a serious shove to make it lift its mass enough to tip over one of its corners and plonk down on its next side. A circle will roll on a flat surface; a square has to be constantly driven.

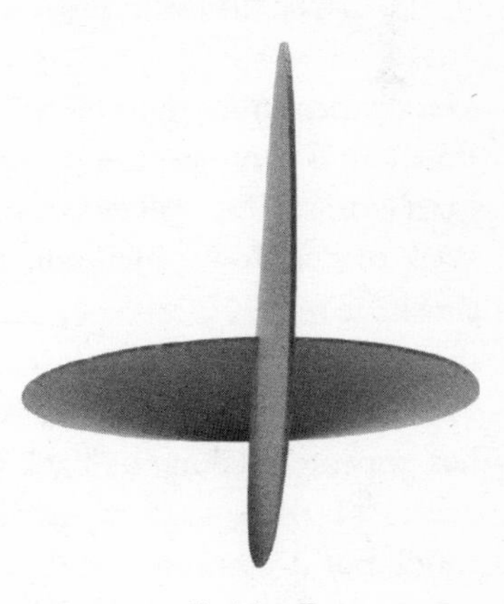

Position B

You can demonstrate that the Wobbler's centre of mass is

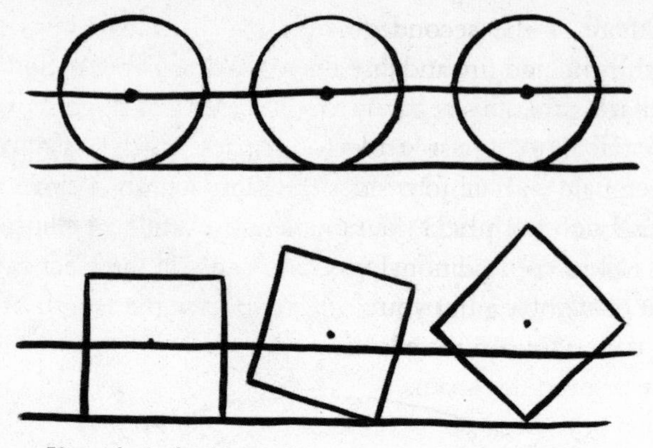

If you look closely enough, you can see the circle is rolling.

exactly the same height off the ground in Position A as it is in Position B with nothing more complicated than a bit of Pythagoras. The full solution is in 'The Answers at the Back of the Book', including the step which shows why the slits have to be 29.2893 per cent of the radius (briefly, so that the separation of the two centres of the circles is exactly $\sqrt{2} \times$ the radius). Once again, that mysterious number $\sqrt{2}$ has appeared, taking us back to our roots. Proving that the centre of mass stays at this height is much more complicated, but it was successfully done by mathematician David Singmaster back in 1990. Even though its height does not change, the centre of mass does swing from side to side, which is why the Wobbler moves forward unevenly, with a kind of lurching action, like a circle which has had one round of drinks too many.

The earliest reference I can find to the Wobbler is a 1966 note in the *American Journal of Physics* by A. T. Stewart on what he calls a Two-circle Roller. Humans have used circles to roll around on for millennia; the invention of the wheel is up there as the second greatest of all human inventions (slightly behind fire and just ahead of sliced bread). So I find it amazing that this new way to roll circles was not discovered until the 1960s. It also works for ellipses. Much as a square is a rectangle with all four sides the same length, a circle is an ellipse with both radii the same length. While an ellipse will not roll in the traditional way, you can still intersect two to form a Wobbler, but you do have to vary the length of the slit, depending on the ellipse.

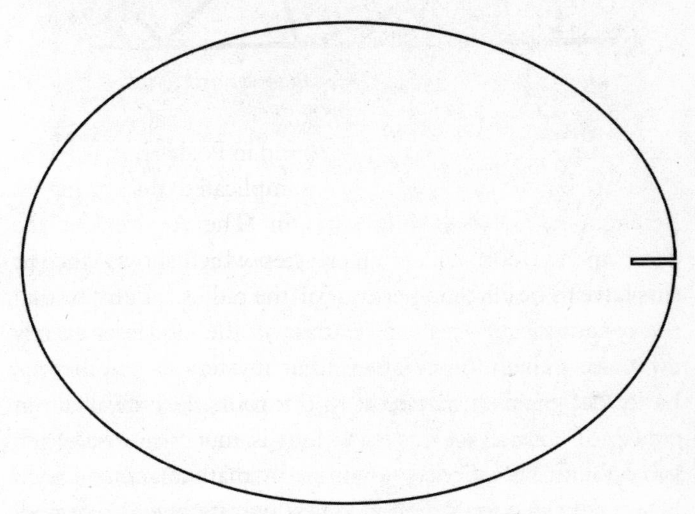

Instead of $\sqrt{2} \times$ *radius, the separation for an ellipse is* $\sqrt{2} \times \sqrt{2(\text{horizontal radius})^2 - (\text{vertical radius})^2}$.

However, we shouldn't be too quick to dismiss the square as a perfectly valid wheel. A square will happily roll, provided it is supplied with a custom-made surface, i.e. a road made of curved bumps so that, as the height of the square's centre of mass goes up and down, it is compensated by the road going

Now that *is the face of a man having a comfortable ride.*

down and up to match the change. For this delicate dance to work, there is only one perfect partner for the square wheel: a very specific curve called a catenary (so named because it is the shape a chain forms when hung between two points; the Latin for 'chain' is *catena*). If you hold a piece of chain at either end you can see this curve for yourself (string is slightly too light to be weighed down into quite the same shape). When the catenary is flipped up the other way and repeated, it forms the perfect road for a square wheel. This isn't only true of square wheels either: almost any shape can roll, given its own strange-bumps road surface.

Shapes of Constant Width

Sitting on my desk in front of me as I type is a triangular coin from Bermuda (you can see what they did there), which is a limited-edition $3 coin made by the Bermuda Mint back in

1998. And limiting it was surely for the best: a triangular coin could never be practical. Historically, coins have usually been circular because they are easier to make that way and have no sharp corners which would be disproportionately subject to wear and tear (and would rip holes in your pocket and purse linings, rather than just burning them). In modern times, having coins of different diameters makes it easy for robots like vending machines to tell them apart. Circular coins can roll down the tracks within a vending machine and, regardless of their orientation, be checked for diameter. A square coin would present all sorts of problems for vending machines because its width (the equivalent of its diameter) changes as it turns.

Unsure how a triangular coin would work, I managed to hunt one down online (for substantially more than three Bermudian dollars) so I could take a closer look at it. Listed on the certificate of authenticity are all the official stats. Apparently, it is made of 20g of 0.925 silver (which made me feel slightly less ripped off), is one of only 6,500 made and has a diameter of 35mm. But how can a triangle have a diameter? Triangles are different distances across, depending on where you measure them. Or are they? I've just taken out a ruler and measured the coin in several directions and it always comes out at 35mm across, seemingly close enough to justify the certificate's claim (and level of precision). This is because the Bermuda $3 coin is no ordinary triangle.

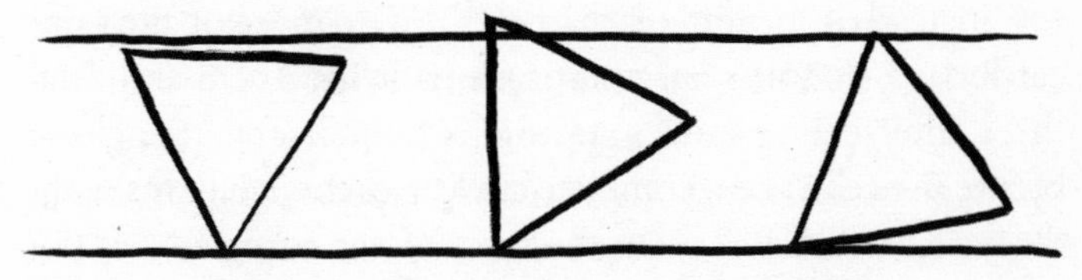

Triangles can pretend to be taller by rotating slightly.

The rounded-off triangle and the Bermuda coin.

It is in fact a shape we've already looked at. Flick back to the diagram in the Making Shapes chapter, to the bit where a series of circles were used to draw an equilateral triangle. The shape of this coin is sitting there behind it, hidden in the oft-overlooked construction lines. The three corners are joined together by straight lines to form a triangle, but they are also joined together by three circular arcs, and it is these lines that make the rounded-off equilateral on which the $3 coin is based. It's the shape that lies behind our solution to the pizza problem.

As well as being a great way to slice up a pizza, this three-sided shape also has the remarkable property of being the same height any which way it's positioned. Which means it can roll. If you take the two circles of cardboard from the Wobbler and place them both flat on a table, they will roll between two parallel rulers because circles have a constant width (for some reason, 'width' is used to describe this property instead of 'height' or 'diameter'). With more of the same cardboard, double your compass's radius to 10cm and draw three intersecting circles around an equilateral triangle as before (it needn't be a complete circle, just enough to make the three arcs which will form the sides of the triangle). Cut this shape out, then repeat so you have a second identical shape. If you place them between the two parallel rulers they will happily roll around. You can swap them in and out with the circles and the rulers will always remain exactly 10cm apart.

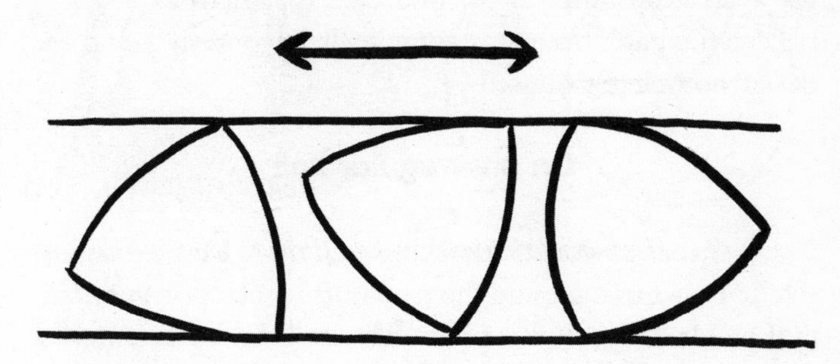

If all of this sounds like too much effort, find some twenty-pence or fifty-pence coins. Both are regular heptagons, but all the sides have been rounded off. You can make this shape out of cardboard, sticking a compass in each corner and using it to draw an arc around the opposite side. This works for any odd-sided regular shape. Your minted fifty-pence coin and the one you have crafted with a compass and some cardboard will each roll perfectly between two parallel rulers, and therefore your odd-sided heptagonal fifty- and twenty-pence pieces will still work perfectly in vending machines, despite not being circular. That said, Australia has had twelve-sided coins with flat edges since 1969, and the UK is introducing a flat-sided coin in 2017, and vending machines can handle a decent amount of variation in width, so the rounded-off sides are slight overkill.

The rounded-off equilateral triangle has so many amazing properties it has earned a special name: the Reuleaux triangle (after nineteenth-century German engineer Franz Reuleaux). Beyond coinage, you can use this shape to make a drill-bit that can drill nearly square holes. This is actually less useful than it sounds, because of the scarcity of square pegs and the accepted incompatibility between round pegs and square holes. However, a patent was granted in the US in 1978 to use a Reuleaux drill-bit to drill coal (patent #4074778). I believe the patent remains completely uncontested, and the design completely unused.

On an Irregular Roll

These shapes of constant width get far more exotic than just the Reuleaux triangle and the rest of its regular polygon family. Our challenge now is to see how irregular we can make a shape that still will have exactly the same width regardless of how you turn it. Any such shapes should still happily roll between two rulers, exactly like a circle would.

Round 1: Different angles

The first challenge is to make shapes with an odd number of sides all of the same length, but with different angles. You can draw semi-regular pentagrams fairly easily. For these sorts of constructions it's easier to work with the stellated version of the shape: in this case, a pentagram instead of a pentagon. So, choose your side length (again, 10cm is a good size and will make the shape backwards-compatible with your earlier creations), then draw your first three sides wherever you want, as long as the third one crosses back over the first. The free-styling is now over; the last two sides can meet in only one location in order to complete the pentagram with 10cm sides and link back to where you started.

Turning this deformed pentagram into a shape of constant width is as easy as it was with the Reuleaux triangle.

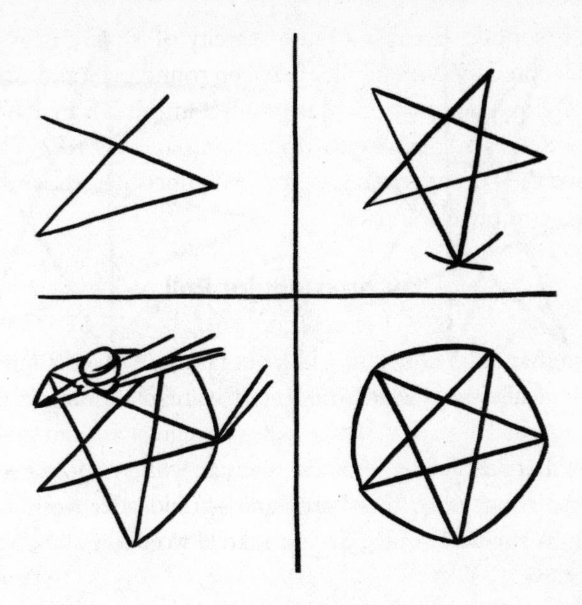

Stab the compass into each point of the star, and draw an arc for the opposite edge. Because all sides are the same length, the arc will line up neatly with the corners. Whatever strange and distorted shape you end up with, the one positive thing you can say about it is that it will have constant width. You can cut it out and test it with any other 10cm-wide shapes, or make two different ones, if you really want to push the boundaries.

Should you want not only to push the boundaries but to go over the wire, the same method works for any odd-sided shape, so you can try a seven-sided or nine-sided distorted stellated shape and round it off. The difficulty here is making sure the corners are in the right order. If they aren't, the arcs will intersect the sides and the shape won't work. Once you've mastered the right order, though, there's

Only one of these heptagons will work. Guess which.

no limit to how many sides you can add. Unfortunately, beyond about nine, the shapes start to look like nothing so much as slightly wonky circles. Like a wheel on a shopping trolley.

Round 2: Edges of different lengths

OK, now we're ready to level up and make shapes of constant width in which the odd number of sides are all of different lengths. This is more complicated, because the arced edges will no longer line up with the corners, but the problem can be solved by extending the original sides to be as long as required and patching over the gaps. A triangle with sides of 9cm, 6cm and 5cm is a good starting shape to try. If you follow the instructions carefully, you'll end up with a new deformed shape of constant width that is also always 10cm across.

If you try constructing a few of these, you'll see that the distance across these triangular shapes of constant width is always equal to the sum of the two longest sides minus the

Construct an Irregular Shape of Constant Width

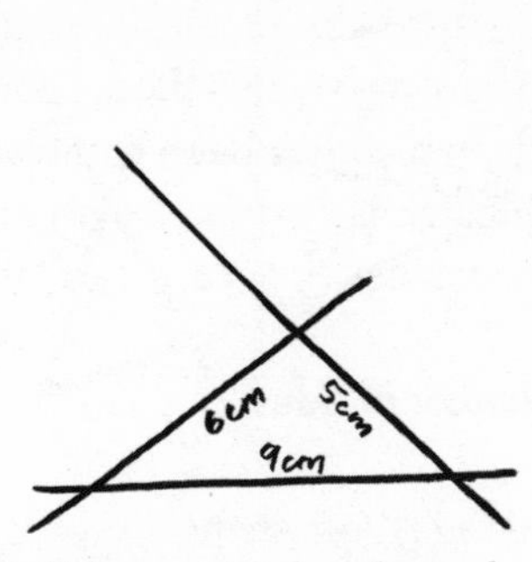

1. Take any triangle and extend the sides.

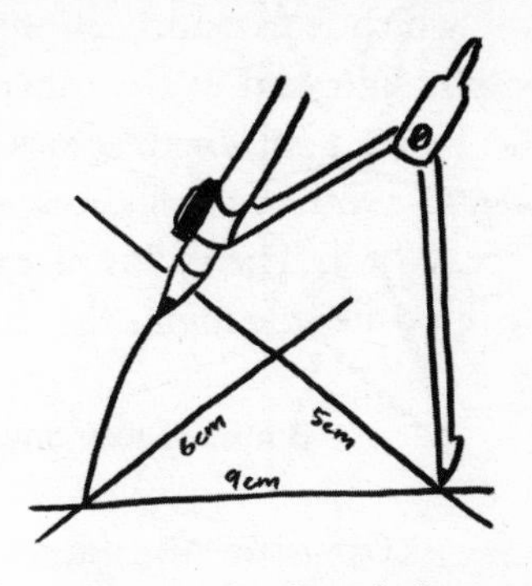

2. Draw an arc from the longest side around to the shortest.

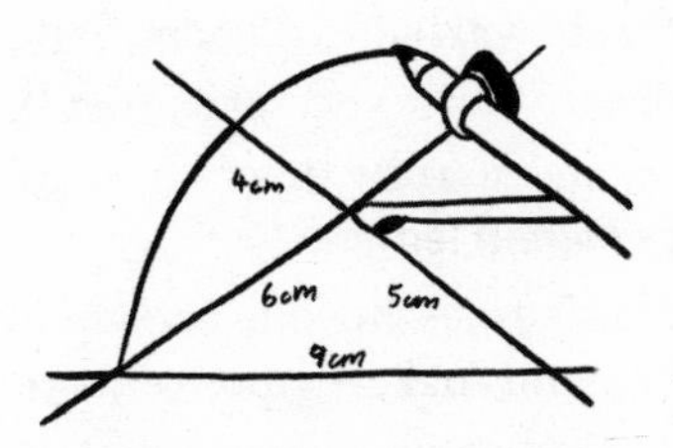

3. Move your compass to the near corner and draw a 'patching arc' around to the next side.

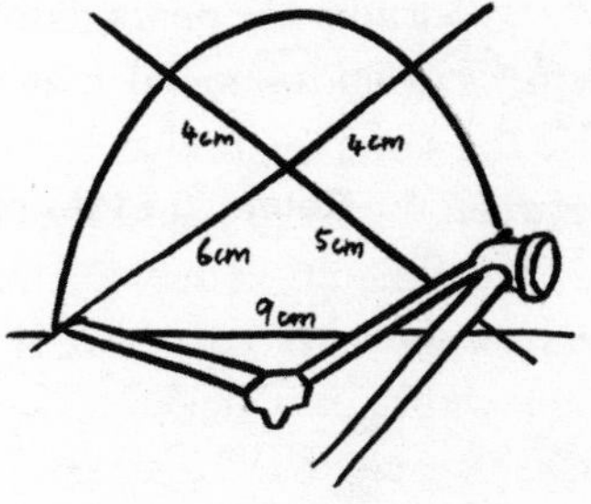

4. Bring your compass back out to draw the next arced edge.

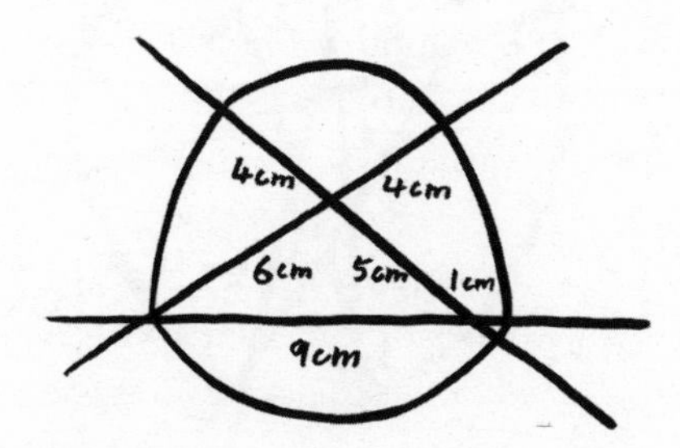

5. Continue around the triangle until you hit the starting point.

shortest. This is most obvious if you look at the shortest side, which is extended in one direction to be equal to the longest side, and in the other to equal the second-longest side. So the total length across is the longest, plus the second longest, subtract the shortest side, which has been counted twice. Got it? The radius of each patch is also the length of the opposite side minus the shortest side.

Bonus round: any number of sides

Now it's time for the premier league of shape drawing: shapes of constant width with *any* number of sides, odd or even. It is possible to extend the previous method to work for any shape with an odd number of different-length sides. My favourite is the pentagram made with sides of 8cm, 7cm, 6cm, 5cm and 4cm (in that order: that's very important!). Use the same method as before, and start by drawing an arc between the longest and shortest sides.

One difficulty in drawing these shapes is making sure there are no bits with 'curve in'. If a shape has edges which curl

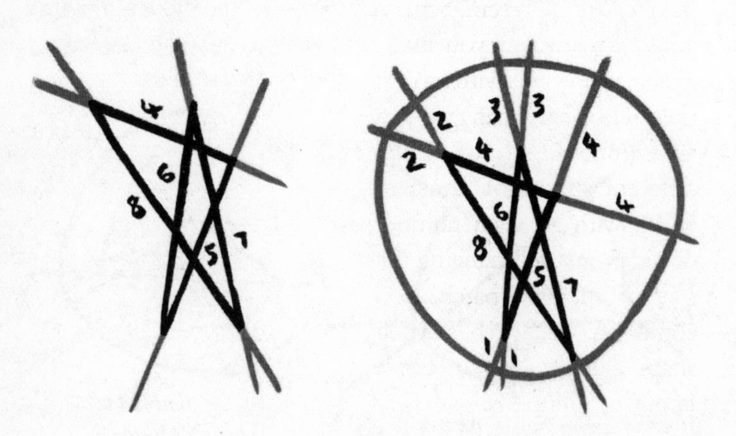

back inside, it's become a concave shape (easy to remember, because the edges bend inwards to form a small cave). Concave shapes do not make rollable shapes of constant width, as some of the edges cannot reach the bottom. Once you master drawing convex shapes (aka not concave), you can scale up to as many odd sides and different lengths as you want.

The real challenge here is how to get shapes of constant width with an *even* number of edges. All the odd-sided polygons also have an odd number of corners, so when you turn one into a shape of constant width you get an arced edge for every side, then a patch for all the corners but one. This always ends up being an odd number of edges in total, because of the one corner which doesn't have a patch. We can move things up to make an even-sided shape of constant width by adding one more patch – which is simply done by making all the patches slightly bigger.

The 9cm × 6cm × 5cm triangle earlier had patches of radius 0cm, 4cm and 1cm. You can increase all these by the same amount to create a new shape. An easy option is to increase the radius of your compass by 1cm, so that the patches are now 1cm, 5cm and 2cm. Or – go wild! – increase it by any amount you like: the options are endless. Every irregular polygon with an odd number of sides can give you an infinite supply of subtly different shapes of constant width with an even number of sides, just by changing the size of the first patch. You could say that each starting shape can produce an entire family of solutions.

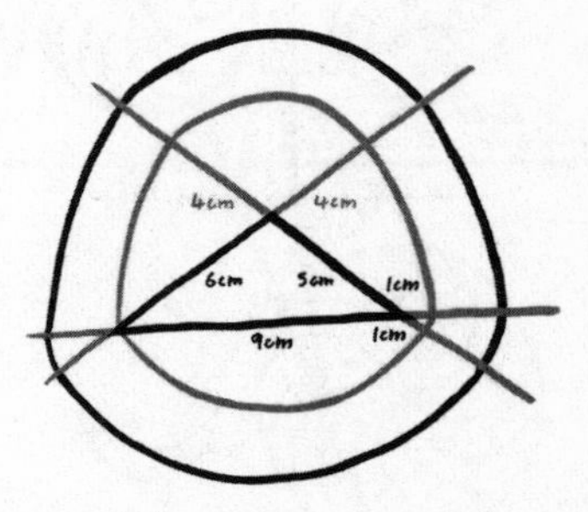

A shape of constant width with six edges.

Pizza of Constant Width

These exotic families of irregular shapes of constant width cannot help us with the pizza problem, but the regular versions can. Every regular shape of constant width with an odd number of sides produces its own family of solutions to the pizza problem. To try one, get a fifty-pence coin and draw around four of the sides on a piece of paper. Then move the coin around so that you can draw around three sides, starting and finishing at the same points as the earlier four (it will fit perfectly), then overshoot for a fourth side. (See diagram below.) If you repeat this again and again, drawing fourteen lines in total, you will have drawn a complete circle: a perfect circle divided into fourteen identical grin-shaped wedges, all of which can be cut in half so only one side touches the middle.

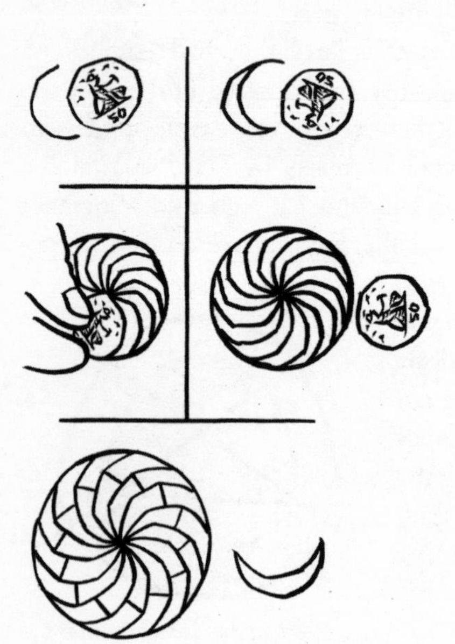

Let's take a closer look at the symmetrical shape into which the pizza is divided. It's a bit like a heptagon, but three of the sides have been flipped inside to make a grin shape. This is known as the 'heptagrin'. Honest. For any regular polygon with an odd number of sides you can flip just under half the edges inside to make the corresponding polygrin. Our original solution was based on the 'trigrin' both cut in half with a straight line and with a curved line. Those curved lines can be used to cut each piece into more than two pieces. These grin-shaped slices of pizza can also be divided into thirds, quarters, fifths, and so on, infinitely, using the same curve method.

To recap, every regular shape of constant width with an odd number of sides gives you a solution to the pizza problem, which means there are infinitely many ways to solve it, as you can always have another shape with two more sides. And beyond that, every different solution can have its pieces cut up into different numbers of pieces: two, three, four . . . and so on, infinitely. So each of the infinite possible solutions has its own infinite family of sub-solutions. That's a lot

of ways to slice a pizza! In the table opposite you can see where there are two different ways to get forty-two pieces (take your pick: heptagrins cut into thirds, or trigrins cut into sevenths).

With a bit of effort, you can see why there are only five ways to get 180 absolutely identical slices but ten ways to get 630 slices where not all of them touch the centre. You can also try to find ways to cut a pizza into identical pieces such that all but one of them touches the centre (useful if you want to maximize the number of pieces which have some of whatever the central topping is). The opposite cannot be done: you cannot cut a pizza so that the entire centre ends up on just a single piece. We think.

Flexagons

There is one last way that shapes can move and roll, and it's a truly strange one. These bizarre shapes do not roll around on the ground as such; rather, they roll into themselves, exposing new faces. For example, it's possible to make a square which rolls through six different faces. What are we waiting for? Start with a reasonably large square piece of paper. Divide it up into sixteen smaller squares, cut out the central four and follow the folding instructions on the next page carefully. What you are left with is known as a tetraflexagon.

To keep track of which face is which, put a figure 1 in all four squares on the front and a figure 2 in the squares on the back. (Once you get the hang of flexagons, you'll find yourself susceptible to far more creative labelling of these faces . . .) When you fold the square in half, you'll see that there are two ways to open it back up. One way goes back to where you started; the other gives you a new face. Label these four new squares with the figure 3. And open and close again.

Folding a tetraflexagon

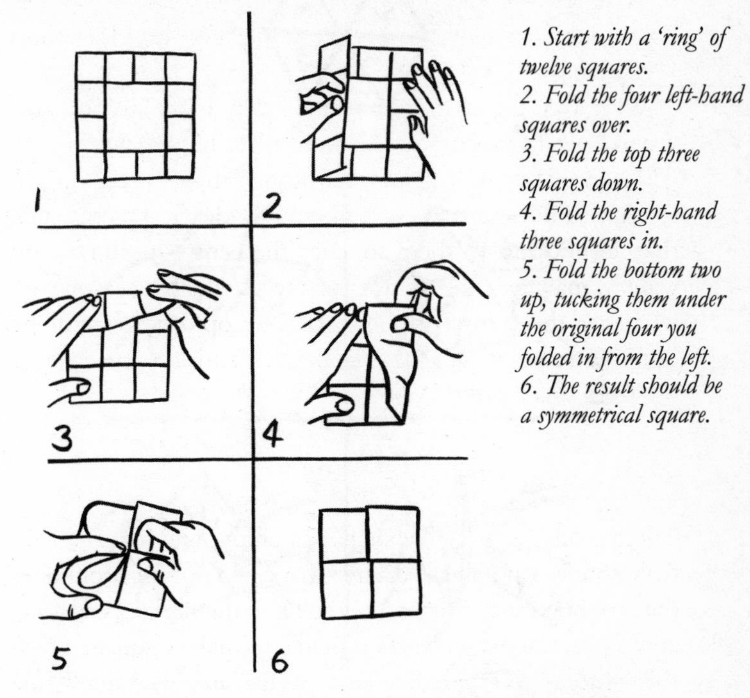

1. Start with a 'ring' of twelve squares.
2. Fold the four left-hand squares over.
3. Fold the top three squares down.
4. Fold the right-hand three squares in.
5. Fold the bottom two up, tucking them under the original four you folded in from the left.
6. The result should be a symmetrical square.

Flexing a tetraflexagon

1. Fold the tetraflexagon over in half, hiding the first face.

2. Fold it back open from the other side to expose a new face.

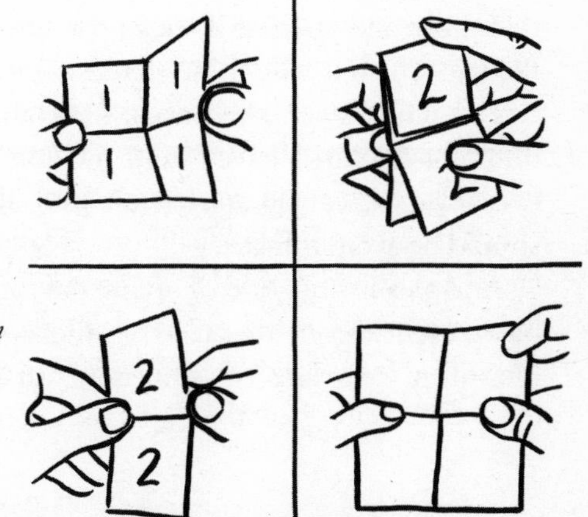

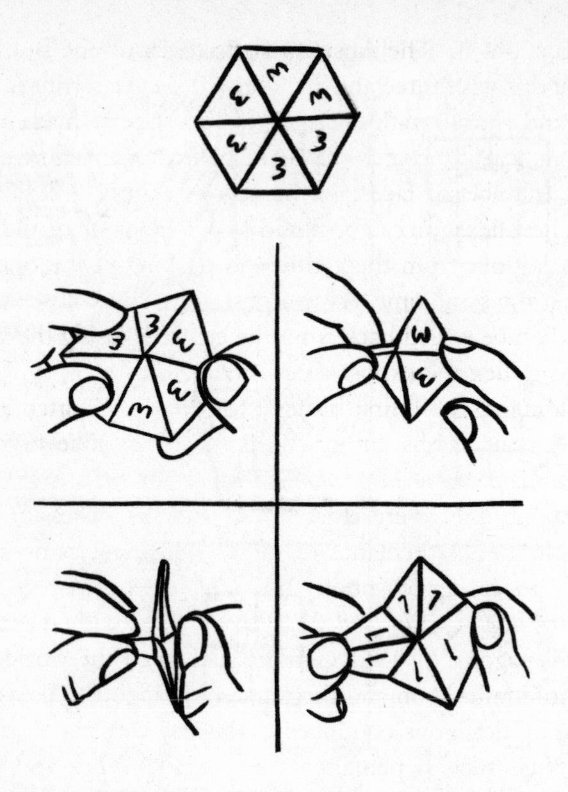

As you go, different faces will appear and disappear, like some sort of over-engineered origami fortune-teller. Folding this shape and opening it back up a different way is known as flexing, which is why these shapes are called flexagons. This one's a tetraflexagon because it has four edges. Keep flexing until you think you have found and labelled all the faces (you can always cheat by completely unfolding it again). There should be six in total.

And this is only one of many flexagon shapes. The ones based on hexagons are called hexaflexagons and can be made out of a long line of equilateral triangles. I've included instructions in 'The Answers at the Back of the Book' for flexagons with three and six faces, so they're a trihexaflexagon and a hexahexaflexagon. The square one we made before is specifically a hexatetraflexagon. (The nomenclature always goes {number of faces} – {number of edges} – flexagon.)

A hexaflexagon can be flexed by pinching one of the folds radiating out from the centre and pushing in the opposite one at the same time. As you push it in, you will see that it can then be pulled back from the centre, causing the whole shape to flex and expose a new side.

Flexagons were first brought to the public's attention by the famous maths author Martin Gardner. The first ever column he wrote in *Scientific American*, in 1956, was about flexagons; they were what introduced his obsession with recreational mathematics to the world. I first came across them in his 1959 book *The Scientific American Book of Mathematical Puzzles and Diversions*. Along with describing hexaflexagons, by 1961 Gardner had given the world both the tritetraflexagon and the tetratetraflexagon. This exploration of flexagons continues to this day with the discovery of things such as pentaflexagons, octaflexagons and dodecaflexagons. All of which move in strange and unexpected ways. At a recent maths conference someone excitedly showed me a decadecaflexagon. That alone made the whole trip worthwhile.

They're not that old either. The trihexaflexagon was the first flexagon discovered, and that was only back in 1939. The English mathematician Arthur Stone (a colleague of Paul Erdős) was working in the USA and had to cut strips off American paper so it would fit into his English foolscap binder. While absent-mindedly folding some of these strips of paper into equilateral triangles – hey! before he knew it, he had made a trihexaflexagon and, before long, several of his friends (including Nobel Prize-winning physicist Richard Feynman) had joined him to form the Princeton Flexagon Committee. One of their first discoveries was the Tuckerman traverse (named after committee member Bryant Tuckerman), in which, if you repeatedly flex the same place on a hexaflexagon until you can flex no more, then move to an adjacent corner, you will definitely work your way through all the possible six faces without fail. (If you flex about randomly, some of them can be quite hard to find.)

Your challenge is slightly more complicated. On the hexatetraflexagon it's possible to find all six faces easily enough. There are the front and back faces you see to start with (which I often label F and B for convenience). You can flex each face either horizontally or vertically, and these four flexings produce all four of the hidden faces. (These can be systematically labelled F_V, F_H, B_V and B_H.) What you need to do now is to see which of these six faces you can have exposed at the same time. You can obviously begin with F and B at the same time, but can you get F_V on one side with B_H on the other? Some pairs can be placed opposite each other, while others cannot. (More in 'The Answer at the Back of the Book'.)

This is where the creative labelling comes in. Mark the faces with a combination of colleagues' names and office chores; the combinations they come up in look elaborate and random, but you can control them. 'Sorry, Mavis, I have no idea why you're making the tea again, but the flexagon has spoken.' Or prepare a bespoke flexagon for the office Christmas party and wow that special someone from Accounts Receivable.

As for me, I made a hexahexaflexagon beer coaster,

with different drinks on the faces and a record of whose round it is. The hexatetraflexagon can also be generalized to work for any rectangle (aka squaroid) should you require an overly elaborate business card with three times the surface area for no increase in size. I know I do. Thankfully, I have a second business card to explain how to use the first one.

Five

SHAPES: NOW IN 3D

I told you that, for me, circles are one of the best shapes in mathematics, so I've drawn you another one. This one's exactly the same size as a five-pence coin. Please cut it out of the book (or copy it on to a piece of paper, if your copy of this book has already suffered enough), leaving as perfectly circular a hole as possible. Your challenge now is to fit a two-pence coin through that hole without ripping the paper. Yes, a standard UK two-pence piece coin is 25.9mm across, making it substantially bigger than the 18mm-diameter five-pence hole you have cut out, so it is entirely reasonable to expect that you cannot fit this larger coin through this smaller hole without ripping the piece of paper – but it can be done. I'll leave you to struggle alone with that for a while. (Top tip: there's a substantial hint in the fact that this is flagged as the chapter on 3D shapes.)

Five-pence piece, to scale.

I cannot remember who it was who first showed me the two-pence-through-five-pence puzzle: that knowledge is lost deep in the deficiencies of my memory. Apart from the very few bits of mathematics I can claim to have discovered myself, everything else I have, by default, learned from someone else. Which makes mathematics a very sociable, sharing kinda subject. While much of this interaction may be with long-dead mathematicians via the printed page, the truly surprising bits tend to come from living people. From this coin trick to cutting cuboid cakes, the amount of mathematics I learned in my maths degree is probably rivalled by the amount that's been passed on by people who've wanted to 'show me something interesting'.

The loophole you can fit the two-pence piece through is the third dimension. Instead of leaving the piece of paper flat, you can curve it into a 3D shape and the coin will easily pass through. Once it has, you can flatten the piece of paper back into 2D. To put it very simply, we consider a piece of paper to be two-dimensional because it is completely flat: you can only move left–right and backwards–forwards. For it to remain 2D, no part of a flat piece of paper must lift up off that flat surface: moving up–down is the third dimension. And strange things start to happen when you do mathematics on a surface which has been curved into the third dimension. Larger objects pass through the eye of smaller objects, and straight lines bend.

Big coin through a small hole

1. Fold the paper in half so that the hole looks like a semicircle from the side.

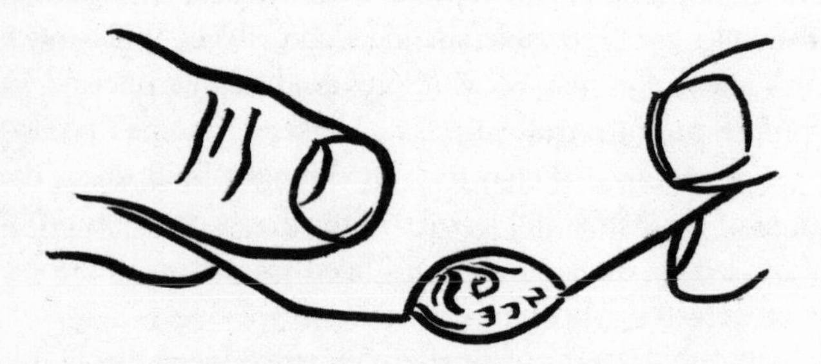

2. Grasp each of the folded corners, one in each hand, and move them in together so that the semicircular hole is stretched out.

3. The two-pence piece will drop easily through the hole.

A few years ago I was on a road trip in the USA with some friends, driving down a fantastically straight Nevada highway. Given that the desert is almost completely obstacle free, there is no reason why the roads can't be completely straight, taking the shortest path possible with nothing to go around. Suddenly, though, the Nevada highway would bend to the right, then back to the left, resuming its dead-straight path as if nothing had ever happened. It was as if the road were avoiding obstacles invisible to the non-highway eye.

No doubt like many road trippers before us, we tried to guess why the highway kept swerving like this – had there been a tree or a building there to avoid when it was constructed? Then it dawned on me: the road planners would have been working from flat maps and, in the real world, those straight lines had to cope with the fact that the surface of the Earth is not flat but a curved sphere. Straight lines act differently on a sphere than on a flat page; those kinks in the road represent the difference between flat and spherical geometry, and the dog-leg corners are a way of patching the problem. My friends told me to save my maths for Vegas.

You can re-create the geometry of a spherical curved surface by drawing on a balloon. Although not strictly spherical, balloons are readily available and, if you don't over-inflate them, they're sufficiently spherical for our purposes. So, get a balloon, a felt-tip and a piece of string. Draw a line all around the very centre of the balloon to represent its equator. Now draw a line parallel to that line. Now this is kinda tricky. Start by marking two dots the same distance above the equator but about a third of the distance around the balloon from each other. In mathematics we define a straight line as the shortest distance between two points. To find this shortest distance, use a piece of string to join the two dots and pull it as taut as you can, without distorting the balloon. However you do it, the string will not lie parallel to the equator. It will curve up and away from it and then back down. If this is not evident on your balloon, a ball (larger, firmer) may do the trick.

To move beyond our balloon, the lines of latitude marked on the Earth are *not* straight lines. You'll have seen this effect before. For example, when you look at flight paths around the globe, the aeroplanes seem to follow curved paths instead of taking the more obvious straight distance between two points. This is because they are following the shortest distance between two points *on a sphere*. Straight lines on a sphere do not look straight to our eyes, because they are not straight from an external point of view; they curve with the surface. But, if you're confined within that surface, they *are* the shortest paths. My theory is that straight lines on the Nevada road planners' map didn't translate into straight lines on a sphere.

This curved line is the shortest distance between two points.

We grow up assuming that lines can be parallel to each other, and railway lines, for example, fool us into thinking that it's possible, but it turns out to be much more complicated. Euclid himself discovered this when writing *The Elements*. I said that Euclid made two starting assumptions:

that you can draw lines with a ruler and circles with a compass. In fact, he phrased this as four different postulates. That was what he called them anyway; today's mathematicians call their starting assumptions axioms. Whatever you call them, these are things so simple and self-evident that we can safely assume they are true without requiring any further proof. Euclid's postulates/axioms were, in his order but my words:

- Any two dots can be joined by a straight line.
- Any straight line can be extended to be as long as you want.
- You can draw a circle of any size you fancy, wherever you want.
- All right angles are the same (that is, a quarter of a full revolution).

When writing *The Elements*, Euclid set about proving the existence of parallel lines using these four axioms. Only, he didn't succeed. And it wasn't only him having a problem. Try as they could, none of the ancient mathematicians could prove that parallel lines exist using only Euclid's axioms. It is a giant leap missing in *The Elements*.

Euclid's solution was simply to take parallel lines for granted and add them in as a fifth postulate. But even he must have thought that was a bit of a cheat. The truth of that statement isn't self-evident like his first four axioms – the existence of parallel lines was a more complex and nuanced concept and feels like it needs proving. In *The Elements*, the first four postulates are stated in simple and obvious statements, together amounting to only thirty-four words. The fifth postulate alone requires thirty-five words. This was the ancient Greek equivalent of 'citation needed'.

It's a massive 'get out of Euclid free' card. As we know,

maths is a game in which you choose your rules and then play within them. It turns out that Euclid had not chosen the *only* set of rules possible . . . it just took mathematicians two millennia to come up with some equally good alternatives. In 1823, János Bolyai and Nicolai Lobachevsky each – separately – tried to use the same initial four axioms as Euclid but leave out the fifth one about the parallel lines. To their surprise, geometry continued to work just fine without it. Only now, it was strangely different.

As they continued through the same proofs as Euclid, they realized they were producing subtly different results. Even things like triangles were behaving differently. In the first book of *The Elements* Euclid proved some standard facts about triangles, including that old chestnut about the sum of all the angles in a triangle adding up to 180°. However, in order to do so, he used the postulate assuming that parallel lines exist. If you try to draw triangles on a surface where there are no parallel lines, then his proof no long holds: the angles in a triangle do not have to add up to 180°. Through the simple act of drawing on a balloon, you are playing within different mathematical rules.

On a sphere, if a triangle is small enough, then it looks pretty normal: the sum of its angles will still be around 180°. But, if you draw a bigger triangle, the sum of its angles gets bigger too. Let's prove it. Balloon and felt-tip at the ready. Start with your pen at a point either at the top or the bottom of the balloon, draw a line all the way around a quarter of the balloon's surface so you hit the equator, turn 90° in one direction and proceed along the equator another quarter of the balloon's circumference, turn 90° again and go straight back up to the pole, making, magically, another 90° with the line descending from where you started. This is a truly right-angled triangle, one with *three* 90° corners . . . adding to 270°

(*not* 180° – here's lookin' at Euclid!). The sum of the angles in any spherical triangle will be somewhere between 180° and 540°, depending on how much of the surface area of the sphere they cover.

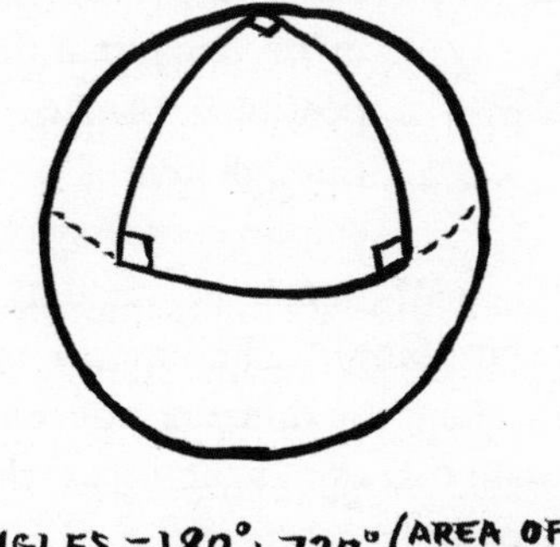

A triangle with angles which add to 270°.

You can also create a whole new shape on a sphere, one with even fewer sides than a triangle. If you have two dots on a flat piece of paper, there is only one way to join them with a straight line, and even though I'm a big fan of one-liners, a one-line shape does come up a little short. On a sphere, however, there are two different straight lines which can unite the same pair of dots. Get that balloon, that felt-tip and that piece of string. Put two dots on your balloon exactly opposite each other and try to link them together with the string. You'll find it does not matter which direction the string goes, it is always the shortest path. Any two of these straight lines form a shape called a lune. (The two-sided lune does not exist in normal Euclidean geometry, which is starting to look rather flat and dreary.)

There are actually two different non-Euclidean geometries, and they each match one of the two different ways you can ignore the fifth postulate. The fifth postulate insists that if you

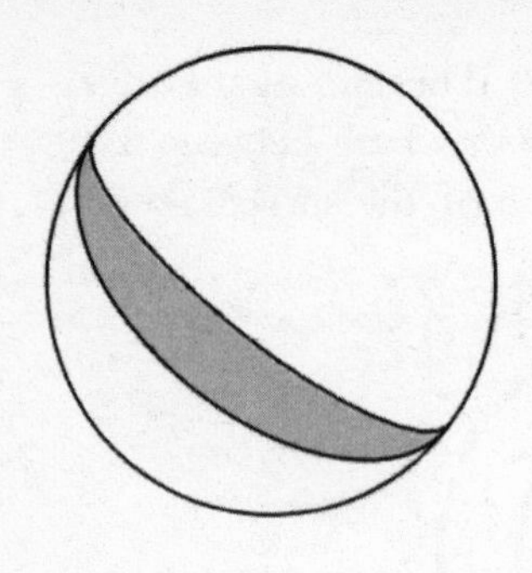

The two-sided lune.

have a line with a dot nearby, there is one way, and one way only, to draw a line parallel to the first going through the dot. The first way you can disagree with this is by declaring that there can be no possible parallel lines going through that dot. This gives us elliptic geometry (which is pretty much the same as the spherical geometry we've been doing).* The other option is to argue that there are more than one different possible parallel lines which can go through that dot . . . and that takes us into the bizarre world of hyperbolic geometry.

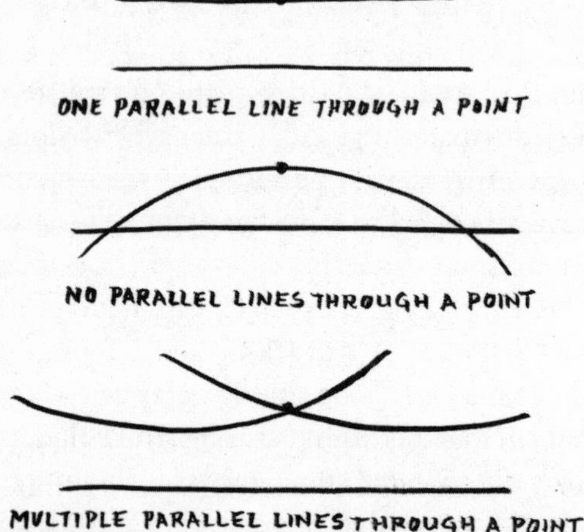

The three options for parallel lines through a point.

* The difference between the two is that elliptic geometry retains Euclid's second postulate that lines can be as long as you want. This is not true on a sphere: eventually, a line will go all the way around, and hit where it started.

Hyperbolic geometry is a real curve ball. The rule-of-thumb difference is that, as you move around on a spherical surface, you get less space than you expected; on a hyperbolic surface, you get more. If you've ever tried to gift-wrap a ball, you'll have seen this effect: there isn't enough room on the surface of the ball for all the paper, so you end up crumpling it all up in ugly lumps. Conversely, if you were to gift-wrap a hyperbolic surface, the paper would stretch and rip, as the hyperbolic surface expands as you move across it. Mercifully, it is possible to make a hyperbolic surface by crocheting shapes from wool, so you can re-create this effect by crocheting a circle but continually adding extra loops as you go. (Or find someone who can crochet and get them to make it for you in exchange for love/money/their crochet hook back.) This produces a hyperbolic surface which has to bow up and down in order to fit all the extra area in.

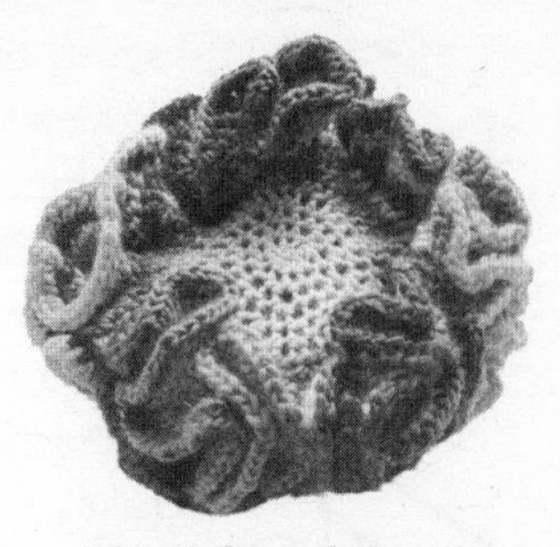

Hyperbolic crocheting: the surface expands as it goes out.

Cubism

Is it possible to fit an object through itself? Well, not literally through *itself*, but through an exact copy of itself. If you had two identical shapes, could you make a hole big enough in one to fit the other through it? In the seventeenth century, Prince Rupert wagered that it was possible to fit a cube through a hole in a cube of the same size. And he was right! It is even possible to fit a cube through a hole in a smaller cube. Prince Rupert was so right that a cube with such a hole

Ways to slice a cube.

sliced through it is now called a Prince Rupert's Cube. (Unlike my unsuccessful attempts to christen 90,525,801,730 the Matt Parker Number.) See if you can guess what Rupe's Cube looks like, or even try to make such a hole in a cube yourself. It's a perfect use for surplus cube cakes.

The Prince Rupert's Cube Trick uses, unexpectedly, the cross section of a cube. If you cut a cube perfectly in half, you can either give the new surface of each half the traditional boring square (cut from the middle of one edge straight across), the halfway more interesting rectangle (cut diagonally from one corner to another), or the completely interesting hexagon . . . Go on: try it at home. Failing left over cube cake, buy the most cuboid loaf of bread and take up your bread-knife. Cut a cube off the end. If you can make the right sloping cut which goes through all six faces, the exposed surface will be an exact hexagon. (You can produce the same shape and without getting covered in a thin layer of breadcrumbs by taking any old, not necessarily edible, cube and holding it against a light source in a certain way: the shadow cast will be a hexagon.)

This hexagon has a surface area bigger than that of the square-faced cube, so, if

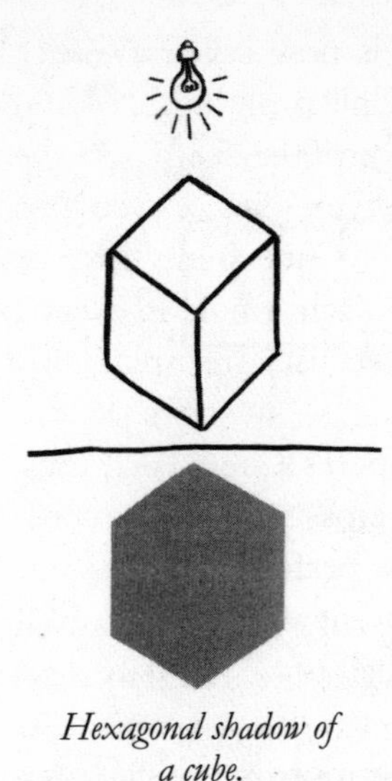

Hexagonal shadow of a cube.

you cut a square hole straight through the centre of the hexagon cross section, the cube will fit through it. There's even a bit of leeway: the hexagon is large enough to fit a cube 3.5 per cent bigger than the original cube through. And if you tweak where the square hole is and move it so it's at a slight angle to the hexagon, you can fit a 6 per cent bigger cube through. When making a Prince Rupert's Cube, this spare capacity is normally used to reinforce the edges of the hexagon hole/cube (particularly if the cube is big, it may easily snap and break). If you want your own Prince Rupert's Cube, it can fairly easily be assembled from cardboard.

The cube is a great 3D shape, and it's so great partly because it's so regular. In 2D, if you remember, we had our regular polygons which had all angles and edges the same, and now we can extend this to 3D. The 3D equivalent of a polygon is a polyhedron, and it's made by joining polygons

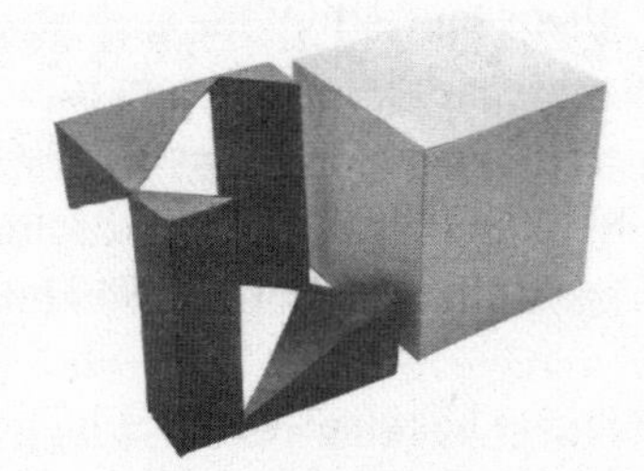

The hole through Prince Rupert's Cube.

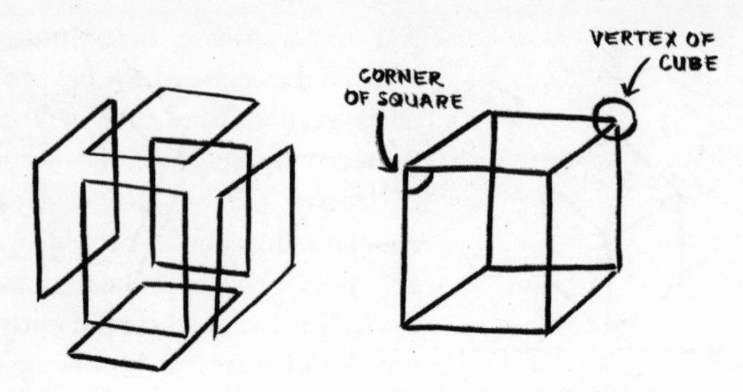

Six squares maketh the cube.

together in the third dimension. A cube is a polyhedron made by joining six square polygons together. A tetrahedron is a polyhedron made from four triangle polygons. Whereas polygons have only corners and edges, a polyhedron also has vertices where the polygon corners meet. A regular polyhedron is one made only from identical regular polygons and in which all the vertices are also exactly the same.

Regular polyhedrons were described by the Greek philosopher Plato around 350BCE and so are commonly referred to as Platonic solids (3D shapes are often called solids, and I've just told you where the 'Platonic' bit comes from). For unknown reasons, Platonic solids are such lovely and pleasing shapes that humans across time have granted them almost mystical significance. Plato himself thought that the indivisible 'atoms' of nature must be Platonic-solid shaped. The German astronomer and mathematician Johannes Kepler said he had an epiphany on 19 July 1595 that the ratios of the planets' orbits must be based on the Platonic solids. There have even been regular solids unearthed in Scotland which date back to the Neolithic period, severely pre-dating Plato.

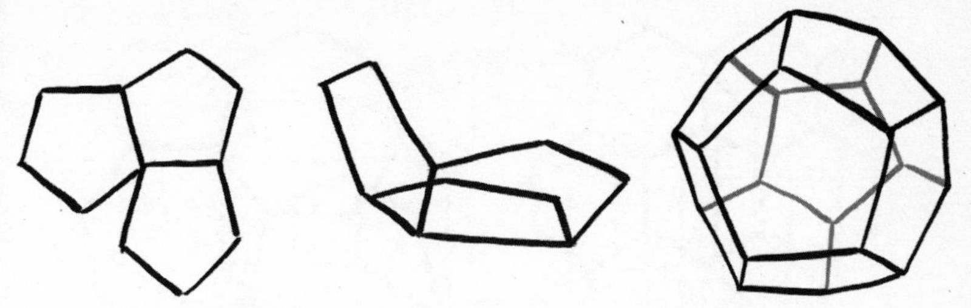

Three pentagons make the 3D vertex of a dodecahedron.

We had better make some of these amazing solids. The cube is easy enough: cut six squares out of cardboard and tape them together. Four cardboard triangles will quickly give you a tetrahedron. For the dodecahedron, start by cutting twelve pentagons out of fairly thick card, then stick three of them together to make a corner. You'll see that, when they lie flat, these three pentagons together leave a small gap. Closing this gap causes the pentagons to bow out and form a 3D vertex. Join twelve pentagons, three to each vertex, and you will have your very own dodecahedron. At last, you will not have to use the communal one.

Sadly, this method does not work with hexagons; three regular hexagons fit together perfectly with no gap at all. Joining loads of cardboard hexagons together will just cover a flat surface in hexagon-shaped tiles. Heptagons onwards, it gets even worse: none of these shapes fit together in threes so you can't even use them to tile a 2D surface. There are no regular polygons beyond the pentagon that can be used to make regular polyhedrons. There are, however, two other 3D shapes you can make with equilateral triangles.

Instead of joining three triangles at each vertex, you can join four, and there will be a 120° gap. Putting them together forms a slightly less pointy vertex than in the tetrahedron.

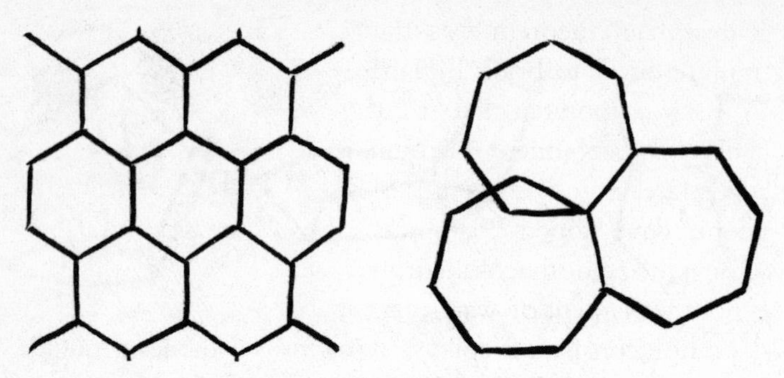

Hexagons fit together too well; heptagons don't fit at all.

Keep going, with four triangles at each vertex and, eight triangles later, you'll have an octahedron. Going up to five triangles at each vertex will leave a 60° gap, and an incredible twenty triangles will join in this way to give you an icosahedron. A sixth triangle on a vertex fits perfectly, leaving no gap, so that's the end of the road for Platonic solids.

That only five Platonic solids exist is one of the most famous proofs in mathematics. Euclid himself used his very last proof (Book 13, Proposition 18) to demonstrate this. It is the pinnacle of *The Elements*. And it's not as if it didn't have some exciting moments along the way. The proof of

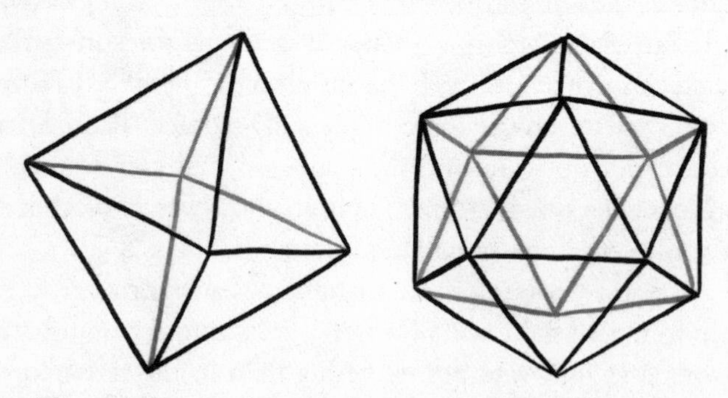

The octahedron has four triangles per vertex; the icosahedron five.

Neolithic Platonic solids . . . yeah, right.

Pythagorean theorem was the gripping finale to Book 1. Hard to beat a moment like that. Sadly, however, these Platonic solids are not as mystical as people have hoped. Plato was wrong: the smooth icosahedron is not the element of water; fire does not consist of spiky tetrahedrons (shame, though). Kepler's model of the solar system as a 'nest' of Platonic solids was proved incorrect (to his credit, the actual ratios of the orbits are, coincidentally, about right). I would also argue that Neolithic people made a lot of differently shaped rocks with bumps on them, and so of course some of them happened to look like Platonic solids.

Platonic solids do exist in nature, though, and in a far more personal and intimate way than you might expect. Many viruses are shaped like icosahedrons. If you have ever caught certain viruses (such as herpes), then you've been invaded by icosahedrons. Platonic solids are the simplest 3D shapes, with the fewest parts, so it's really easy for a virus to

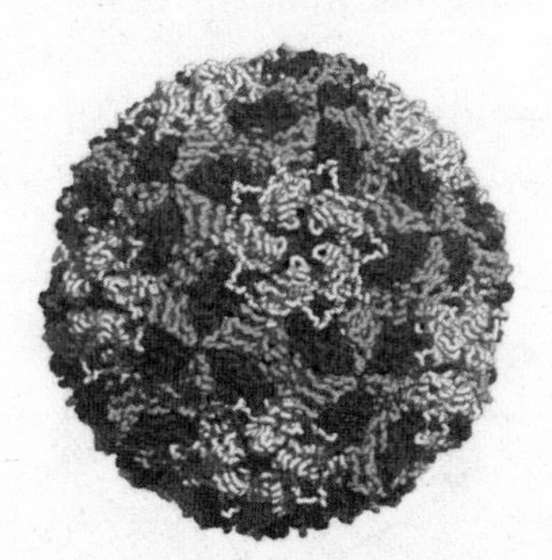

The icosahedron-shaped Foot and Mouth virus.

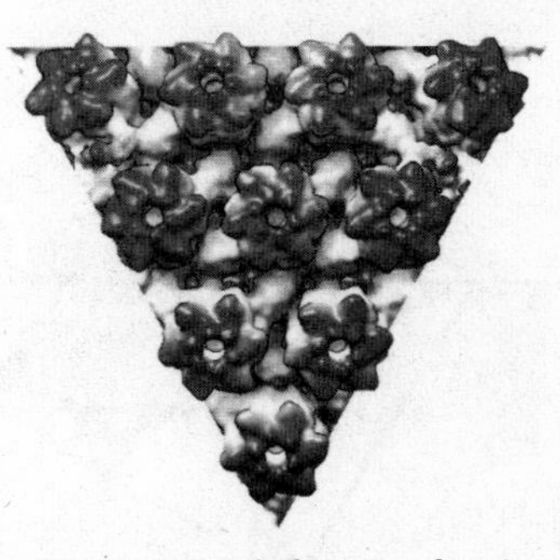

Print twenty of this triangle, join them into an icosahedron, and give a friend or loved one herpes.

build them: it takes very little information to describe an icosahedron. Basically, their genetic code only has to say, 'Make loads of bits and join them five to a vertex.' Which is exactly what these viruses do.

The Forgotten Platonic Solids

Despite the fame of those limited-edition *five* Platonic solids, there are in fact four more, which were completely unknown to the ancients. (Honestly, first parallel lines and now this.) Just as we produced bonus regular 2D shapes by letting the edges pass through one other without forming corners (the stellated polygons), we can make more Platonic solids in which the faces intersect each other without forming edges. Euclid assumed that faces could not intersect; let's assume otherwise. Twelve pentagons form a Platonic solid (the great dodecahedron), as do twenty triangles (the great icosahedron). Next up is the bonus round: stellated polyhedrons made by joining together polygons which are themselves stellated. It's stellation squared. Twelve pentagrams can be joined three to a vertex to give the great stellated dodecahedron; five to a vertex will give you the small stellated dodecahedron.

These stellated Platonic solids have been discovered by

Great dodecahedron, great icosahedron, great stellated dodecahedron, small stellated dodecahedron.

various people over the centuries. The earliest example seems to have been found in a cathedral in Venice: in a mosaic from around 1430 the artist Paolo Uccello depicts what is undoubtedly a small stellated dodecahedron. In 1813 the French mathematician Baron Augustin-Louis Cauchy 'did a Euclid' and tidied everything up, proving that the stellated dodecahedrons and icosahedrons great and small are the only four bonus Platonic solids.* It is possible to make these shapes out of cardboard, but the intersecting faces do make it a bit tricky.

Mosaic by Paolo Uccello in Saint Mark's Cathedral, Venice

All the Platonic solids, stellated or otherwise, share one final, fantastic property. For every Platonic solid, there is always another one that fits perfectly inside it. For a final polyhedron party trick: if you draw a dot in the centre of each of the 'faces' (the polygons which make them) of a Platonic solid, another Platonic solid will fit inside in such a way that its vertices touch all those dots. This second solid is called the dual of the first. The twelve faces of a dodecahedron match the twelve vertices of an icosahedron inside it; the twenty faces of an icosahedron line up with the twenty vertices of a dodecahedron. In a similar fashion, the cube and the octahedron pair up; and the tetrahedron is its own dual. This continues nicely with the great stellated dodecahedron

* These shapes are now called the Kepler–Poinsot solids, because Kepler described the great and small stellated dodecahedrons in 1619 and Louis Poinsot the great icosahedron and great dodecahedron in 1809.

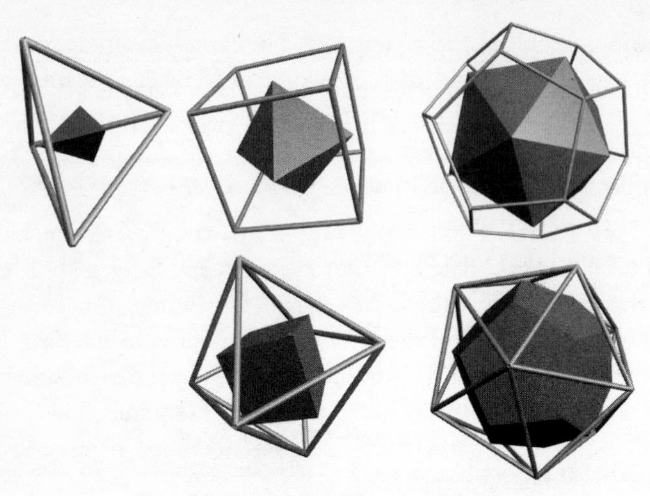

Some Platonic solid on Platonic solid action.

and the great icosahedron being a dual pair, as are the small stellated dodecahedron and the great dodecahedron.

Back to Balloons

And now for more magic in the third dimension. For this trick, take an inflated balloon and draw any 3D shape you want on it. We're not using the balloon as our surface, as we were for the spherical-geometry stuff, but rather as a handy way to position the edges of a shape in 3D. You could imagine that, if the balloon were removed, you would be left with a 3D wire frame of the shape you have chosen. Draw lots of dots on the balloon to represent vertices, then join them up with edges. They don't even have to be straight. Draw whatever crazy shape you want. The only rule is that the edges cannot intersect each other.

Having done this, you now have a shape which has probably never been drawn before. Yet I can say with certainty that, if you count the number of faces you have drawn, add the number of vertices, then subtract how many edges there are, the result is two. Ta-da. Hushed awe. Whispered 'How does he *do* that?' The minimum number of people asking for their money back.

The magic of this is that any shape you draw on the balloon will always have two more vertices and faces than there are edges. Try as you might – without cheating – you cannot break this rule. And the only way to cheat is to have edges that cross each other (or faces that intersect each other). And that is forbidden by the rules of the game. This was noticed by the mathematician Leonhard Euler (pronounced 'oil-er'; mispronouncing it will lose you several nerd points), who mentioned it in a letter he was writing to fellow mathematician Christian Goldbach in 1750. The value of *faces* + *vertices* – *edges* is now known as the Euler characteristic, and for any shape drawn on a normal balloon it must equal 2.

There is one other way to cheat: if you draw a shape with a hole in it, the Euler characteristic is no longer 2. If you were to find a doughnut-shaped balloon with a hole through the middle (they do exist, and are ideal for maths parties) and draw a shape on its surface, the number of vertices and faces would not add up to be two less than the edges. In fact, the number of vertices and faces would be the same.

The relationship of *faces* + *vertices* – *edges* = 0 holds for any shape with a hole through the middle (with no holes through any of the faces themselves). This means that the Euler characteristic for all one-holed shapes is 0. This doughnut-shaped surface is called a torus. If you have a torus with two holes, the Euler characteristic is –2, and it continues to drop by 2 for each additional hole. If you know what the Euler characteristic is for a polyhedron, you can calculate how many holes it must have by using the formula: *Euler characteristic* = 2 – (2 × *number of holes*).

Shapes of Constant Width: Now in 3D

We have gone from the flat 2D triangle to the 3D tetrahedron, and from the flat 2D square to the 3D cube. Can we do the same with our flat 2D shapes of constant width? It turns out we can: there are solids of constant width which are not spheres. They are non-spherical balls which are the same height however they are placed on a flat surface. If you place three of these under a flat book and roll it around, it moves so smoothly you cannot tell there are not three perfect spheres under it.

There are in fact two different ways to bring a 2D Reuleaux triangle into 3D. The first is to take the Reuleaux triangle and spin it around, then make a 3D object the same shape as the space this spinning Reuleaux triangle moves through. These sorts of shapes are called solids of revolution. Personally, I consider this a bit of a cheat: it's not really that different from the Reuleaux triangle. If you slice through the centre of this solid, the cross section will be a Reuleaux triangle. Which is why they work: whichever way you put them down, they are effectively a 2D shape of constant width balanced on its edge. Nothing new is introduced into the maths that is an advance on the 2D version.

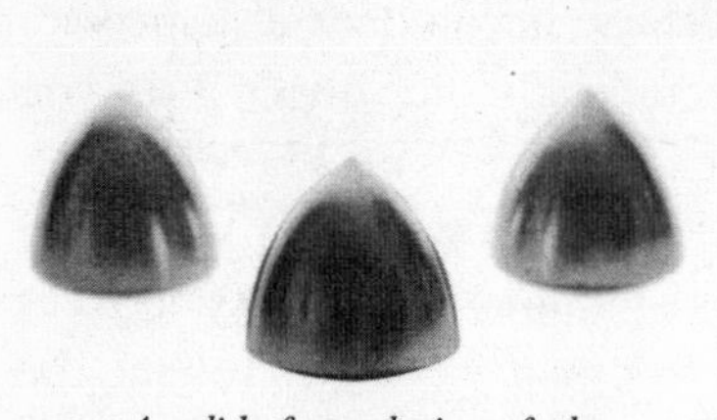

A solid of revolution of the Reuleaux triangle, aka 'cheating'.

To go truly 3D, you need to start with a tetrahedron and replace the flat faces with sections of spheres. This is the same as when we used a compass to draw arced edges, except, instead of using bits of circles as the arcs, we're using sections of spheres. Then you need to round off three of the edges as well (with three which meet at a vertex, or three which go around the edge of a face), and you get what is called a Meissner tetrahedron (one of two versions, depending on which edges you rounded). First mentioned in 1911, it is named after Swiss mathematician Ernst Meissner, who, coincidentally, had the same highschool maths teacher as Albert Einstein. Meissner tetrahedrons are incredibly hard to make, but some friends and I managed to get them injection moulded. Behold:

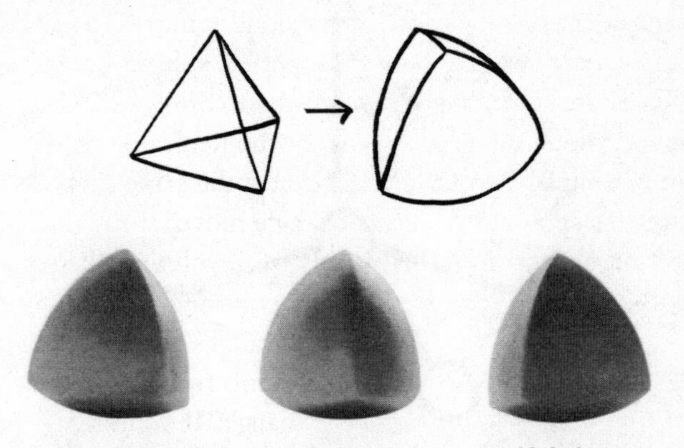

Meissner tetrahedrons made from injection-moulded plastic.

You know you want one. Or three.

Six

PACK IT UP, PACK IT IN

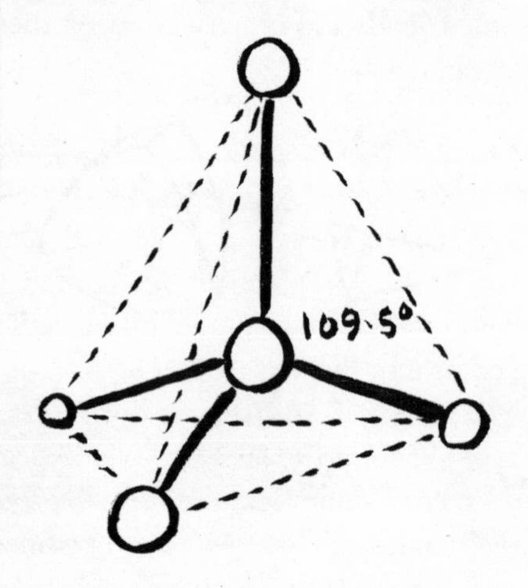

Let me begin with a challenge: find seven two-pence pieces and try to fit them all into the square on the next page. It *is* possible. And it's possible because mathematicians know beyond a doubt that this box of 74.25mm by 74.25mm is the smallest square possible into which seven two-pence pieces can fit. This wasn't worked out only for two-pence pieces: mathematicians also know that, if you have seven of any circle, the smallest box they can pack into is 2.866 times the diameter of the circle. The diameter of a two-pence piece is 25.9mm, so

This square is exactly 74.25mm × 74.25mm.

25.9mm × 2.866 = 74.23mm. With a bit of jiggling, you *can* get all those coins into that square.*

Too easy? OK, see if you can fit thirty-one two-pence pieces into a 144.9546mm square. Or try and beat that. The current world record stands at thirty-one two-pence pieces in a square with sides of 144.9546mm, but one thing that mathematicians haven't yet figured out is whether that is the best that can be done in fitting round objects into a square hole. No one has yet found the most efficient way to pack thirty-one circles into a square. That record could be yours for the taking. In fact, records have only been finalized up to thirty coins, and, in one extra case, for thirty-six coins. For

* OK, so the box of 74.25mm is bigger than the 74.23mm you actually need, but that's below the accuracy of a printed line, and the precision we know the size of the coin to, so it's fine. Consider the extra 2 per cent of a millimetre a gift from me to you.

all other options, there could still be wriggle-room to do better.

N	RATIO	SIZE FOR 2p PIECES
7	2.866025404	74.23 mm
10	3.373720762	87.38mm
13	3.731523914	96.65mm
31	5.596701676	144.9546mm
42	6.426611073	166.4492 mm
101	9.788881942	253.532042mm

If you up the ante and choose to spend your time trying to fit 101 coins into a square, you might start to wonder – after several hours, perhaps – if there is a quicker method. And there is! Program a computer to do it for you. (I'd argue that's totally within the rules.) Computers have found pretty good solutions for everything from one coin up to ten thousand coins.* However, they do run into the same problem you would if you found a better arrangement for thirty-one coins: how do you know if you have reached the ultimate: the greatest number of coins possible in the smallest amount of square space? There will always be the

* You can download all ten thousand current record-holding arrangements of circles in squares from Eckard Specht's website: http://packomania.com/csq

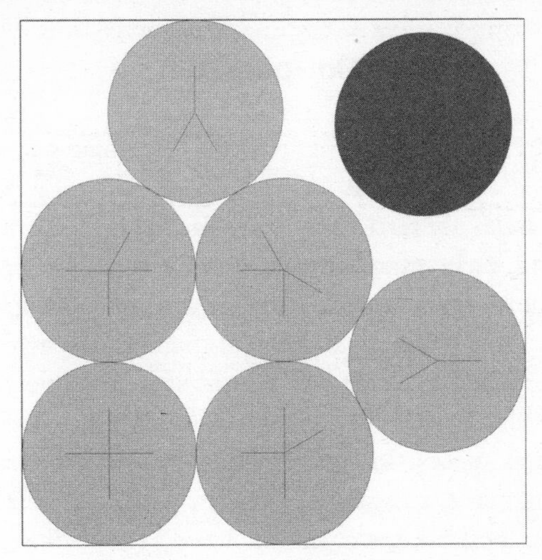

Seven coins in a box. The shaded one has wriggle-room to rattle around.

irksome possibility that someone, sometime, will break your record.

I find this whole area of packing problems hugely exciting. Humans know very little about the mathematics of packing objects into spaces. It seems like a simple thing to do: arrange objects around each other to take up as little space as possible. But we just can't seem to get it right. The really fascinating thing, though, is that the subject of packing, mathematically speaking, is tantalizingly open: it's in the same kind of disarray in which other mathematical areas historically found themselves before a revolutionary breakthrough propelled them up to the next level of understanding. You can sometimes feel that much of maths was all done and dusted way back in the distant past, but here's an area where the best maths is yet to come.

Flat-packed

We'll start at the beginning: packing flat objects on flat 2D surfaces. Even without being limited to fitting within a square, circles are terrible at being packed because they can't fit around each other without leaving gaps. Other shapes such as hexagons can pack together without leaving any, and of course this is the most efficient method. As anyone who has remodelled a bathroom knows, it is possible to fit all sorts of tiles together to cover a flat wall with no unsightly in-between spaces. In mathematics, this act of covering a flat, blank plane with shapes is called tiling. It differs slightly from the better-known maths word 'tessellation', which requires there to be some order to the tiled shapes and their positions. Tiling includes tessellations as well as a complete free-for-all on other arrangements, so I'll use it as a more all-encompassing term.

We do know of some great ways to tile the plane, but we're a long way from finding all of them. There are the boring options, such as using all squares or triangles, but we can do much better. One of my favourites is an irregular nonagon, which completely tiles a surface in a spiral pattern. We can also use more than one shape. Squares and triangles can team up to give snub square tiling, and the dream team of hexagons, squares and triangles produces rhombitrihexagonal tiling (a great word, yet worth only thirty-six points in Scrabble). With all these amazing options, whenever I see a bathroom wall with square tiles, I'm secretly very disappointed.

There is one packing puzzle which, I believe, holds the record for the longest time between the answer being discovered and proof being found that this answer is definitely the correct answer. A solution was found before humans

This spiral is the same nonagon repeated over and over.

Snub square tiling and rhombitrihexagonal tiling.

even existed, but it wasn't proved until 2001. The puzzle is: what is the best shape to pack in 2D? By 'best', we mean that this is a shape that fits together leaving no gaps and covers the maximum area possible for the length of its edges. The winner is the hexagon, as was discovered by bees long before humans even conceived of mathematics. Known as the honeycomb conjecture, it was not until millions of years later that the American mathematician Thomas Hales proved that, even if you resort to crazy shapes with strange curved sides, you'll never flat-pack better than you can with a hexagon.

If hexagons are the best 2D packing shape, what about the worst? Unfortunately, this is one more thing that humans don't yet know. We do know that, even though circles are pretty rubbish (at best, tightly packed circles can cover 90.7 per cent of a surface), there is something even worse. It's called the smoothed octagon. If you take an octagon and replace the corners with rounded hyperbola curves (hyperbola are from the same family as circles and ellipses, at about the cousin level), the resulting shape can never cover more

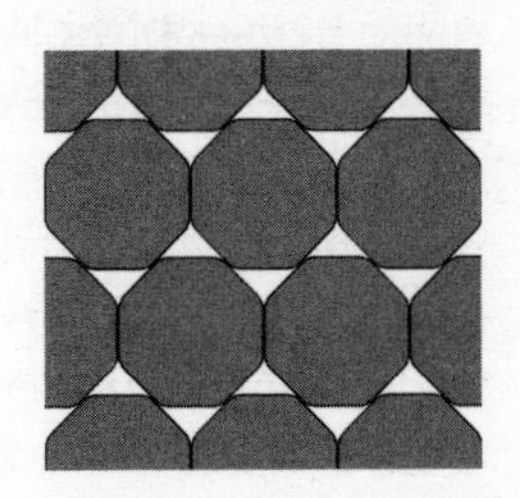

Smoothed octagons leave bigger gaps than circles.

than 90.24 per cent of a surface. There could, however, be a worse-still 2D packing shape out there – we've just not found it.

Space to Fill

Humankind's understanding of how polyhedrons tile in 3D space is a joke. And not a very funny one. To look at the different ways single polyhedrons could fit together, I went to the online resource of all things mathematical: Wolfram MathWorld. Normally an authoritative guide to mathematical topics, the entry on space-filling polyhedrons exhibited an unusual air of despair. Apparently, the mathematician Michael Goldberg spent the years from 1974 to 1980 trying to categorize all space-filling polyhedrons, including the twenty-seven hexahedrons, sixteen heptahedrons, forty hendecahedrons, sixteen dodecahedrons, four 13-hedrons, eight 14-hedrons, one 16-hedron (there are no space-filling 15-hedrons, in case you're wondering), two 17-hedrons, one 18-hedron, six icosahedrons, two 21-hedrons, five 22-hedrons, two 23-hedrons, one 24-hedron, and what he believed to be the biggest possible -hedron: a single 26-hedron. Phew. I'm impressed he found that much space in his schedule. And then filled it. With polyhedrons.

Then, in 1980, Peter Engel alone found another 172 new space-filling polyhedrons of between seventeen and thirty-seven faces. And more have been added. The MathWorld entry currently ends with the quiet plea: 'A modern survey would be welcome.' All we seem to be doing at the moment is listing ways that polyhedrons can fill space but without any systematic understanding of what is really going on. What is missing is a decent theory. But, as we saw with the honeycomb conjecture, you can't rush these things.

Even if we limit our search to the Platonic solids, mathematicians are still learning new things. In 2D, of the regular polygons, the triangle, square and hexagon can completely tile a surface. Of the Platonic solids, only the cube can fill a 3D space. People often claim that the tetrahedron is also space-filling (including Aristotle, in his work *On the Heavens*) but, despite the fact that this, er, 'fact', is repeated again and again, it isn't, in fact, a fact: it isn't true. If feels like it should work but, if you stack tetrahedrons, there will always be some small gaps between them.

I personally consider even the cube to be a bit of a cheating solution, as it's just an extension of the square tiling the plane. When a 2D shape is extended into 3D with parallel sides, it's called a prism. As well as the cube (a square prism), the triangular prism and the hexagonal prism work equally well. None of the Platonic solids fills 3D space in a way any different to what already works in 2D. It's a bit like the way in which spinning a 2D Reuleaux triangle *technically* gives you a solid of constant width, but the Meissner tetrahedron is the first truly 3D solution.

In fairness, the tetrahedron can stack nicely, but only when it partners up with the octahedron. There is a repeating pattern of two tetrahedrons paired with one octahedron (with the same-length edge as the tetrahedrons) which *is* space-filling. And it's not the only option. As recently as 2011, the then 73-year-old mathematician John Conway discovered that you can fill space with one octahedron combined with six smaller tetrahedrons. This is still using only Platonic solids, and we're still finding out new things!

If you want to find the best polyhedron to stack in 3D, you can try making a slight tweak to a Platonic solid. If you cut the corners off an octahedron, it exposes new square surfaces and gives a solid called the truncated octahedron. This

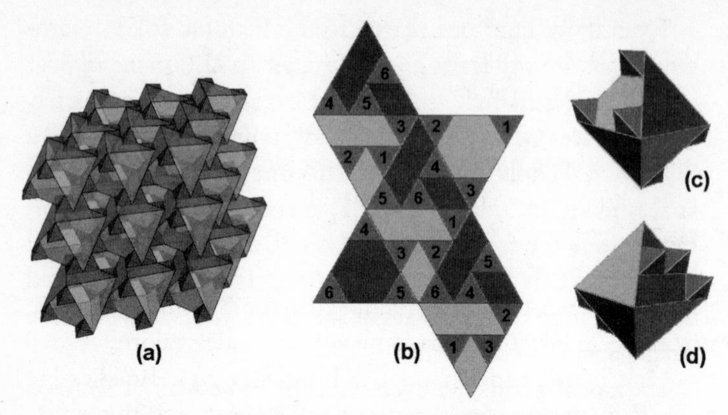

An octahedron with some tetrahedrons attached can tile perfectly.

was the candidate for the 'Best 3D-Space Filler' gong proposed by Lord Kelvin in his 1887 paper 'On the Division of Space with Minimum Partitional Area'. When mathematicians want to share their ideas or findings with each other, they write up a detailed description of their concept in papers, which are then published for everyone else to read. Lord Kelvin wrote this one extolling the virtues of the truncated octahedron. Over a century later, mathematicians would find he'd actually got it wrong. Still, he lived the glory.

There's No Space Like Foam

The proof that Kelvin was wrong was right in front of everyone's nose the whole time. There is a way to fill 3D space better than with a truncated octahedron and, once again, nature got there first. The structure of these superior shapes had already been seen in crystals. It even appears as a diagram in the 1960 edition of *The Nature of the Chemical Bond* by Linus Pauling as the structure of the chlorine hydrate crystal.

The problem is that no one bothered to see how efficiently this structure tiles in 3D.

And it wasn't for lack of looking. When Lord Kelvin conjectured that the truncated octahedron should win the trophy, other mathematicians immediately started trying either to prove he was correct, or to find a better shape as a counterexample to prove he was wrong. As the decades passed and the problem remained unsolved, it became more and more notorious. Anyone who could settle the Kelvin Conjecture would gain instant maths fame and accolades galore.

One mathematician on the hunt was Ken Brakke. He was ultimately unsuccessful (someone else found this structure before him), despite his father owning a copy of *The Nature of the Chemical Bond*. Unfortunately for him, he never happened to flick through it and find the chlorine hydrate diagram. 'This book was sitting on my father's bookshelf 10 feet from me as I was trying to beat Kelvin,' he later lamented. But Brakke did still make a major contribution to solving the Kelvin Conjecture: he created the modern computer equivalent of blowing bubbles.

Bubbles, and the soap film they're made of, have played a vital role in investigating space-filling shapes, because of one very important intrinsic property: soap film will always try to contract in order to have the smallest surface area possible; this is limited only by the volume of air inside it. If you blow a bubble in the air, it forms a sphere, because that is the minimum surface area possible to surround a volume of air. If a bubble touches anything else, it will change its shape to whatever the new best solution is. Which means it's possible to blow a cube bubble.

To blow a cube-shaped bubble, you'll need to make a wireframe cube and dip it into some bubble mixture. If you get this just right, you'll have six bubbles attached one to each

'face' of the wire cube and a single cube bubble suspended right in the very middle. If it's not quite right, dip a straw into the bubble mixture and add or remove bubbles with that. The only problem is, if you look at the cube bubble very closely, you may notice that it isn't quite an exact cube: the faces are not quite flat. This is because of the subtle ways in which soap film behaves.

A foam made of soap film may look like a very complicated structure, but it all comes down to just a few simple mathematical rules. These were described by the Belgian physicist Joseph Plateau in 1873, and they are known today as Plateau's four laws. The first two cover the fact that soap film is always nice and smooth and that a bubble will always curve by the same amount anywhere on its surface. The final two deal with bubble-on-bubble action: faces always meet in threes to form edges, and edges always meet in fours to form vertices. The next time you're washing up, or drinking a beer, or both simultaneously, look closely at the foam. Regardless of how complicated it may appear, you will only ever see faces in groups of three and edges in groups of four.

Beyond that, faces and edges always meet at exactly the same angles. All faces meet with 120° between them, so they are always equally spaced. Dividing a 360° full circle into thirds gives you the 120° equal separation in 2D. Equally spacing things in 3D gets slightly more complicated, because there are now more directions to deal with (the up–down, as well as the left–right and backwards–forwards from our first venture into 3D), but this is a problem already solved by the tetrahedron. If you join all the four vertices (which are perfectly spaced out) to the exact centre of the tetrahedron, those lines all meet at angles of 109.5°. This is called the tetrahedral angle, and all edges in foam meet at this angle.

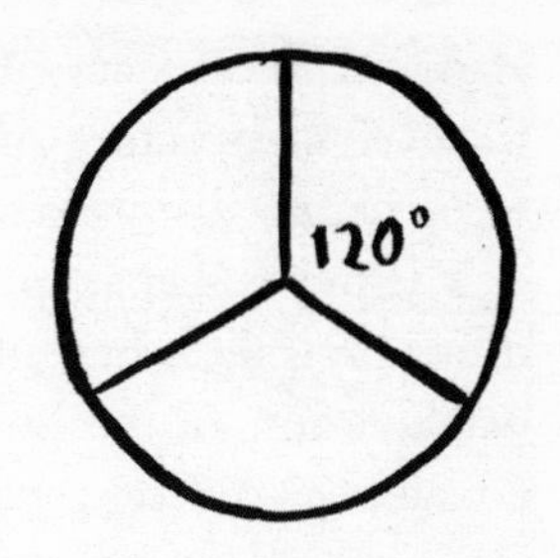

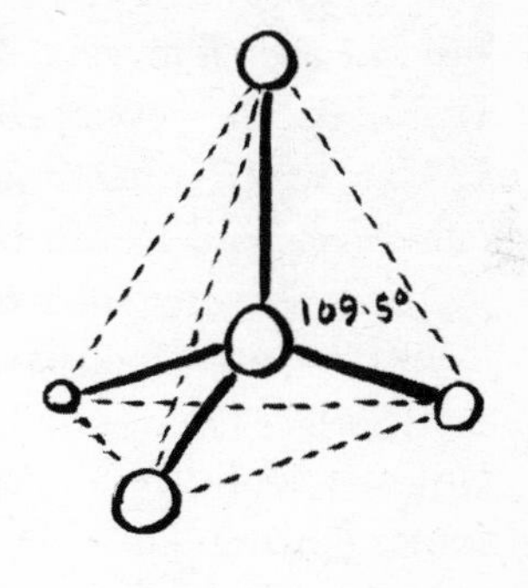

Lord Kelvin's 1887 paper focuses heavily on using soap film as a way to find the most efficient way to fill space. For an antiquated science paper, it's a good read. A decent amount at the beginning is him talking about playing with a cube bubble like the one we made just now. He calls it the Plateau cube, because Plateau used the same cube wire frame to do his experiments. Lord Kelvin described how he would blow on the Plateau cube to distort it and then stop to watch it spring back to its optimum reduced-surface position.

Lord Kelvin's theory was that if you stacked equal-volume bubbles, they would take the shape of truncated octahedrons (or tetrakaidecahedrons, as he called the 14-sided polyhedrons). Or, rather, they would take the shape of the bubble equivalent. The cube we made before was not a perfect cube, as I mentioned, because the faces are slightly curved, meeting at 120°, and the edges meet at 109.5°, instead of the standard 90°. A stack of bubbles should form what is called the Kelvin structure of truncated octahedrons, with their faces slightly distorted to match Plateau's laws.

The pity is that soap film will form this truncated octahedron only if the other bubbles it rests against are also truncated octahedrons. If the boundary surface the foam eventually contacts is a different shape, this will have ramifications across the entire foam structure and it will not form the perfect shape. However, if you get close enough to the correct structure, the soap film will relax into that shape. This is what Ken Brakke did: he developed a software package called Surface Evolver which could simulate soap-film surfaces and do digitally what Lord Kelvin had to do manually. Fun fact: both methods are susceptible to getting bugs in them.

In 1994, two physicists working at Trinity College, Dublin, used Surface Evolver to find a structure better than the Kelvin structure. Known as the Weaire–Phelan structure (after Denis

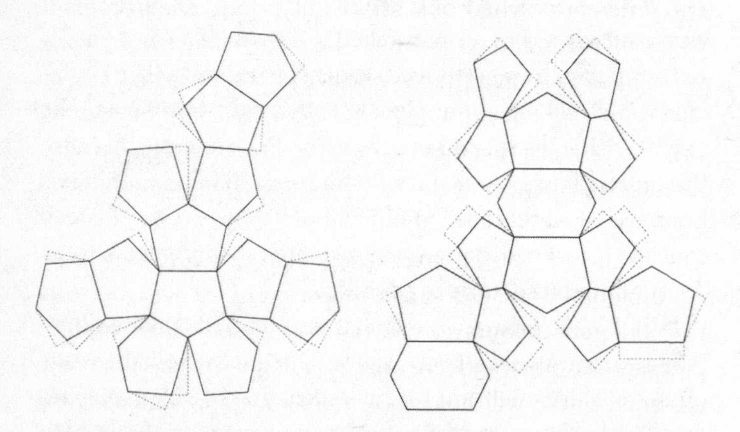

Dodecahedron made from non-regular pentagons (one side is 130.9% of the length of the other four), and 14-hedron built from twelve non-regular pentagons (two sides 86% and one 57.6%) but with an extra two distorted hexagons squeezed in (two sides 152.3% of the other four).

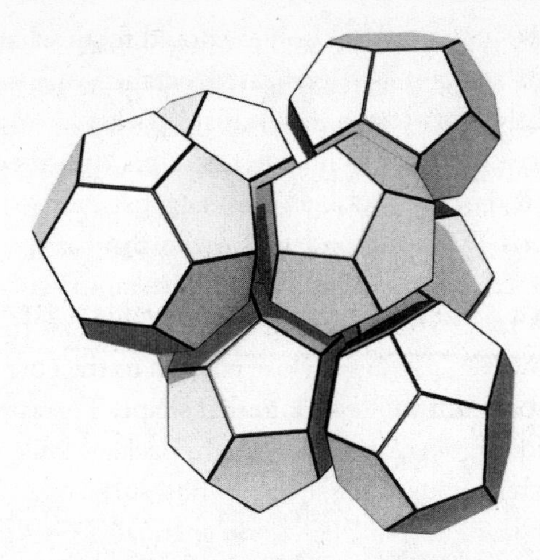

The repeating unit of the Weaire–Phelan structure made from six 14-hedrons and two dodecahedrons.

Weaire and his PhD student Robert Phelan), it consists of two different polyhedrons: an irregular dodecahedron made of twelve pentagons and a 14-hedron made from twelve pentagons with an extra two hexagons squeezed in. If you cut the patterns out and make these models from cardboard, they can be stuck together to form the structure. The cheating method is to make only the 14-hedrons and, as you stick them together, the voids left behind will be the dodecahedrons.

Even better, you can make the whole structure as a frame of drinking straws. Get a massive packet of straws of at least four different colours so you can colour-code sides of four different lengths, and an equally massive packet of pipe cleaners. These will hold the straws together. Cut the pipe cleaners down to a length of 10cm, twist two together in the middle and fold each end back to make a four-pointed star. Now slide a straw over each point. Thanks to Plateau, we know that this is the only situation we need ever face. The straws will not all meet at the exact centre of the vertex because of their thickness, so they need to be cut slightly shorter. Repeat the different vertex options shown below and, after a bit of fiddlesome fingerwork, you will have created an amazing Weaire–Phelan structure which you can look right into, getting a good view of its inner workings. Strawsome.

EDGE	RATIO	FOR 6mm-DIAMETER STRAWS
A	2.272	7.3 cm
B	1.736	5.5 cm
C	1.492	4.7 cm
D	1.000	3.0 cm

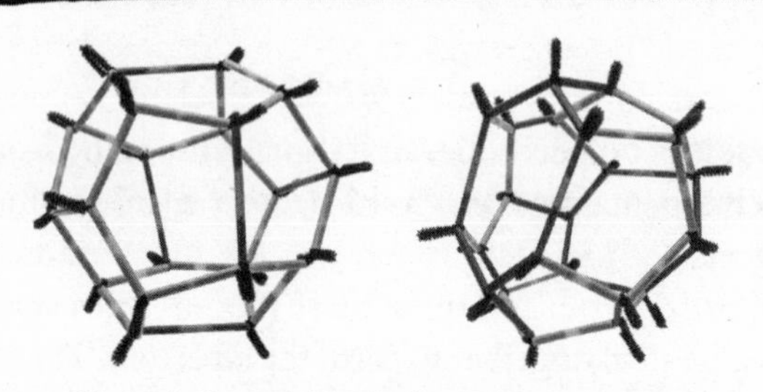

Dodecahedron and 14-hedron made from straws.
You're going to need a lot of these.

The average surface area required for the Weaire–Phelan structure is 0.3 per cent less than that required by the Kelvin structure. This may not be much area saved, but it had a huge impact on the world of maths. After a century of searching, mathematicians couldn't believe that someone had found a better structure than Lord Kelvin's. They were ecstatic that such a structure had been unearthed, yet it received very little attention in the mass media. Such a monumental discovery deserved to have as its result an international architectural firm designing and constructing a model of the Weaire–Phelan structure the size of a large building (say, requiring about £90 million from wealthy private investors) and then have it featured in the biggest televised international event possible. And that's exactly what China did during the 2008 Beijing Olympics.

To complement the circular Bird's Nest stadium, the Beijing Olympics wanted a square aquatics centre, and the competition to design it was won by the Australian firm PTW Architects (in collaboration with various engineering firms). They wanted their design to have the feel of natural, bubbling water, and their research led them to the Weaire–Phelan arrangement of stacked bubbles. To make it look less regular, they have effectively sliced through the foam on angles that do not align with the polyhedrons, giving it a random, chaotic look.

Named the Water Cube (despite, strictly, being a cuboid), it was a huge success, and billions of people around the globe got to see a fantastic cut-through of the Weaire–Phelan structure. After the Olympics it was turned into a swimming

centre for the public. I made a maths pilgrimage to see it for myself, but was not allowed to go for a swim in it as I didn't have the necessary official health certificates (true story). But, inside, you can still see some of the huge steel beams forming some of the distorted pentagons of the Weaire–Phelan structure. If you're ever in Beijing, the glowing mathematical architecture remains there to be marvelled at. It's a monument to why we do maths.

Oh, Balls

Moving on . . . Get an orange. Now see how many other oranges you can make touch it at the same time. No time to waste: I'll give you the answer. If you take a sphere, you should be able to get twelve other identical spheres to touch it simultaneously. In maths, having two shapes touch each other like this has the rather sweet, technical name of 'kissing', so the kissing number of a sphere is at least twelve. But could it be thirteen? If you do manoeuvre twelve oranges into kissing position, you will see there is a decent amount of wiggle-room in which to shift them around. It feels like there should be space to get a thirteenth in there somewhere.

Such luminaries as Isaac Newton conjectured that there was no room for a thirteenth sphere to join in the fun, but no one could rule it out for sure. There are so many ways to arrange the spheres it's hard to calculate all the options. Proofs that the thirteenth sphere could not fit finally arrived in the late nineteenth century, but there were still plenty of other unanswered questions about arranging spheres. For a start, no one knew the best way to stack them.

It is in fact extremely annoying that many varieties of fruit are spherical, because in 3D, as far as we know, a sphere is the absolute worst way to fill space. It's not like in 2D, where the circle is dreadful but there are, technically, worse options; in 3D, we have not found a less efficiently stackable shape than a sphere. The opposite of an optimal shape, the sphere is known as the pessimal shape. However, we may *know* it is the worst out of all possible symmetric shapes, but we have not ruled out the slim chance of some incredible exotic shape turning up and under-trumping it. But it seems that oranges could be literally any other shape and they would stack better. For centuries, greengrocers have made do with what they believe to be the best way to stack oranges. Kepler agreed with them, and in 1611 he conjectured that the greengrocer option was the absolute best option when it came to stacking spheres.

When you were building pyramids of oranges in Chapter 3, you will have quickly discovered that square-based pyramids are next to useless. Unless you are stacking something as heavy and immobile as cannonballs, it's hard to keep spheres in a square shape as they stack. Oranges will, like all spheres, slip into a different* arrangement in which each row is offset from the previous one and each piece of fruit sits in the dips. If you continue this triangle arrangement of fruit, you will build a tetrahedron of spheres, where the spheres fill 74.048 per cent of the available space – which Kepler and

* The arrangement of oranges in a square pyramid is the same as in a tetrahedron, only on a different orientation. It's the orientation that makes it much harder to stack.

greengrocers claim is the best you'll ever do. When life gives you oranges, make a tetrahedron.

Proving this, though, was near impossible. As we saw when we put the coins in a square, sometimes a jumbled arrangement is more efficient than an ordered one. When you push a thirteenth sphere into the twelve others kissing the central orange, for example, you can move some of the others away slightly so they are no longer quite kissing and still get a fairly dense arrangement. The problem is, as I said above, there are so many ways to arrange the spheres when doing this, no one could ever check them all and see if any are better than Kepler conjecture.

In steps Thomas Hales once again. Having defeated the honeycomb conjecture, he took on Kepler conjecture. What gave Hales the advantage to succeed where everyone else since 1611 had failed was that he was not afraid to let computers do some of the work for him. To be fair, he also has the advantage of living after the invention of the computer. Hales managed to narrow down all the possible irregular arrangements of spheres to around five thousand different situations. He then projected these down to their 2D shadows to make calculation easier and split them into individual problems which could be individually checked by a computer.

Even using computers to do as much of the work as possible, this was still almost impossible. When he began work on this in the 1990s, computers were only barely powerful enough. The final paper he wrote (published in 2005, after revisions) ran to 250 pages along with several gigabytes of vital computer code and data. In the paper, Hales makes a summary list of all the computer programs he used, and the last one listed is a program that was used to keep track of what all the other programs were doing. Hales thus used computers as an aid in using other computers. As he wrote,

'The organization of the few gigabytes of code and data that enter into the proof is in itself a non-trivial undertaking.'

But, somehow, it worked. The computers checked every possible arrangement of spheres, and none was better: Hales proved that Kepler and greengrocers had been right all along. He also writes with a level of lucidity rarely seen in academic mathematics papers (making him a hero of mine), so I recommend reading the first few pages of his paper 'Historical Overview of the Kepler conjecture'. It sets out the Kepler conjecture better than I ever could, with much more detail. Sadly, as readable as the start is, no one has ever read and checked the entire proof. Before a mathematics paper can be published, it needs to be checked by other mathematicians (it's called peer review) to make sure there are no mistakes. Hales's paper was given to a panel of twelve experts but, because so much of the work was done by computer, after four years checking it, they could only say they were 99 per cent confident that it was all correct. And that just won't do.

Hales's response to this is genius. He is going to use computers to double-check and peer review the work done by computers. Genius, as I say. His proof used computers to do the heavy lifting, meaning that something too complicated for humans to calculate themselves was achieved but the resulting proof was too involved for humans to check. Hales's new adventure is called the Flyspeck Project (because 'flyspeck' is one of the few English words with the letters FPK – 'Formal Proof of Kepler' – in that order), and if you wish to join in and help, it's still ongoing,* at http://code.google.com/p/flyspeck/.

* Days before this book went to print, Hales completed the Flyspeck Project, giving me barely enough time to add this footnote. Updates on everything in this book after it has been set in paper will be online at www.makeanddo4D.com.

I love all these packing problems because humankind is on the verge of being able to crack open so many answers. We obviously know so little about how to fill 2D and 3D space; only in the last few decades – a blink in mathematical time – have we even begun to understand the most simple and regular of cases. Should we meet the Hypotheticals, this is one area in which they could have much to teach us. And if we don't meet them soon, it's OK – I suspect humans cannot be more than a century or two away from mastering this facet of mathematical logic.

Seven

PRIME TIME

Prime numbers are the A-list celebrities of maths. They may not own mansions or have unexpected breakdowns, but they do have two other very significant properties. Firstly, they are numbers which no other numbers divide into. Secondly, they are chased by the maths paparazzi: there are cash prizes for finding them. An anonymous sponsor has put up the cash via an organization called the Electronic Frontier Foundation as a bounty on prime numbers: $100,000 for the first prime with over 10 million digits, $150,000 for one over 100 million digits and $250,000 for a billion-digit monster. The first prize has already been claimed, but the other two are still up for grabs.

Anyone can join in the search for large prime numbers and have a chance to share in the prize money. Because finding prime numbers requires so much computing power, the Great Internet Mersenne Prime Search (GIMPS – a somewhat unfortunate acronym) was started in 1996 as a 'grassroots supercomputing' project. Any Tom, Dick or Harriet can take part, using their own computer as part of a distributed virtual supercomputer, currently made up of 9,700,000 processors performing around 164 trillion calculations every second. It was GIMPS who found a prime with over 10 million digits in 2009 and received the first prize. As far as newspaper headlines go, 'PRIME NUMBER DISCOVERED BY GIMPS' is up there with the best of them.

The problem with primes is that you cannot accurately predict where they are. Attempts such as those by GIMPS to locate large primes pretty much consist in guessing a big number which may be a prime and then checking through possible numbers which could divide into it (its possible factors) to make sure it really is a prime. If you join the GIMPS network you'll be assigned one of these large numbers for your computer to check. The vast majority of the time, the numbers end up having a factor. To stack the odds slightly in their favour, GIMPS does not hand out completely random numbers to be checked: they deliberately use only numbers which are one less than a power of 2.

It was the ancient Greeks who first noticed that, even though not all numbers which are one below a power of two are primes, these numbers do have a propensity for primality. Up to the thirteenth power of two, five of them sit above a prime number. The chances get much lower as the numbers get bigger, but the odds are still better than choosing a large number arbitrarily. So if you join GIMPS your computer will be given a power of two so large the answer has millions of digits, and it will check if the number one below it is a prime. The winning 2009 prime was one less than $2^{43,112,609}$.

$2^2 - 1 = 3$ YES
$2^3 - 1 = 7$ YES
$2^4 - 1 = 15$ NO 3×5
$2^5 - 1 = 31$ YES
$2^6 - 1 = 63$ NO $3 \times 3 \times 7$
$2^7 - 1 = 127$ YES
$2^8 - 1 = 225$ NO $3 \times 5 \times 17$
$2^9 - 1 = 511$ NO 7×73
$2^{10} - 1 = 1023$ NO $3 \times 11 \times 31$
$2^{11} - 1 = 2047$ NO 23×89
$2^{12} - 1 = 4095$ NO $3 \times 3 \times 5 \times 7 \times 13$
$2^{13} - 1 = 8191$ YES

Despite being studied since at least 300BCE, numbers one less than a power of 2 to a certain power are named after the seventeenth-century monk Marin Mersenne who popularized them. He published a list of all the Mersenne numbers which are primes – which was, unfortunately, riddled with errors. However, there's nothing like doing something wrong to spur other people into doing it correctly. Mersenne claimed that $2^n - 1$ is a prime for $n = 2, 3, 5, 7, 13, 17, 19, 31, 67, 127$ and 257, and no other values below 257. Other people had other ideas.

The low values of 2, 3, 5, 7, 13, 17 and 19 had all been checked, and they stand as prime numbers. The ancient Greeks knew about 2, 3, 5 and 7, and both 17 and 19 had been confirmed by Italian mathematician Pietro Cataldi by 1604.

Everything beyond these was just Mersenne guessing, although he did state these stabs in the dark with such certainty it was really quite convincing. Mersenne's motto was: conjecture hard or go home. Cataldi had conjectured his own list of $n = 23, 29, 31$ and 37, but as Mersenne's guesses were slightly less wrong than Cataldi's, we remember him and forget the Italian. (Also, the Frenchman was well known for other scientific work, so he had a rolling start on prime-number notoriety.)

It was over a century later that Euler (think 3D shapes and balloons) was able to confirm that the first guess on the list was correct: $2^{31} - 1 = 2{,}147{,}483{,}647$ is indeed a prime. This was the eighth Mersenne prime and, at over 2 billion, everyone assumed that humans would never do any better. Numbers above that just seemed too big to be checked by human brain alone (even with a steady supply of paper, pencils and maybe even an abacus). Mathematician Peter Barlow, writing in 1811, said that the prime number 2,147,483,647 'is the greatest that ever will be discovered; for as they are merely curious, without being useful, it is not likely that any person will ever attempt to find one beyond it'.

He was proved wrong in 1876 by Édouard Lucas (think cannonballs), who proved that $2^{127} - 1 = 170{,}141{,}183{,}460{,}469{,}231{,}731{,}687{,}303{,}715{,}884{,}105{,}727$ is a prime. Lucas found some cunning short cuts to check if a number had any factors, but it still involved an excruciating amount of working out. To this day, it holds the record for the largest prime number found by hand (i.e. brain/paper/pencil). Lucas also removed an incorrect entry on Mersenne's list, showing that $2^{67} - 1$ is not a prime. In the years that followed, people spotted a few primes that Mersenne had overlooked (in 1883, it was discovered that $2^{61} - 1$ was a prime and, by 1914, $n = 89$ and $n = 107$ had joined the list), as well as crossing a few off his list (by 1933, $2^{257} - 1$ was shown to be not a prime). However, none of these surpassed Lucas's mammoth effort of finding $2^{127} - 1$.

Interestingly, even though Lucas was able to prove that $2^{67} - 1$ had factors, he didn't know what they were. This is called an existence proof: you can prove something exists without discovering exactly what it is. For example, mathematicians had long known that, if the power of 2 was itself not a prime, then the matching Mersenne number would not be a prime either. So we know that $2^{14} - 1$ has factors, because 14 is not a prime. What Lucas did was enough to prove that $2^{67} - 1$ must also have factors; other mathematicians then leapt into action to try to find out what those factors were.

The mathematician Frank Nelson Cole was the first to locate them, and he presented them in 1903 to the American Mathematical Society in a lecture during which he didn't say a single word. Chalk in hand, he wrote a 1 on the blackboard and doubled it to 2, then to 4, 8, 16, 32, and so on, until, after sixty-seven doublings, he had shown that $2^{67} - 1 = 147,573,952,589,676,412,927$. He then moved over to a second blackboard and began multiplying 193,707,721 by 761,838,257,287 and showed that it gave exactly the same answer. He returned silently to his seat as the crowd went wild. It had taken 259 years since Mersenne claimed $2^{67} - 1$ was a prime for someone finally to find its factors.

Then, in 1952, everything changed. One of the first computers in the world was switched on and the next five Mersenne primes were found at once, with $n = 521$, 607, 1,279; 2,203; and 2,281. The last one, $2^{2,281} - 1$, has 687 digits. Ever since then, computers have continued to find primes bigger than pre-twentieth-century mathematicians could have ever dreamed of. There are now forty-eight known Mersenne primes, over half of which have been found since 1970. People have been obsessing over these numbers since

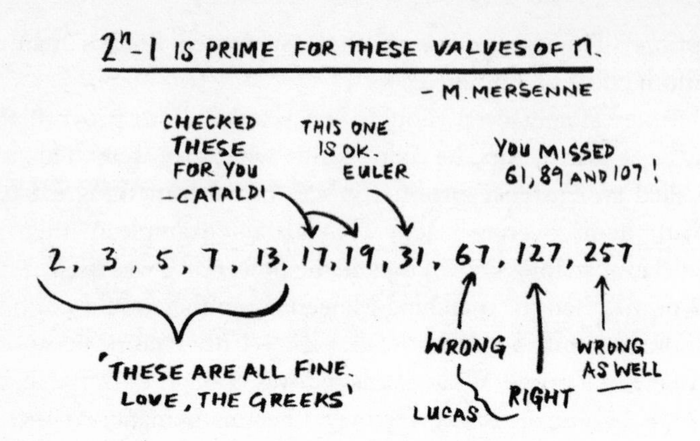

the time of the ancient Greeks, yet in a human lifetime we've more than doubled the list.

It is GIMPS who have found the most recent fourteen Mersenne primes, including all ten of the biggest primes known to humankind. The latest was found on 25 January 2013: with a value of $2^{57,885,161} - 1$, it has 17,425,170 digits. If you were to print all those digits out at a size of 1mm each, they would form a line over 17 million millimetres long. That's over 17km. That's longer than a line of 562 large blue whales (to use the official unit, over half a kilo-whale). The number is so long, double-checking its primality took a powerful 32-core server six days. If you'd like to join the hunt, you can look GIMPS up online, but perhaps avoid their not-very-search-engine-friendly acronym and just type in 'Mersenne prime search' instead.

Critical Factors

Imagine a mathematical gaol with a hundred cells numbered one to a hundred stretching down a single, long corridor. The staff consists of a hundred guards, all of whom have a key which can lock and unlock any of the cell doors. At the start of the evening, all the cells are locked. The first guard (let's assume they're also numbered, so s/he's number 1) walks down the corridor and uses their key to unlock every cell door. Realizing this is a mistake, the second guard (number 2) goes along the corridor and locks every second door. The third guard uses their key in every third cell door but, now, the key can either lock or unlock the cells. The fourth guard uses their key in every fourth door, either locking or unlocking; and so on through all hundred guards. In the morning, which of the cell doors are unlocked?

Given that this is a chapter on primes and factors, it may come as no surprise to you that the answer depends on what factors the number of each individual cell is divisible by. This is a recurring theme in maths: the properties of a number depend on what factors it has, and primes are the only numbers that have no factors. In fact, every number which is not a prime can be split apart into prime numbers multiplied together, and those prime factors explain most of the ways in which that number behaves: they dictate its personality in terms of what properties and patterns it will have. For example, to a mathematician, the number 28 is really $2 \times 2 \times 7$, which is known as the prime decomposition of 28. Prime numbers are, in a way, the atoms of maths, the components that make up all other numbers. The non-prime numbers are known as composite numbers.

It turns out that in our maths gaol it is the cells with square numbers that are found unlocked in the morning. Each cell has a key turned in its lock by every guard whose number (position in the order of key-turning duties) is a factor of that cell. To be unlocked, the lock needs to be turned an odd number of times, and square numbers are the only numbers with an odd number of factors. If you list all the factors of any composite number, they form pairs which multiply to give you back the original number, but square numbers have an extra lonely factor which has to pair with itself. To make a complete list of all of the factors of a number, you need to note down every combination of its prime factors.

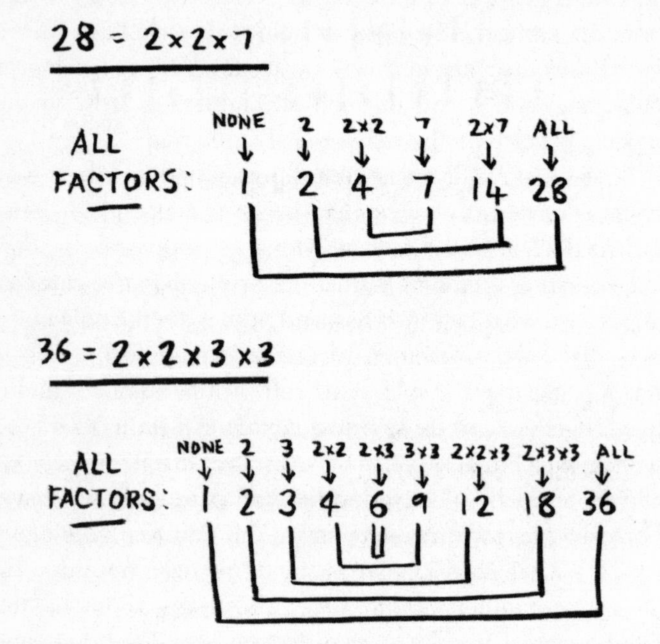

But why should factors be so important? Instead of splitting 28 into numbers which multiply, we could just as easily write 28 as 21 + 7 or 13 + 15, or any of the many other equally valid ways. Who mounted the coup on addition and made multiplication king? The problem is that if you break the addition of a number down to the simplest numbers, you get a long string of 1s being added, which is really only a tally

$$28 = 2^2 \times 7$$

VS

$$28 = 1+1+1+1+1+1+1+1+1$$
$$+1+1+1+1+1+1+1+1+1$$
$$+1+1+1+1+1+1+1+1+1+1$$

of how big the number is. Splitting a number into its factors is more interesting, as you hit dead-ends which cannot be made simpler, and these cul-de-sacs are the prime factors. Mathematicians hold multiplication in higher regard than addition because addition predictably tells you how big a number is, whereas multiplication tells you a bit about its personality.

Factors explain some of the maths patterns we have already come across. When we were drawing the regular 2D shapes that had self-intersecting sides, there were five different options for eleven-sided shapes, but only two versions of the regular twelve-sided shape. This is because eleven is a prime and twelve has four factors. You only get a new stellated regular polygon if the number of corners between each edge is not a factor of the total number of corners. As you draw around the shape, you can see how some numbers mean you have to avoid the corner you started at until the absolute last moment. So, for eleven sides, you can have a side for corners 1, 2, 3, 4 and 5, and none of them is a factor of eleven (hence five options); for twelve sides, only 1 and 5 are not factors (hence two).

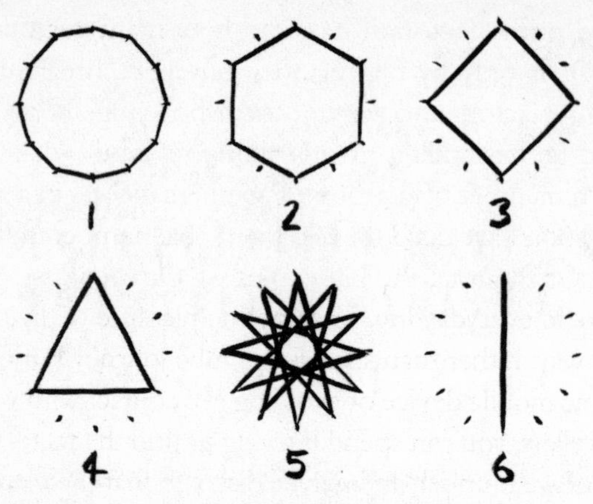

Only the edges between one and five corners hit all twelve corners.

These factors also explain why Dozenalists want to use base-12 numbers: 12 has more factors than 10. In general, the more factors a number has, the better it is as a base system. Because 10 only has 2 and 5 as factors, we can write ½ = 0.5 and ⅕ = 0.2 nice and neatly, but something like ⅓ has digits which never end: ⅓ = 0.3333. . . In base-12 we would be able to write digit versions of ½, ⅓, ¼, ⅙, ⅛, ⅑ and many more. Still not worth swapping, in my opinion.

Even cutting up a pizza links back to factors. We saw that there are two different ways to cut a pizza into forty-two pieces so that not all the pieces touch the centre. This is because of the factors of 42. If you look at how the pizza is divided, you get an initial number of pieces that is double the number of sides the shape of constant width has, and each of these can then be subdivided into two or more pieces. If you pull apart the number 630, for example, you can see why it gives you ten different ways to cut a pizza.

There is thus a delicate and subtle interplay between a number, how many distinct prime factors it has, how many repeated prime factors it has, and how many factors it has overall. It is only by playing with numbers, breaking them down into factors and putting them back together that you build an understanding of and intuition about what's going on. Each member of a group of mathematicians in a restaurant will soon have decomposed their table number into prime factors. Or the total the bill comes to. Or even just random numbers in everyday life. Imagine the pleasure of living your life this way: rather than spending a tube journey buried in an electronic mobile device or avoiding eye contact with your fellow travellers, you can spend it trying to find the factors in the train carriage number, and maybe discover that there are none, it's a prime and you're travelling in an A-list compartment.

One is the Loneliest Prime Number

It's worth clearing up an age-old debate here: is 1 a prime number? It certainly doesn't have any factors, which means that it's ripe to be a prime. Unfortunately, it isn't considered so. At least, not any more. For a long time, mathematicians were happy with 1 being a prime number, but from the eighteenth century on it gradually fell out of favour. Its downfall was that mathematicians found themselves constantly having to add the clause 'except 1' when talking about the things that prime numbers did. It just became easier to relegate it to the B-list.

These days, no one considers 1 to be a prime, but it was still in the running right up until the middle of the twentieth century. Before computers became prevalent, you could buy reference books which listed prime numbers. As late as the 1956 edition of *Factor Tables for the First Ten Millions* was printed, 1 was still listed as a prime number. This was not only the opinion of some fringe mathematician either: the author was an American mathematician named Derrick Lehmer. It was his son (also named Derrick Lehmer – classic Americans) who in 1934 was the first person to improve on Lucas's method for checking prime numbers. Even in 2013, the number $2^{57,885,161} - 1$ was checked using a Lucas–Lehmer primality test.

Modern students of mathematics are taught by rote to claim that 1 is not prime because primes need to have two different, distinct factors (themselves *and* 1), and for the number 1 this doesn't work because, $1 = 1 \times 1$, and 1 is itself 1; there's no 'and' in it. This argument feels unsatisfactory somehow: more like a legal argument hanging on the technicalities of how a definition is phrased. A better reason would give us some insight into how 1 would actually behave as a prime number.

The argument students should learn to use against 1 being a prime is not to recite some technical definition but to point out that 1 is completely incompetent at being a factor. Because multiplying by 1 leaves a number unchanged, you can scatter them wherever you fancy in a prime decomposition. For example, 28 could be $1 \times 1 \times 2 \times 1 \times 2 \times 7 \times 1 \ldots$ and so on. The uniqueness is ruined; the relationship between a number and its prime factors is no longer exclusive, and this is why 1 has been voted out of the prime club.

I personally don't think that mathematicians have gone far enough in voting out prime numbers. Both 2 and 3 are pretty lousy primes. In my opinion, 2 and 3 are only primes because there aren't any numbers smaller than them to be their factors. They're primes by default, not through any property of their own. I consider 5 to be the first true prime number, because it is greater than 4, the smallest composite number. If I had my way, we could reclassify 2 and 3 as subprime numbers, but that is unlikely to happen, as all other mathematicians disagree with me.

Love by Numbers

It is possible to buy split-heart keyrings and jewellery where the heart breaks into two pieces, one with '220' engraved on it, the other with '284'. People buy them, keep one half for themselves and give the other half to a loved one. I may have done this myself. The numbers 220 and 284 have been a sign of friendship and romance since the ancient Greeks and continue to be used by nerds in love to this very day.

The factors of 220 are 1, 2, 4, 5, 10, 11, 20, 22, 44, 55 and 110, which may not seem very special – until you realize that, if you add them all together, the total is 284. Still not that special? Add the factors of 284 (1, 2, 4, 71 and 142), and they come to 220. Because adding up the factors of each of the numbers gives you the other number, 220 and 284 are considered to be intimately linked, and have earned the name 'amicable numbers'.

These two are not the only options available. Fermat found the next pair of amicable numbers – 17,296 and 18,416 – in 1636, but to fit all the digits on, you're going to need a bigger keyring/piece of jewellery, and for those discovered by René Descartes in 1638 – 9,363,584 and 9,437,056 – it's going to be bordering on bling. By 1747, Euler had joined the game and was just showing off, finding about another sixty pairs. All of them missed the second-smallest pair – 1,184 and

$a^3wh = aaawh$.

1,210 – which was found by sixteen-year-old secondary-school student B. Nicolò I. Paganini in 1866. We don't know if he was motivated by love or maths.

You can find a list of all known amicable pairs and who found them here: http://amicable.homepage.dk/knwnc2.htm should you need a lot of them.

220 & 284	PYTHAGORAS	ANCIENT TIMES
1,184 & 1,210	PAGANINI	1866
2,620 & 2,924	EULER	1747
5,020 & 5,564	EULER	1747
6,232 & 6,368	EULER	1750
10,744 & 10,856	EULER	1747
12,285 & 14,595	BROWN	1939
17,296 & 18,416	FERMAT	1636

We still don't understand amicable numbers very well. For a long time, it was conjectured that every amicable number was a multiple of 2 or 3 – that is, until 42,262,694,537,514,864,075,544,955,198,125 and 42,405,817,271,188,606,697,466,971,841,875 were found in 1988. The conjecture was then upgraded to become that every amicable number was a multiple of 2, 3 or 5 – but a 193-digit counter-example was found in 1997. It is also conjectured that there is an infinite supply of amicable numbers, but even though there have been at least 11,994,387 pairs found to date, to be honest, I just don't know who to trust any more.

Numbers such as 12,496 are a variation on this theme. If

you add the factors of 12,496, they total 14,288, and the factors of 14,288 add up to 15,472. But if you continue adding up the factors, 15,472 gives you 14,536; 14,536 gives you 14,264; and 14,264 gives you 12,496, so we're right back where we started. But hey, it was a hell of a ride. Adding factors gives a chain of five numbers which loop around together, and these are called sociable numbers. There are sociable numbers in groups far bigger than five. Not so intimate as the amicables perhaps, but we're open-minded about these things. (You'll notice we're not using the original number as one of its own factors, which means we're adding the so-called 'proper factors', i.e. all the factors including 1 but excluding the number itself.)

But the best is yet to come! There are a few rare numbers which, when you add up their factors, come to themselves. The smallest such number is 6, because its factors are 1, 2 and 3 and 1 + 2 + 3 = 6; then comes 28, because 1 + 2 + 4 + 7 + 14 = 28. These were called perfect numbers by the ancient Greeks. The next perfect number is 496, then there's an even bigger jump, to 8,128. After that, it gets silly. The next one is 33,550,336; the one after that 8,589,869,056; far from hot on its heels, 137,438,691,328; and dragging its feet behind, 2,305,843,008,139,952,128.

The ancient Greeks had found only the first four, up to 8,128. The first appearance of 33,550,336 was in 1456, and then a further seven perfect numbers were found over the next half a millennium, the largest seventy-seven digits long. Since 1952, computers have allowed us to find a further thirty-six perfect numbers. The largest so far was discovered in 2013, and it weighs in at 34,850,340 digits long (the last digit is a 6). Amazingly, all its 115,770,321 proper factors add up to equal itself.

Finding this perfect number in 2013 was actually an offshoot of something we've already talked about: finding Mersenne primes. Every single perfect number that has ever been found is a multiple of a Mersenne prime. As long ago as Book 7 of *The Elements*, Euclid defined a perfect number as 'that which is equal to its own parts', then proceeded to prove that all Mersenne primes are a factor of a perfect number. Euler later proved the (subtly different) fact that all even perfect numbers have a Mersenne prime as a factor. (Euclid and Euler together at last, and not just because they're in alphabetical order). So, whenever we discover a new Mersenne prime, we get a new perfect number thrown in for free.

There is only one missing piece to perfect numbers: odd numbers. So far, we have only ever found *even* perfect numbers, but there is no reason there couldn't be odd perfect numbers out there somewhere. If they do exist, we know that they won't have Mersenne prime factors: they'll behave in a whole new way we haven't yet thought of. While it's generally conjectured that odd perfect numbers do not exist, the hunt is still on to try to find one. As always, this involves a huge amount of computational power, so there is, naturally, a distributed-computing project to try to find them. You can join in at www.oddperfect.org.

Pretty Prime Patterns

Because primes, as the building-block factors of all other numbers, are so significant to mathematicians, understanding them is of huge importance. Frustratingly, there's no good way to predict prime numbers: they appear to be randomly distributed among the other numbers. Only, they're not. They may be unpredictably distributed, but it's certainly not random.

The square of every prime number is one more than a multiple of 24.

That sentence can freak out even the most balanced of mathematicians. It feels as if the prime numbers shouldn't be that well behaved, yet if you pick any prime number from 5 onwards (it doesn't work for the subprimes, 2 and 3) and square it, the answer is always one more than a multiple of 24. How can that *be*? It turns out that the primes follow all sorts of amazing patterns; the problem is that none of them can be used to predict what the next prime number is. The 24 pattern doesn't work in reverse: not all numbers which are one more than a multiple of 24 are the square of a prime number.

A proof of how the 24 pattern works is in 'The Answers at the Back of the Book', but we can look at some more straightforward patterns here. Some of these can be nicely visualized by colouring in each prime in a number grid. In most grid-widths, the primes appear to be distributed haphazardly, as most people would suspect. However, when the grid happens to be a multiple of 6 wide, suddenly all the primes snap into dead-straight lines. They form an incredibly regular pattern, with only the subprimes breaking rank. All prime numbers are one more or less than a multiple of 6.

With some thought, you can see why the primes have to have this type of pattern: they simply cannot be in any random position. For a start, no prime number can be even, because all even numbers are divisible by 2. This is widely known and accepted; people are happy to exclude prime numbers from half of all possible number positions. The next step is to exclude all the spots that are divisible by 3; and 1, as discussed above, is not a prime, so the only remaining spots to survive are the numbers directly one above or below each multiple of 6.

Prime numbers in grids in various widths

1	**2**	**3**	4	**5**	6	**7**	8	9
10	**11**	12	**13**	14	15	16	**17**	18
19	20	21	22	**23**	24	25	26	27
28	**29**	30	**31**	32	33	34	35	36
37	38	39	40	**41**	42	**43**	44	45
46	**47**	48	49	50	51	52	**53**	54
55	56	57	58	**59**	60	**61**	62	63
64	65	66	**67**	68	69	70	**71**	72
73	74	75	76	77	78	**79**	80	81
82	**83**	84	85	86	87	88	**89**	90
91	92	93	94	95	96	**97**	98	99
100	**101**	102	**103**	104	105	106	**107**	108
109	110	111	112	**113**	114	115	116	117
118	119	120	121	122	123	124	125	126
127	128	129	130	**131**	132	133	134	135

Width = 9: fairly random looking.

1	**2**	**3**	4	**5**	6	**7**	8	9	10	**11**
12	**13**	14	15	16	**17**	18	**19**	20	21	22
23	24	25	26	27	28	**29**	30	**31**	32	33
34	35	36	**37**	38	39	40	**41**	42	**43**	44
45	46	**47**	48	49	50	51	52	**53**	54	55
56	57	58	**59**	60	**61**	62	63	64	65	66
67	68	69	70	**71**	72	**73**	74	75	76	77
78	**79**	80	81	82	**83**	84	85	86	87	88
89	90	91	92	93	94	95	96	**97**	98	99
100	**101**	102	**103**	104	105	106	**107**	108	**109**	110
111	112	**113**	114	115	116	117	118	119	120	121
122	123	124	125	126	**127**	128	129	130	**131**	132
133	134	135	136	**137**	138	**139**	140	141	142	143
144	145	146	147	148	**149**	150	**151**	152	153	154
155	156	**157**	158	159	160	161	162	**163**	164	165

Width = 11: still looks chaotic.

1	**2**	**3**	4	**5**	6	**7**	8	9	10	**11**	12
13	14	15	16	**17**	18	**19**	20	21	22	**23**	24
25	26	27	28	**29**	30	**31**	32	33	34	35	36
37	38	39	40	**41**	42	**43**	44	45	46	**47**	48
49	50	51	52	**53**	54	55	56	57	58	**59**	60
61	62	63	64	65	66	**67**	68	69	70	**71**	72
73	74	75	76	77	78	**79**	80	81	82	**83**	84
85	86	87	88	**89**	90	91	92	93	94	95	96
97	98	99	100	**101**	102	**103**	104	105	106	**107**	108
109	110	111	112	**113**	114	115	116	117	118	119	120
121	122	123	124	125	126	**127**	128	129	130	**131**	132
133	134	135	136	**137**	138	**139**	140	141	142	143	144
145	146	147	148	**149**	150	**151**	152	153	154	155	156
157	158	159	160	161	162	**163**	164	165	166	**167**	168
169	170	171	172	**173**	174	175	176	177	178	**179**	180

Width = 12: Bam! Straight lines!

There's no reason why we should stop there. If you block off all the multiples of 5, the newly excluded positions will give rise to a more complicated rule – something like 'all prime numbers are one more or one less than a multiple of 6, except not five above or below a multiple of 30' (this strikes out 5, 25 & 35, 55 & 65, 85 & 95, etc.). While describing the patterns can become unrealistically complicated, you can continue to strike out the multiples of 7, then of 11, 13, and so on, leaving the prime numbers in their wake. Called the Sieve of Eratosthenes, this method of filtering out composite numbers is a great way to find smallish primes.

An even better pattern involves putting numbers in a spiral, and it was first discovered by none other than someone who was just trying to pass the time in a boring lecture. The next time you're bored in a class or meeting, remember that your doodlings could lead to a new maths breakthrough. In this case, the doodler was the Polish-American mathematician Stanisław Ulam. Famous for his maths skills as well as his involvement in the Manhattan Project (he worked on the 'explosive lens' effect, which compresses plutonium to its critical density), what he discovered is now called the Ulam spiral.

Ulam began by writing numbers in a spiral and circling the primes. To his surprise, they started forming patterns. In his words, the prime spiral 'appears to exhibit a strongly non-random appearance'. The primes seem to form straight diagonal lines, both when there are only a few numbers and also when you zoom out to look at thousands of numbers at once. There are other patterns as well, my favourite being the blank horizontal straight line which starts 8, 9, 10 and continues 27, 52, and so on. No matter how far you zoom out, that line stands out for its stark lack of primes. The Ulam spiral is a fantastic reminder of the structure behind the prime numbers.

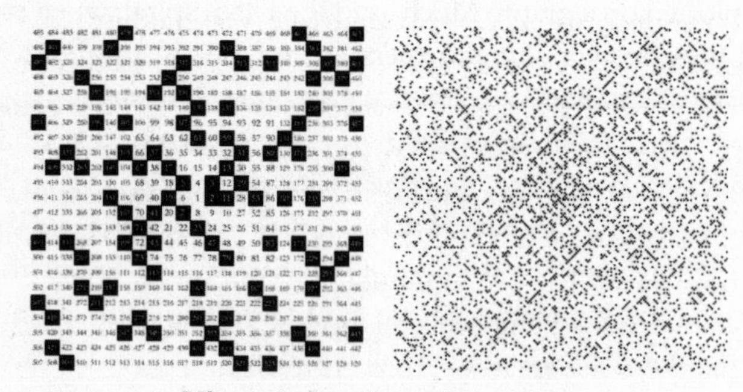

Ulam spiral up close and zoomed out.

Yet all these patterns pale in comparison to the most famous of prime patterns. There is another alignment of the primes, which has mathematicians obsessed. This is one which threatens to offer real insight into the prime numbers; it may allow us finally to understand why these numbers fall the way they do in the numerical order. First suggested in the 1859 paper 'On the Number of Primes Less than a Given Magnitude' by Bernhard Riemann, it is now called the Riemann Hypothesis. Why it did not get the correct name, Riemann Conjecture, is itself open to conjecture.

The Riemann Hypothesis relates to how many primes there are below a given limit. Even though the prime numbers appear to turn up in random locations, their density, on average, seems to be very predictable. Mathematicians think they have a way to calculate this density but, for us to know for sure it's giving us the correct answers, the Riemann Hypothesis needs to be proved as well. Sadly, the Riemann Hypothesis has resisted all attempts. There's even a $1 million bounty on its head.

The Riemann Hypothesis itself involves something more complicated than putting numbers in a spiral; instead, the numbers are put through a complex process and the results are plotted on a graph. Much as in the Ulam spiral, when you plot these results, some of them line up. But, in another way, this is not like the Ulam spiral, with its few glimpses of alignments: here, there is one dead-straight line of points with *no* exceptions. The Riemann Hypothesis is now infamous as one of the most difficult problems in maths: no one knows why that alignment is there. However, we do know that cracking this problem will give us an insight into the primes, and therefore an insight into all numbers.

Twins

There's something else that has baffled mathematicians about prime numbers, and only in 2013 have we found a possible path to the answer. This mystery is how many twin primes there are. Twin primes are prime numbers that are two numbers apart, such as 5 and 7; and 29 and 31. Unlike normal prime numbers, no one has ever managed to show that there are an infinite number of twin primes. Another way to state the problem is: can you prove that there's a highest pair of twin primes? And we simply don't know the answer to that question. They could stop, or they could go on for ever.

There have been some breakthroughs. One of them was made in 1915, by Norwegian mathematician Viggo Brun. He knew that if you start adding all the prime fractions ($\frac{1}{2} + \frac{1}{3} + \frac{1}{5} + \frac{1}{7} \ldots$) the answer will keep getting bigger as you add more. Brun then managed to show that, if you add the twin prime fractions together, the answer does not keep getting bigger. It approaches a set value, which we currently estimate to be 1.902160583. . . (this figure doesn't end here, but I thought nine decimal places were enough; more are, however, available on request), no matter how many you keep adding. This figure is known as Brun's constant, and although it doesn't solve the problem, it does provide us with a pretty hefty clue: it means that either the twin primes stop at some point, or that they become very scarce indeed.

The big breakthrough came in 2013, which was an amazing year for primes. On 13 May a rumour spread across the Internet that mathematician Yitang (Tom) Zhang had managed to prove that there are infinitely many primes only 70 million numbers apart. Yes, 70 million is a lot bigger than 2, but it was the first time anyone had managed to put any upper bound on how far apart primes could be. If Zhang was right, then no matter how big numbers become, there will always be primes within 70 million numbers of each other. At 3 p.m. that day Zhang presented his work at Harvard, and it was accepted as correct.

The breakthrough here is that there is any limit at all, regardless of how high it may be. Zhang's actual limit was 63,374,611, but he rounded it up to 70 million, because the details didn't matter. At least, they didn't . . . to start with. In maths, once base camp has been established, the challenge is on to regroup and complete the climb. A few people took some early easy wins, with Tim Trudgian bringing the limit down to 59,874,594 and Scott Morrison claiming it to be

59,470,640 in late May 2013. Commenting on his slight edge over Trudgian, Morrison said, 'I just couldn't resist momentarily "claiming the crown" for the smallest upper bound on gap size.'

Morrison managed to tighten his grip on that crown to the tune of 42,342,946 by 31 May – until Terence Tao took it a day later with 42,342,924. Australian-born Tao was a maths prodigy, having completed a maths degree by the age of sixteen and then moved to the USA to do his maths PhD. In 2006, he won a Fields Medal, which is the highest award possible in mathematics; often described as the Nobel Prize of mathematics, it is actually harder to get than a Nobel Prize and therefore superior. Like many maths luminaries past and present, Tao could accidentally give the impression that you need to be a hyper-genius to partake in maths. But that is not the case.

Tao started an open project as part of Polymath in which anyone could contribute in bringing down the 'bounded gaps between primes'. As of 20 July 2013, the bound has come crashing down to 5,414 through an incredibly collaborative and chaotic effort. The Polymath project itself was started by another Fields Medal-winning mathematician Tim Gowers to encourage massively collaborative mathematics. When I asked him how the project was going, he was impressed by the level of amateur mathematician involvement, both in writing and running computer code for some of the projects, as well as contributing interesting remarks. It isn't just the heavy maths-hitters working on these things; all sorts of people can contribute and, in this case at least, plenty were keen to. In Gowers's words, 'They had never lost their love of mathematics, but not being part of an academic set-up did not have a way of pursuing it.'

Of course, not all of us can contribute at such a high level, but we can still have a go. All around the world, people meet in pubs once a month for MathsJams, where they enjoy playing maths games and solving puzzles. You may well be one of them. Regardless of how advanced mathematics may become, it's still driven by that same motivation to play around with patterns and explore things for the sake of it – even though the results may end up being unreasonably useful. Google, a company which excels at finding pragmatic ways to apply pure mathematics, hasn't forgotten the playful origins of the maths now driving their company. In a 2011 auction for a suite of patents, their competitors were scratching their heads at Google's bid of $1,902,160,540. It was a nod to Brun and his prime numbers.

This is the great irony of prime numbers: for millennia, they were the poster-child of pointless maths but now they are vital to our modern lives in very practical ways. As late as 1940, G. H. Hardy used prime numbers as a perfect example of useless but harmless mathematics. All that changed with the rise of digital communication, and now everything from the Internet to smartphones would not function without our understanding of the properties of numbers.

Prime numbers are used to encrypt information because they can be almost impossible to find without secret information. If you multiply two large enough prime numbers together and give the answer to someone, if they don't know one of the secret primes you started with, they'll find it nearly impossible to work backwards and calculate the other one. This is why it took twenty-seven years to find the factors of $2^{67} - 1$, even though people just knew they were in there somewhere. Using secret prime numbers in difficult-to-reverse calculations like this is the foundation of modern cryptography.

And this means that not only are mathematicians trying to understand prime numbers to sate their curiosity about numbers, but that other people want to find the same patterns for various nefarious reasons. The coming decades of research into prime numbers are going to be very interesting indeed . . . Mathematics is nearly in its prime.

Eight

KNOT A PROBLEM

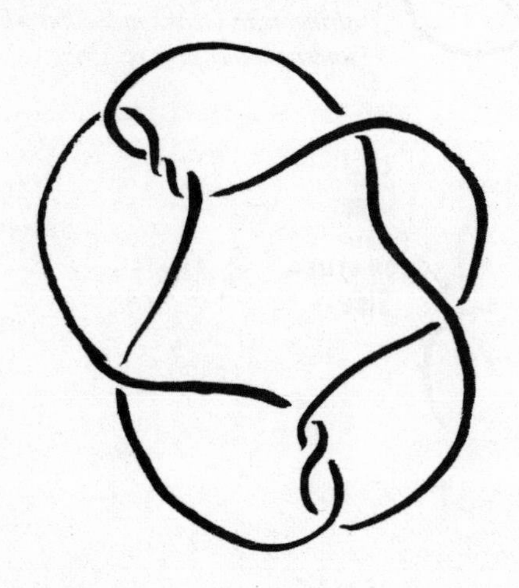

Bacteria are better than humans in one very important area of maths: undoing knots. When bacteria reproduce, their DNA becomes tangled and, if they are unable to unknot it, they will cease to function and will die. All living things use enzymes called topoisomerase to untangle DNA, but in bacteria they have a different variation, Type II topoisomerase. This enzyme is incredibly good at finding the perfect place where a long strand of

DNA crosses itself in order to make a snip in one piece and join it back together on the opposite side of the other piece.

A knot. To avoid any confusion right from the get-go . . . All maths knots are drawn in a closed loop, with the two ends joined together, so you can move them around without an end slipping out and the knot coming undone. Also, knots are drawn as a single line for the string, with breaks in the line when it goes under another piece of string.

Switching this one bit of the string to go over instead of under completely undoes the knot.

So Type II topoisomerase is able to find the optimum places to cut and perform switches in DNA: these chemical molecules inside cells can de-knot a knot far more efficiently than top mathematicians at the top of their game. We have some catching up to do.

The rewards of catching up are great. Because humans and bacteria have subtly different topoisomerase, it is theoretically possible to target only those undoing knots in bacterial DNA. If bacteria cannot multiply because their DNA

Microscope image of knotted DNA.

is tangled, then the infection they are causing cannot spread. In a very similar way, when human cells run amok and become cancerous, tying their DNA in a knot could be a way to stop a tumour from growing. A future wave of anti-bacterials and medical therapies therefore depends on us expanding our mathematical understanding of knots.

This is Knot Maths

Knots may not seem strictly to belong in maths but, then again, any situation is fair game for an attack using mathematical methods. For a long time, knots were only really studied by sailors and, up until very recently, the authoritative guide was 1944's *The Ashley Book of Knots*. This book contains over two thousand different knots, all categorized by type or usage. This is a very practical book, intended to be a guide for people actually tying knots. Mathematicians, however, became interested in knots because they wanted to see if it was possible to calculate the best way to *undo* a knot.

It turns out that understanding knots mathematically is much harder than we expected. Even a knot as simple-looking as this one has mathematicians baffled. If you

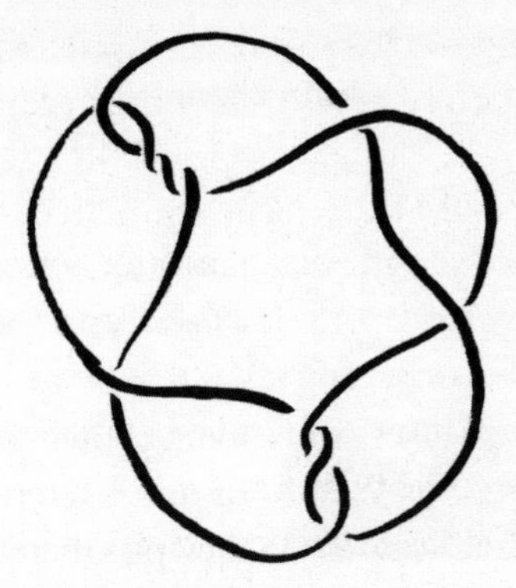

Can you undo this knot in fewer than three crossing switches?

were to make this knot out of string, you could try to untangle it by cutting and rejoining bits, exactly like topoisomerase enzymes do to DNA. The place where a knot crosses over itself is, sensibly enough, called a crossing, and if you cut it and move one piece to the other side, this is a crossing switch. We know that you can definitely untangle this knot with three crossing switches, but no one has managed to find a way to do it with only two.

Let's Get Physical

Before we get into the abstract maths of knots, we can have a look at physical methods to solve knot (knotty?) problems. One perpetual source of tangles is headphones. It seems whenever I get my headphones out they have managed to snarl themselves up, and I would love to find a way to avoid this happening. Mathematicians thought they had a way to reduce the likelihood of something becoming tangled: to join the ends together to form a loop. The loop conjecture proposed that if something is a loop, it is much less likely to become tangled than if it were an open long string.

Instructions

1. Find a piece of string.
2. Jumble it around in your hands for ten seconds.
3. Find the two ends and pull them apart to see if the string has become tangled.
4. Keep track of how often this happens and repeat until you become bored.
5. Now tie the string in a loop and repeat the experiment.

One way to check if the loop conjecture is on the right lines is to put it to the test. Jumbling up pieces of looped and unlooped string to see how often they become tangled will not prove or disprove it for certain, but it may provide some pretty good evidence in one direction or another. However, doing this repeatedly, physically, with actual string, can become rather tiring. To save yourself the trouble, why not enlist hundreds of school students to do it again and again in their maths lessons and collect all the results together? This is exactly what the Great British Knot Experiment did in 2010.

At one school alone (Coundon Court, a school in Coventry) the students ran over five thousand experiments on eleven different lengths of string, looped and unlooped. They showed that tangles were reduced by a factor of 2.09 when a length of string was looped, delightfully close to what the loop conjecture predicted. Mathematically, we would like to prove this for certain, because no matter how many times you try it with string, the results could just be happening by accident. However, for all practical purposes, the moral of the story is that if you want something to be much less likely to become tangled, make sure it forms a closed loop.

If you look at the end of most headphones you'll see a small clip that can be used to join them together. I don't think anyone is sure what this clip is for. On my latest set of headphones, the instructions made no reference to it or its intended purpose. My theory is that someone at a headphone manufacturer knows about the loop conjecture, or discovered it independently, and this clip is for joining the two earbuds together and forming a loop. Try it: it certainly works . . . or at least it doesn't work half as often.

We can now do our own Great Knot Experiment to try to unmesh the knot on p. 158. With the sheer number of people reading this book, I think we can succeed where mathematicians have failed. Find a piece of string and tie this knot in it.

So, the puzzle is: how many crossing switches does it take to go from this knot to an unknotted loop?

To start with, your knot will look exactly the same as the diagram, and we already know that in this arrangement it cannot be undone in only two crossing switches. For it to be possible, you need to move it around so that different parts of the knot cross each other. It's the huge number of ways to arrange the knot which makes establishing the minimum number of switches required so difficult. But if we all try it, we can check loads of them. Together we can do it!

Instructions

1. Make this knot out of string.
2. Place it flat and take a photo of it.
3. Choose and somehow indicate the two crossing points where you will make switches.
4. Complete the switches and see if you get an unknotted loop.

5. If you do, email me the photo of the arrangement and position of the switches.
6. Wait for mathematical immortality to be yours.

In Theory

Some of the first mathematical theories of knots were developed by physicists in an attempt to understand the nature of the universe. In the 1800s, there was a raging debate in science which seems very odd to us now, because we know the outcome. Most people today do not even know it happened, but for a long time there were two competing theories about what makes up the universe: one arguing that matter was made of atoms, the other that it was made of knots. Yes, for a long time it was argued that the universe was filled with an undetectable ether and that the matter we experience is merely knots tied in that ether. Molecules were formed when these knots became entangled and linked.

The originator of this knot theory of matter was our good friend Lord Kelvin. We can be very dismissive now, because we know that atoms exist, but for a while it looked as if knots could be the answer. In those days, knots in an undetectable ether was as sensible an option to explain what matter is made of as particles too small to be detected (this was long before the discovery of protons and neutrons). Given that they could be holding the universe together, there was a sudden need to understand knots. The idea was that if you could find a way to categorize and tabulate all the different ways knots could form, you'd have a sort of periodic table of knots.

Thankfully, mathematicians didn't have to categorize every single possible knot you can tie in a piece of string: they already knew that there are prime knots. The area of maths dedicated to investigating knots is known as knot theory (with the mathematicians involved called knot theorists) and, much as the area of number theory uses prime numbers to learn about all the other numbers, knot theory is only really worried about prime knots, because, if we understand them, we are within reach of understanding all knots.

A number is split into its prime factors by dividing it until you're down to the smallest numbers possible. A knot is split into its prime knot factors by dividing it into smaller knots until they cannot be divided any further. To do this, take a knot and squeeze two bits of string together in the middle so they form a bottleneck with some of the knot on each side. Then cut through them both and tie each pair of ends together to separate the two sides. You've now divided one knot into two. If you take exactly the same steps in reverse, then what you have are instructions on how to add two knots together.

Combining and separating knots is somewhere between what we normally recognize as addition/subtraction and what we normally recognize as multiplication/division. To simplify things, it's known as knot addition and to indicate its differences to arithmetic addition, the # symbol is used as a kind of double, more complicated + symbol. As I mentioned above, once you hit a knot which cannot be separated

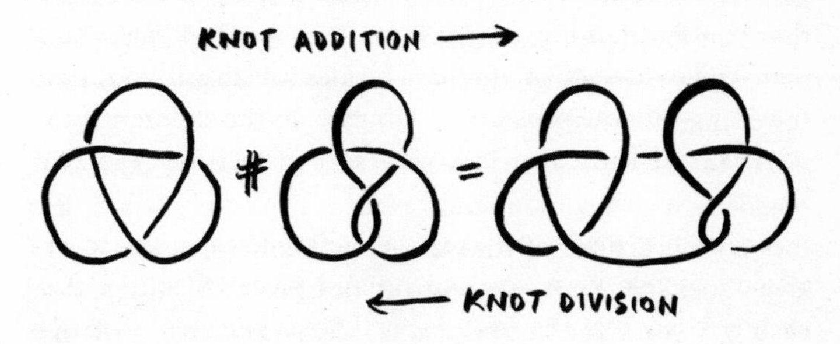

into two smaller knots, you've come to a prime knot. If you do try to separate it further, you'll just end up with the same knot and an unknotted loop. A loop without a knot is known as an 'unknot', and it behaves a bit like the number 1 in that adding the unknot to another knot is the same as multiplying something by 1: nothing changes. And like the number 1, the unknot is still a valid knot, but it is not thought of as a prime knot.

Knots can be simplified one step further because some of them are mirror images of each other. The absolute simplest knot possible is known as a trefoil, and it crosses itself three times. There are two different trefoils, each a reflection of the other, but they behave very similarly, so we need only worry about one of them. If we do want to talk about both, they are described as being the left-handed and right-handed versions of the knot. This 'handedness' in mathematics has been called chirality ever since Lord Kelvin took the Greek word for 'hand', *cheir*, and coined the word in 1894.

It was Scottish scientist Peter Guthrie Tait (also an accomplished sportsman, playing rugby and golf) who first came up with a system for categorizing the prime knots. His method was to look at the minimum number of crossing points a knot could have, such as the three in a trefoil knot. It is of course possible to move a trefoil knot around and give it more than three crossing points, but this is still exactly the same knot, just rearranged a bit. The trefoil therefore has a crossing number of three: it cannot be arranged to have fewer than three crossings.

If you have a knotted piece of string, there are countless positions in which you could place it on a flat surface, but they are all versions of the same knot. Mathematicians like to give things specific names to avoid possible ambiguity, and so each way you put a knot down is called a projection of that knot. If you look at the diagrams of knots in this book, each one is a 2D flat projection of what is actually a knot tied in 3D. Much as a cube can be projected as a hexagon or a square, the same knot can have endless different projections.

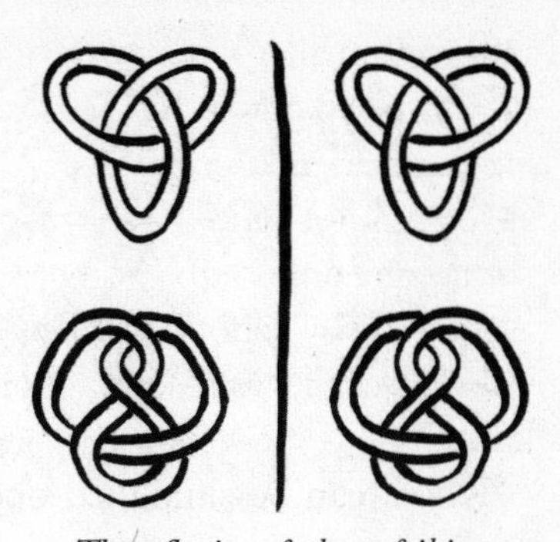

The reflection of the trefoil is a new knot; the lower knot can be rearranged into its own reflection.

If you take the mirror image of a projection of a knot such as the trefoil, it's a projection of the 'other-handed' version of the knot. Interestingly, there are some knots for which the mirror image of any projection is still the same knot. You could pick the original knot up and rearrange it into its own reflection. These ambidextrous knots are known as amphichiral.

As the trefoil is the only knot possible with a crossing number of three, it is given the official name 3_1. There is also only one knot with a crossing number of four (4_1), but there are two five crossing knots (5_1 and 5_2). (The subscripts have no meaning other than being an agreed way to name different knots with the same crossing number.) On the next page are all thirty-five different prime knots (and the one unknot) that have a crossing number of eight or below. As the crossing number gets bigger, the number of different possible knots explodes. To extend this table up to a crossing number of sixteen would require drawing an extra 1,701,900 different knots.

Starting with the trefoil as the simplest knot (the hydrogen of the knot world, as it were), physicists now had a growing table of the prime knots needed to build the other possible knots. Frustratingly, having done all this work on categorization and

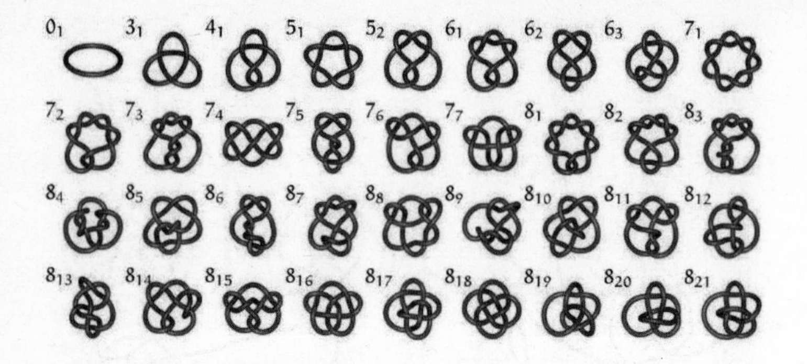

tabulation, it turned out that the universe was not made of knots. Ironically, it was the publication of the chemistry periodic table by 1872 that discredited the knot theory of matter. The numerical relationship between the masses of the different elements in the periodic table hinted at the existence of the proton – and the rest is history. Scientists forgot all about knots. Thankfully, however, mathematicians did not.

Colouring In

Which of these is a knot? They both look like tangled messes of string (or the cables behind my TV) but if you were to pick them both up and shake them around, one is the unknot, whereas the other can be rearranged to a minimum of ten crossings, at best. If the knots were much more complicated and you couldn't make them out of string, it would be incredibly difficult to tell if a projection of a knot could be untangled. With the scientists happily distracted by playing with their atoms, mathematicians took over knot theory and set about trying to find a way to calculate what is a knot and what is not.*

* Technically, the unknot is not not a knot, it's a trivial knot. Glad I could clear that up.

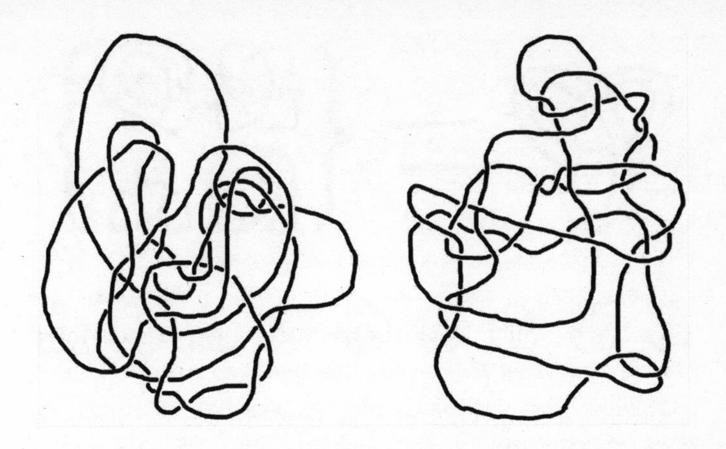

First, however, they stole some words from the scientists. In 1913, on the suggestion of the doctor Margaret Todd, chemist Frederick Soddy had created the word 'isotope' from the Greek root *iso*, meaning 'equal', and the Greek word *topos*, meaning 'place'. He used it to describe atoms in which, even though the number of neutrons varies, the number of protons remains the same and so their chemical properties do not change: they stay in the same place in the periodic table. Moving from one knot projection to another was therefore called an isotopy. The string may move but the knot stays the same.

A knot was now the same as the collection, or class, of all the possible arrangements it could be projected as: its isotopy class. The trefoil knot is, officially, the isotopy class of every possible way it is possible to put a piece of string tied in a trefoil knot down on a flat surface. The first goal of mathematicians was to be able to look at any projection of a knot and calculate if it was a knot at all. But could they work out if any random projection someone gave them was a member of the unknot isotopy class?

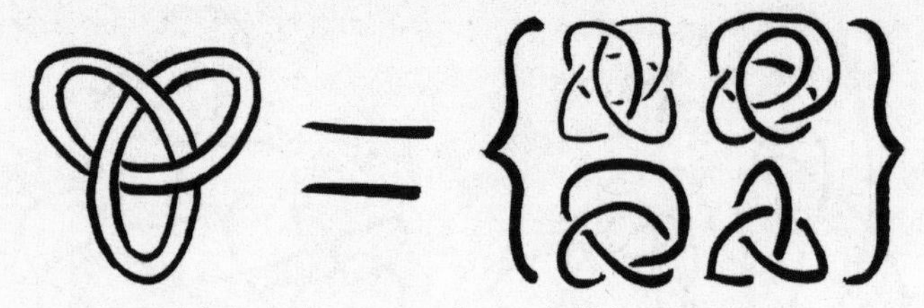

A knot is the same as the set of all its projections.

To do this, the mathematicians had to get their coloured pencils out. What they needed to find was a property of a knot which does not change regardless of how you move the piece of string around. They needed something about a knot which would never vary. They soon found one such invariant property: how the knot could be coloured in. It was discovered that if you can colour a projection of a knot using only three colours, then you can also colour any other projection of the same knot with three colours. Either all the projections in an isotopy class are tricolourable, or none of them is.

There is only one rule for colouring in a knot, and it relates to the crossing points: when the string goes under a different bit of string it must either be all the same colour as the piece on top, or it needs to be two colours different to the piece on top, changing from one to the other as it goes underneath.

Some coloured-in knots, all of which use three different colours.

It's impossible to tricolour the untangled unknot, i.e. no matter how it gets jumbled up, no projection of the unknot will ever be tricolourable. So if you *can* colour a projection of a knot in this way, then you know it *must* be knotted. It cannot be untangled back into an unknotted loop.

This work was started by the mathematician Kurt Reidemeister, who was a lieutenant in the German army during the First World War (in the lead-up to the Second World War, he was firmly against the Nazis, and this cost him his academic position). In 1932, he published a book of only seventy-four pages which revolutionized knot theory (but was not translated into English until 1983, the ninetieth anniversary of Reidemeister's birth). In it he demonstrated that moving between any projections of the same knot could be done in only three different moves, none of which changed the knot's tricolourability. These are known as the Reidemeister moves and form the basis of knot theory to this day.

Unfortunately, even though no tricolourable knot can be unknotted, the opposite is not true: if a projection cannot be tricoloured, it does not automatically mean it can definitely be unknotted. The knot overleaf cannot be tricoloured, but neither can it be turned into the unknot. If you start with the top crossing as either all the same colour, or all three different colours, neither of these options allows the tricolouration of

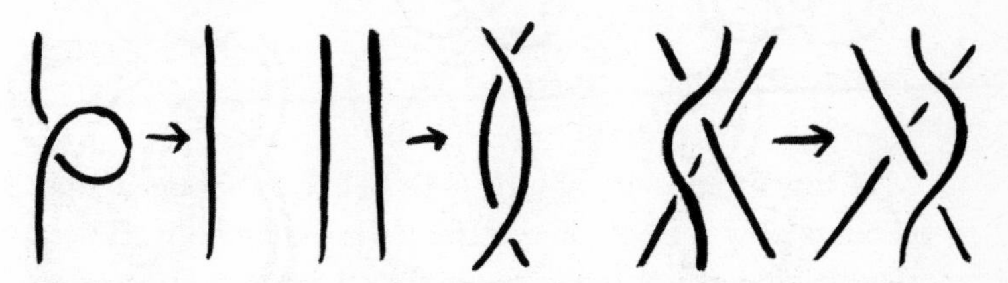

R1: untwisting a kink. *R2: dragging a loop over a different piece of string.* *R3: sliding a piece of string over a crossing point.*

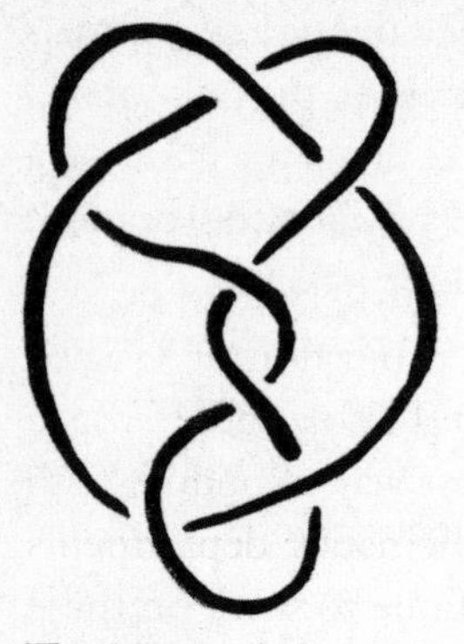

This knot cannot be tricoloured but yet cannot be unknotted.

the rest of the projection. While tricolourability allows us to find some knots, it does not catch them all.

Even if you are able to tricolour a projection and prove it is a knot, however, you still won't always know the best way to unknot it. The minimum number of crossing switches it takes to undo a knot is known as the unknotting number for that knot. Most knots we come across have a reasonably low unknotting number. The first knot to require two crossing switches is the 5_1 knot. The 7_1 is then the first case with an unknotting number of three and the 9_1 is the first requiring four moves. Of all 165 different ten-crossing knots (primes only, of course), forty-four of them can be undone in a single move, ninety-three require two moves, then fifteen need three, and a final four take four moves each.

Knots 5_1, 7_1 and 9_1.

As the more alert among you will have noticed, this accounts only for 156 of the 165 ten-crossing knots. There are still nine ten-crossing knots which humans are yet to work out the best way to untie. The unsolved knot I showed you earlier on p. 158 is the 10_{11} knot, and it is the current lowest-rank unsolved knot. Even if we crack it with our Great Knot Experiment (and I really hope we do), there are eight more ten-crossing knots to go. And that's before we even get to eleven-crossing, or more complicated knots.

The upshot of all this is that mathematicians do not yet have a cover-all method to look at a projection of tangled string or DNA and tell if it is in fact a knot; and, if it is a knot, we cannot definitely find the best way to undo it with minimum crossing switches. In mathematics departments around the world, knot theorists continue to work on these problems. Only now, instead of physicists waiting for the results, it's the biologists who need the answers, to develop a new wave of medical treatments.

Knot theorists have developed an amazing array of techniques and methods to investigate knots, but they're far from finished. Much as it took Lucas to revolutionize the search for Mersenne primes and Lord Kelvin to introduce new ways to look for space-filling shapes, I think we're still waiting for the key mathematical techniques needed to fully understand knots. I suspect a new generation of mathematicians will have to go to university and become knot theorists in order to give us such a breakthrough.

Undoing Our Good Work

Knots are traditionally used to tie things together. The knots in *The Ashley Book of Knots* were developed to be as strong as possible and not to come undone easily. However, as a bit of a challenge, mathematicians sometimes try to design knots and links which come apart really easily. For example, it's possible to link several loops together so that they cannot come undone – unless a single loop is broken, that is, and then the rest completely separate.

Your first challenge is to find a way to link three loops so that if you cut any one of them the other two separate as well. This arrangement is often called the Borromean rings, after an Italian Renaissance family. The rings appeared on their coat of arms, and the theory is that they represent the unity of a family: if you remove any one member, all the others suffer. The Borromean rings have been used to symbolize all sorts of three-aspect unities, including the Christian Holy Trinity. Hard to beat the Ballantine's Ale logo, though, where the rings represent the joint forces of purity, body and flavour required for great beer. Delightfully, a 'ballantine' is now another name for the Borromean rings.

This configuration can be extended to four, five or more rings; still, all of them link together and removing any one of them sets all the others free. If you can't think of any yourself, you can find some examples of these types of links with more than three loops in 'The Answers at the Back of the Book'. The Borromean rings work perfectly with flexible loops but, unfortunately, because of the way they weave through each other, you cannot loop three rigid circular rings in this way. It does, however, work with three rigid elliptical rings.

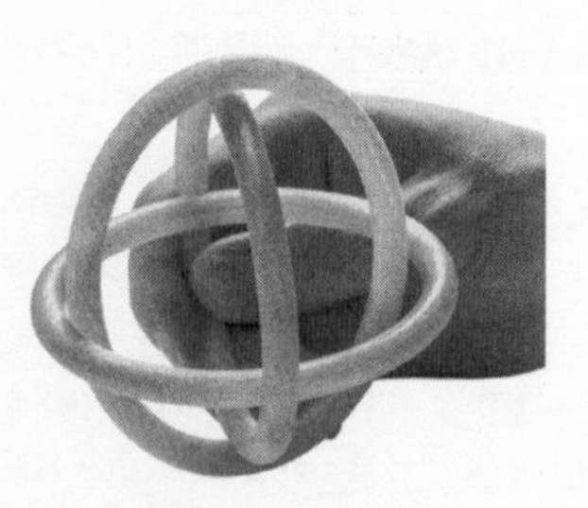

3D-printed elliptical Borromean rings

Finally, we can try to hang a picture in such a way that it can fall off the wall as easily as possible. Not something you'd want to do in your own home, perhaps, but perfect if you need an overly elaborate prank to play on a friend/enemy/combination of the two. A painting suspended by wire over a single hook in the wall is already in a perilous position. If that one hook comes out of the wall, the picture will go crashing to the ground. If, however, there are two hooks and the picture wire goes over both of them, then either one of the hooks could come out and the picture would continue to hang safely on the other one. Perversely, the real challenge is to hang a painting over two hooks so that if either one comes out the picture will fall.

There are sneaky trick ways to do this, the sorts of solutions you'd expect to a physics logic puzzle. You could balance the picture on top of both hooks, so that if either one came out the picture would fall. Or, the hooks could be so far apart that either one coming out would provide enough slack in the wire for the picture to hit the ground. Both of these would work if the end goal was to cause the painting to collide with the floor. Our end goal, though, is to see if it's possible that a wire placed around two points can become completely untangled by removing one of the points.

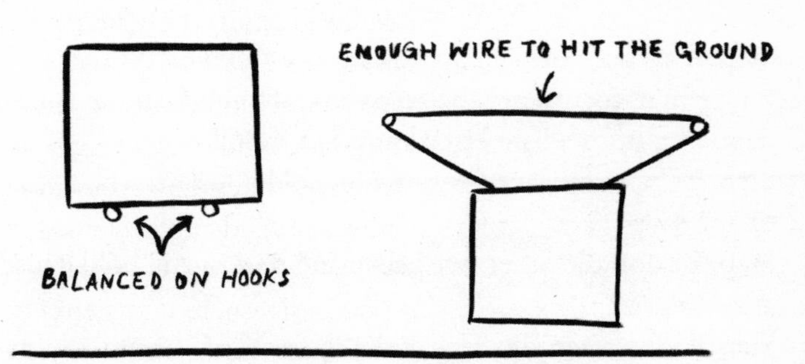

A quick guide to cheating at this puzzle.

There is, in fact, more than one way to do this. Not only that, but it is possible to hang a picture on any number of hooks in such a way that removing any one of them will cause the picture to unravel from all the others and fall to the floor. My record is five hooks. I had five people hold their arms out to represent a hook each, and we wrapped a very long ribbon around the arms to represent the wire. With one arm removed (and, it has to be said, a lot of gentle tugging to help it slip free), the ribbon eventually fell away from the other four arms. Not the most useful application of the maths of knots by a long way, but it kept us in stitches for a while.

Tie Your Shoes the Maths Way

Doing up your shoes (if you're not a member of the cult of Velcro, that is) is a daily chance to practise knot-tying, and I think tying them the same way every time is a wasted opportunity. There is a much quicker and easier way of forming the traditional shoelace knot, but everyone still seems to tie it the long way!

Instructions

Step 0: As per normal, loop the laces around each other and pull tight to form a neat bed for the knot. Fun fact: this foundation knot for all shoelace knots is a trefoil!
Step 1: Loop one lace forward and hold it just after the peak of the loop.
Step 2: Loop the other lace backwards and, again, hold it just after its peak.
Step 3: Place each shoelace under the peak of the other loop.
Step 4: Grasp the laces with the other hand and pull tight.
Step 5: Ta-da!

With a bit of practice, you can do this in a fraction of the time it normally takes to tie your shoes. To anyone watching, it will look as if you are just grabbing the two laces, passing them over each other and magically forming a knot. If they inspect the knot, there is no way to determine whether it was tied the long way or your fast way: it's the same knot. You can try tying one shoe each way and compare the knots; you'll not be able to tell them apart.

Nine

JUST FOR GRAPHS

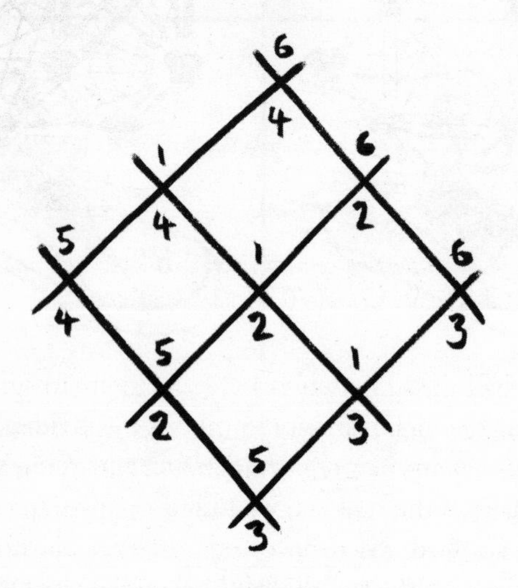

During the 2010 Football World Cup some mathematicians were able to predict correctly the winner of the final match between Spain and the Netherlands. To be fair, as there were only two teams to choose from, around 50 per cent of people who predicted the result got it right. As did an octopus. But it's the way these mathematicians at Queen Mary University of London (QMUL), made their prediction that is so interesting. For every team in the competition, they downloaded all the statistics for how many times the ball had been passed between each pair of players. They then made a network to show which pairs of players in each team had the most passes between them.

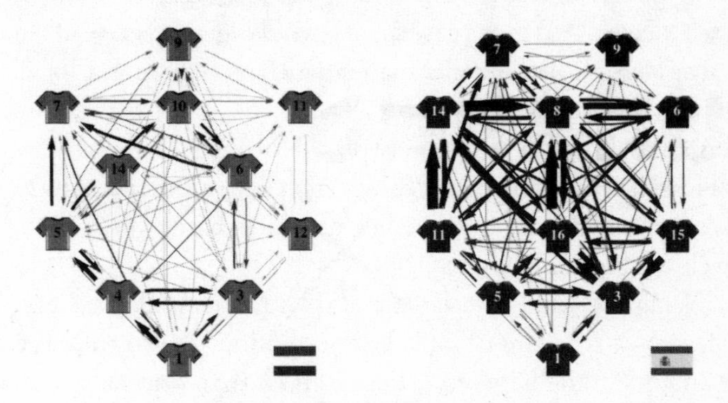

Football-passing networks for the Netherlands and Spanish 2010 World Cup teams.

These ball-passing networks gave them insights into a football team which were not immediately obvious when just watching the team play a game. During a match it's the goal-scoring players (the strikers) who get all the attention, but the networks showed who the other essential players were. A low-key player in the middle of the field may be vital in making sure the ball gets to the striker in the first place. Very few people remember the passes that happen before a goal, let alone recall and spot the patterns that occur in those passes. The networks made the progress of the ball around the pitch easy to see. Even more than this, mathematicians were able to compare the networks of two teams and see how they overlapped. They found this to be an interesting predictor of the outcome of a game, which is how they were able to forecast who would win the final.

If one team were to calculate a ball-passing network for an opposing team it's about to face, it could use that network to plan its strategy. One option is to find the player whose removal from the network would have the biggest impact on everyone else. If the team could then dedicate more defensive players to that player, this would give that team an advantage. Conveniently, mathematicians have put a lot of work into finding ways to calculate the most central and important points in a network. Normally, this kind of research is used to protect things such as computer and infrastructure networks from malicious attack, but it works just as well for football players.

What we need to calculate is then the centrality of each player – a measure of how connected they are to the other elements in the network. There is more than one way to calculate the centrality of a point in a network, and each offers a different type of mathematical insight. The crudest way would be to work out the total number of times each player passes the ball, but this would not take into account how each player connects to the rest of the network. A more nuanced method would be to add up how far away each player is from each of their team mates in terms of passes, a measure known as farness. But, like the England football team, we can do better.

At some point, across many games, most players will pass the ball to each one of their team mates, but for some pairs of players it will be only a few times, whereas for others it will be lots. The networks made by the QMUL mathematicians had thick lines to show strong connections where a lot of passes had been made, and thin lines where they were weak. It's possible to take each player and calculate the combined total of the links required to connect them to a different player; the average value of this for all their team mates is then a measure of their farness. This shows you which players are never very far from any other player in terms of passes, but it doesn't show you what would happen to the network without them.

The better option is to calculate how vital a player is as a link between other pairs of players. If two players want to get the ball from one to the other, then we can see if it has to go via certain other players. This is called betweenness. To find it, you calculate the shortest possible paths between every single pair of players, and then work out what fraction of the passes goes through a given player in the middle. This is what the mathematicians at QMUL did. For each football team of eleven players, mathematicians were able, for all fifty-five possible pairs of players, to calculate the shortest and strongest path of passes between them.*

This revealed the significant players in the middle of the field who had a high betweenness centrality; without them, it would become much harder to pass the ball around the pitch. If the opposing team puts extra players on these midfielders, it can disrupt the other team's play more than if it just targeted its high-profile goal scorers. By aligning the Spanish and Netherlands team networks, the QMUL mathematicians could see which would have the greater impact on the other. This is how they correctly predicted that Spain would beat the Netherlands in the final. That the mathematician Javier López Peña at QMUL who started all this is Spanish of course had no impact on his conclusion.

We can take the maths of networks and use it to solve a problem we've come across before. When you were building

* You can see the original graphs and read the published paper with the results on the site: http://www.maths.qmul.ac.uk/~ht/footballgraphs/.

those flexagons, you undoubtedly spent a while getting lost trying to get from one side to another: the 'number 5' face, for example, would suddenly become strangely hard to find. Working out which of the combinations of sides are possible requires knowledge of how you can move from one pair to another: there is a network behind every flexagon. By simply drawing that network, you can see how to navigate between the sides of a flexagon. Don't forget that convincing co-workers to do your bidding relies on your ability to find sides quickly, and possibly under duress.

So, if you draw the network of which combinations of sides are possible on a flexagon and link the pairs you flex between, you'll see a pattern emerge. If you straighten the links so all the occurrences of the same side sit on the same straight line, the pattern becomes more obvious: the links form a diamond network. This also shows that, even though both a hexatetraflexagon and a hexahexaflexagon have six sides, they behave in different ways. Some of the lines in the hexatetraflexagon network come to an abrupt stop, whereas the same lines in the hexahexaflexagon network carry on to give extra crossing points. The structure of these networks is

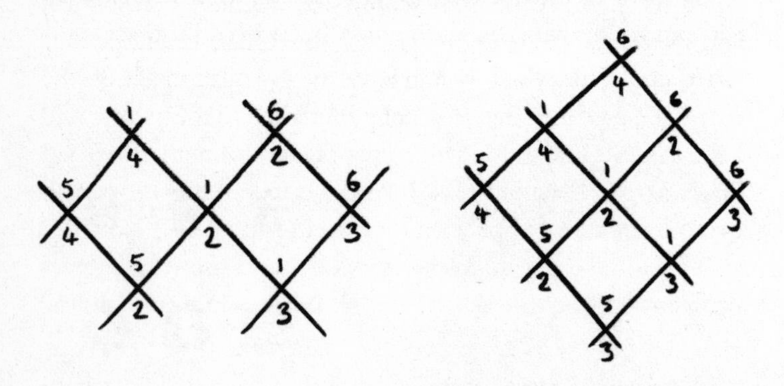

Combinations of sides possible for a hexatetraflexagon and a hexahexaflexagon.

a good predictor of how they behave when you flex them. Colleagues tend to be far more *un*predictable when you flex them.

The two types of networks we have seen so far differ in a few ways. The ball-passing network used lines of different thicknesses to indicate the strength of a link, and the flexagon graphs used only aligned straight lines going in particular directions. Each example shows a certain application of networks designed to solve certain problems. In general, for a network, we need only a set of things which are linked to each other somehow. This is the basis of mathematical networks.

The utilities problem

You have been put in charge of connecting three houses to all the three utilities: power, water and gas. The only problem is that the connections between the houses and the utilities are not allowed to cross each other. This puzzle is known as the utilities problem, and I am such a big fan that I have a mug with the houses and utilities printed on it. The glazed surface of a mug functions just like a whiteboard, so you can try to solve the puzzle over and over using a marker pen. Or you can try drawing on the version here. It's a bit of an odd

puzzle, though, because it's impossible – which is perhaps an extreme variation on the concept of a 'puzzle'.

As we did for football teams and the sides of flexagons, most normal people would call the lines linking these houses to the utilities a network, but for some reason mathematicians have decided to be difficult and also call it a graph. The mathematicians who work on the football networks are in fact graph theorists, and if you want to find their website you'll need to search for 'football graphs'. This area of maths is thus called both network theory and graph theory, depending on the person and the context. To avoid even more confusion, what is normally called a graph at school is called a plot, or chart, by mathematicians, so the word 'graph' refers only to networks.

Back to the whiteboard: the poser in the utilities conundrum is to draw a graph that links all three points (the houses) to another three points (the utilities). The points in a graph are commonly called vertices and the lines joining them edges. In this case, we want to make sure none of the edges cross each other. If a graph can have all its edges in the same plane without them crossing, we call it a planar graph. Unfortunately, the utilities graph is not a planar graph.

So let's leave it for the moment. Here's a different, non-impossible network challenge: link numbers with their factors without any of the lines crossing. For the numbers between 2 and 12, the factor-linking graph is a planar graph. And you can add the numbers 13, 14, 15, and onwards, and it still remains a planar graph. Eventually, however, you hit a limit and the next number you add cannot be linked to its factors while keeping the graph planar. See if you can find out what that limit is.

Most people find they can get up to about 18 on the factor-linking graph before they cannot easily fit any more numbers in. But if you move the numbers into new

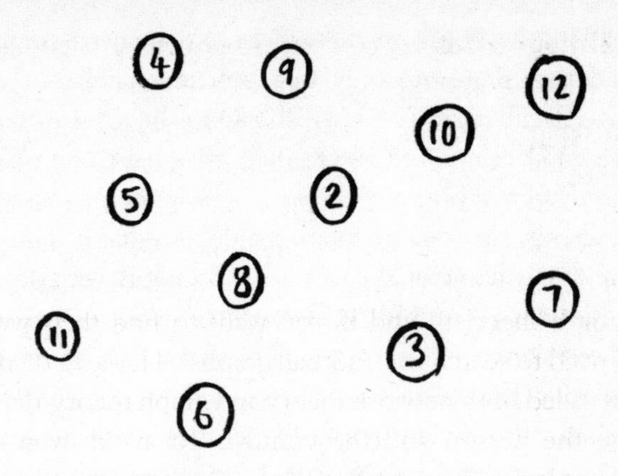

locations, you can find room to fit in 18, 19 and 20. Maybe even more. Solving this puzzle involves a lot of redrawing of the graph, but it's still the same graph. Much as we can have different projections of the same knot, we can have different arrangements of the same network. A graph is just a set of objects and the connections between them, regardless of how you draw it on a piece of paper.

Finding out if a graph is planar is a bit like trying to find the minimum crossing number of a knot: it's hard to find the perfect projection where the crossing number is at its lowest. A graph which is planar may have countless arrangements which contain loads of crossings, but somewhere there is at least one way to draw it with no crossings. (A little note on vocab here: actually *managing* to draw the graph so there are no crossings results in a plane graph; a planar graph has the *potential* to have no crossings.) The factor-linking graph with numbers up to (spoiler alert!) 23 is planar, but finding its plane-graph arrangement is the challenge. When you add 24 to the graph, it feels impossible to maintain it as a plane graph. But how do we know there is definitely *not* a way to do it, a way which we are just not clever enough to find?

In 1930, the Polish mathematician Kazimierz Kuratowski showed that there are only two fundamentally non-planar graphs. All other networks which cannot be drawn on a flat surface without their edges crossing must have one of these two fundamental non-planar graphs inside them somewhere. We've already met one of them: the utilities graph. The other has the catchy name of the Complete 5-Graph, with the nickname K_5, which means it's a graph of five vertices, all of which have been connected to each other. It's the first of the complete graphs (all vertices connected to each other) which is not planar, and all complete graphs bigger than it have it as a subgraph inside them somewhere, and they are all non-planar.

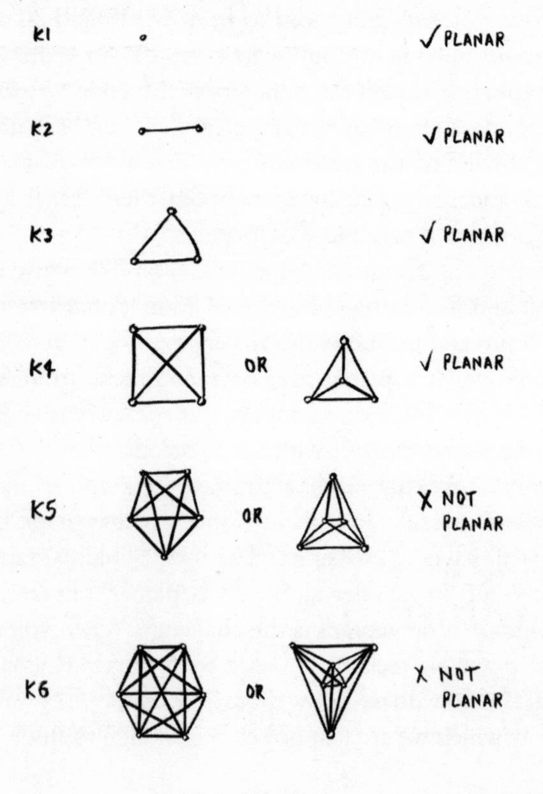

In our factor-linking graph, there must be a K_5 hiding within it (a subgraph) as soon as you add the number 32, because the factor only links between the subset of numbers {32, 16, 8, 4, 2} form a K_5 graph. However, the graph actually becomes non-planar before that: it becomes non-planar when you add the number 24. Subgraphs can be thought of as a little bit like prime factors, in that they combine to give bigger graphs, but there are some very important differences. Importantly, numbers and knots can be split only into the same primes, whereas a graph contains all sorts of different subgraphs. Subgraph decompositions are not unique, but they are still useful.

What we lose in subgraphs being a unique definition is compensated by how flexible they can be. There can be all sorts of interesting subgraphs hidden within a graph, and finding them is quite a challenge. When you add 24 to the factor-linking graph, a utilities graph appears inside it, but it's camouflaged by having extra vertices along some of the edges. We call this a subdivision of the utilities graph, and because it's a subgraph it proves that the factor-linking graph for 24 cannot be drawn without a crossing somewhere. If you're presented with some bizarre graph and you check that it definitely does not have K_5, the utilities graph or any of their subdivisions as subgraphs, then it is definitely planar.

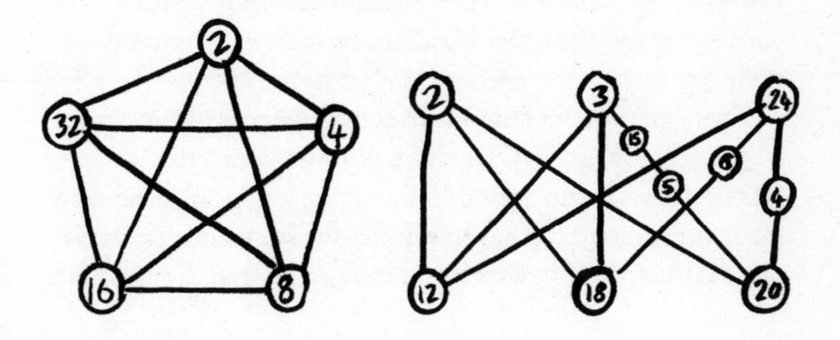

There are only two loose threads to wrap up now. The first is to show why the K_5 graph and the utilities graph are so special and why we know they are definitely not planar. The second is that it is in fact possible to solve the utilities puzzle. (This is why I have it printed on a mug, so I can show people there's a solution over breakfast, or a coffee break.) Importantly, drawing on a mug is not the same as drawing on a flat piece of paper. A hint there of the solution – but to go any further we first need to look at how graphs link to 3D shapes.

Of Shapes and Graphs

Take the dodecahedron you built previously (unless you were so proud of it you gave it away as a present, or had it permanently stuffed and mounted – in which case make a new one). Can you find a way to start at one vertex and follow the edges around so that you pass through every single vertex just once, arriving back at the vertex where you started? This was a puzzle first developed by the Irish mathematician William Hamilton in 1857. His version involved writing the names of major cities on the vertices and seeing if you could plan a 'world tour' to visit each city once. A path that passes through each vertex once is still known as a Hamiltonian path. If you cycle back to the same vertex where you started, on the other hand, that counts as a Hamiltonian cycle.

Trying to find this path on an actual dodecahedron can get a bit annoying, as they are difficult to draw on. But it's OK (particularly if you gifted/framed the dodecahedron you made previously): I've taken a dodecahedron and stretched it out flat into a graph, which you can use instead. It's possible to do this to any polyhedron. If you imagine puncturing a small hole in one of the faces, you can then stretch the whole shape out flat into a plane graph. (By the way, the use of the words 'vertex' and 'edge' for both polyhedrons and graphs is no accident: mathematicians view them both as versions of the same thing.) Ignoring the fact that the shapes of the faces are now distorted, a graph gives you exactly the same structure and linking of the vertices with edges as a polyhedron does.

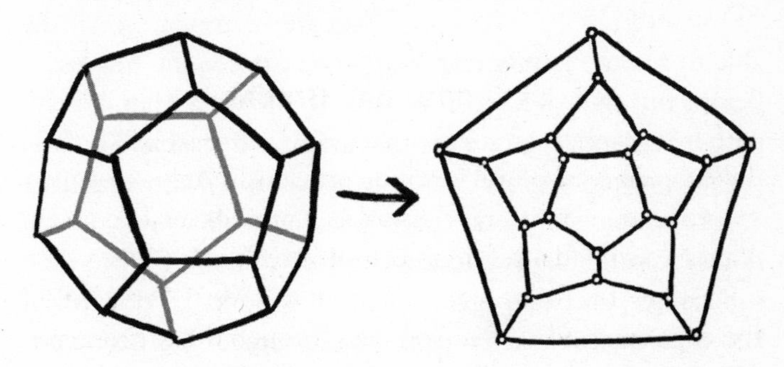

Vocab alert: Any path which does not use the same vertex twice is called, unsurprisingly, a path. If that path ends where it starts, it is known as a cycle. All polyhedrons have cycles, but a graph could have no cycles, or one or more cycles. The longest possible cycle is called the circumference; the shortest cycle is the girth. Mathematicians have a habit of taking synonyms from normal language like this and using them to represent different versions of the same thing. If a graph does not have any cycles at all, it's called a tree, and a collection of tree graphs is called a forest. I'm not kidding.

If you tried the puzzle earlier (see p. 48) to arrange the numbers from 1 to 16 so that if you add two adjacent numbers you always get a square number, what you were actually

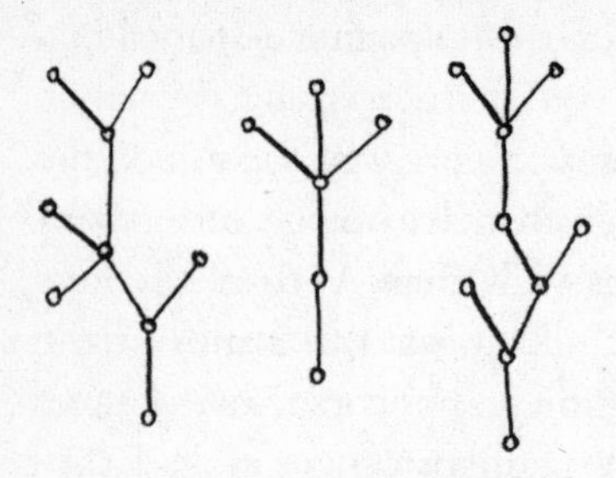

A forest of tree graphs.

doing, although you may not have realized it at the time, was trying to find a Hamiltonian path. If you draw a graph with the numbers 1 to 16 as the vertices and connect an edge between each pair of numbers which add up to give a square number, you'll be able to find a Hamiltonian path. You can see the full graph below, but with node 17 as well. If there's a Hamiltonian path in a graph, then we say that graph is traceable. Much as the factor-linking graph was only planar up to vertex number 23, the square-sums graph becomes traceable when you add node 15 and remains traceable only up to node 17. But then it becomes traceable again once it hits node 23, not for 24 and then traceable again from 25 onwards. It has been verified that a square-sums graph with nodes from 1 to 89 is still

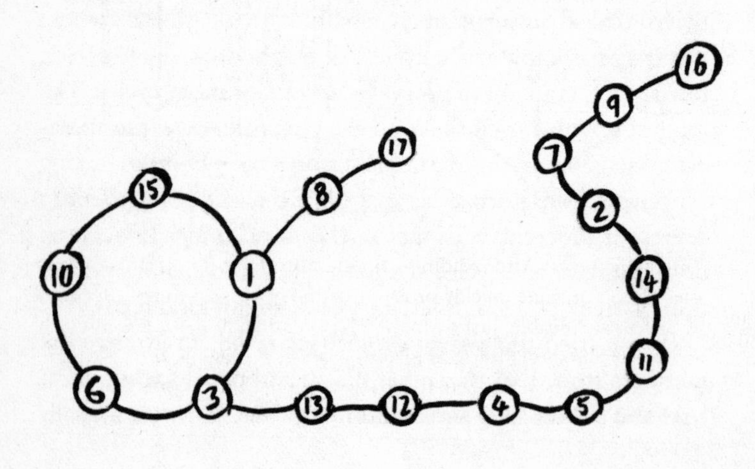

traceable,* and this is conjectured to remain true for all bigger numbers.

The links between graphs and polyhedrons are sometimes rather surprising. Let's say you have three objects on a table – in time-honoured tradition, let's call them A, B and C – and you want to draw a graph of all the ways to remove them from the table, one at a time. The graph would have vertices for all the possible collections of items on the table, linked with edges that represent removing an item. As with the flexagon graph, we can be systematic about the directions edges go in. Starting with ABC at the top of your page and moving diagonally down–left to remove A, directly down for B and diagonally down–right for C, will mean you coincidentally draw a picture of a cube. Maths is just full of these startling twists: who would have thought that the graph for removing three objects is the same as the graph for a cube? You can

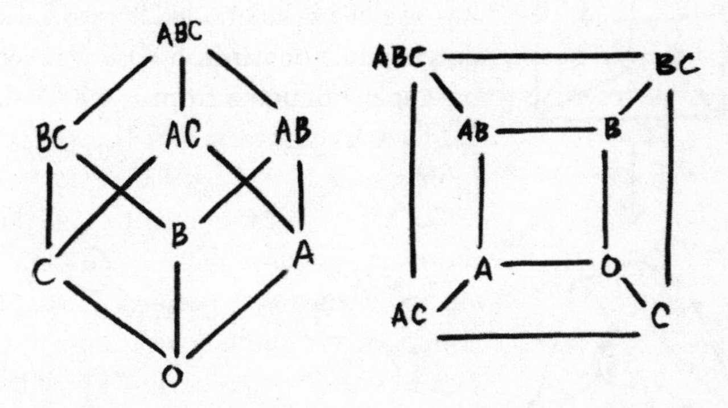

* Breaking news: After reading a final draft of this book, a mathematician friend of mine quickly proved that the graphs up to nodes 90 and 91 are also traceable, just to make what I'd already written technically incorrect. She thought it was hilarious. So the breaking news is: mathematicians can be jerks. Feel free to join in (checking graphs that is, not jerkhood) and check for higher values.

even rearrange it to be planar (A right, B down, C diagonal). The same graph for two objects will give you a square; and the graph for four objects . . . a mess.

This cube graph has a Hamiltonian cycle. In fact, all the graphs of Platonic solids have a Hamiltonian cycle and, as such, are called Hamiltonian graphs. However, this is not true of all polyhedrons. The smallest polyhedron graph which is not a Hamiltonian graph is the Herschel graph, named after the astronomer Alexander Herschel, but not discovered by him (he did work on Hamiltonian paths, just not this shape). You can play with the graph to convince yourself there is no Hamiltonian cycle, but building it as an actual polyhedron is slightly harder. However, the Newcastle-based mathematician Christian Perfect was able to do just that in 2013. He designed the Herschel polyhedron, which not only was as convex as we would hope but also delightfully symmetrical. You can cut out the net on the next page (all of the 2D faces waiting to be joined-up in 3D) to make one.

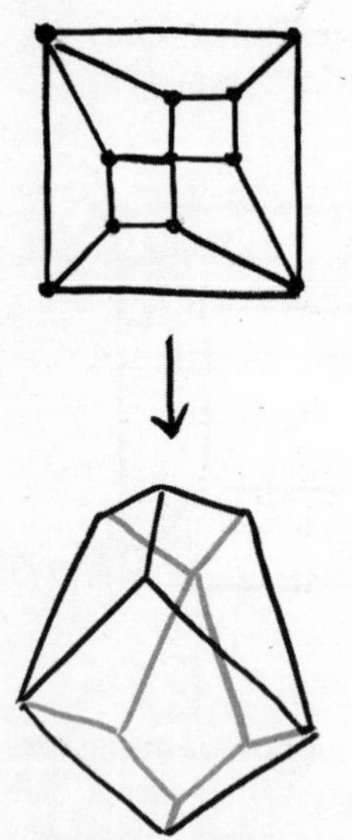

The Herschel graph and the Perfect–Herschel polyhedron.

We have already seen that all polyhedrons are graphs – but not all graphs are polyhedrons (specifically, simple convex ones without holes). It might be nice to find a way to be able to tell if a graph can be turned into a polyhedron and, it turns out, there's a very simple way to tell. It depends whether the graph is planar. All polyhedrons give planar graphs, and all planar graphs can be turned into a polyhedron (provided

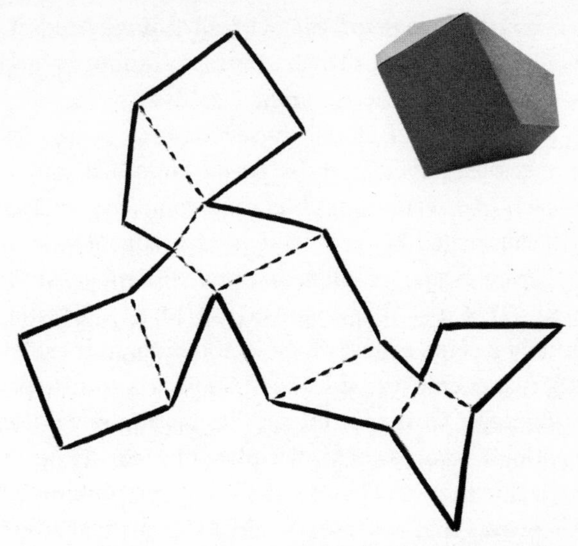

Cut out this net to make your own Perfect–Herschel polyhedron.

they're big enough).* This is exactly what you were doing when you sketched a random polyhedron on a balloon earlier: you were drawing a planar graph. This gives us an amazing new insight into graphs. As we saw before, all polyhedrons sketched on to a balloon have an Euler characteristic of 2, which means all planar graphs have an Euler characteristic of 2. This is exactly what we need in order to prove – or disprove – that K_5 and the utilities graph are planar. Everything comes to she and/or he who waits.

We can still calculate the Euler characteristic of a graph using the formula *faces* + *vertices* – *edges*, but counting the

* 'Big enough' officially means that it is impossible to break the graph apart by removing any two vertices. That said, all planar graphs, even those not big enough to be a polyhedron, still have an Euler characteristic of 2.

faces in a non-plane graph can get really complicated. A better method is to prove that there simply cannot be enough faces. For example, the K_5 graph has five vertices and ten edges so, to have an Euler characteristic of 2, there must be seven faces. But even if those seven faces had only three edges each (the fewest possible) that would require twenty-one edges touched by faces, and as each edge contacts two faces, eleven is the absolute minimum number of edges required ($21 \div 2 = 10.5$, rounded up to 11). But the K_5 graph only has ten edges, which is not enough. It can't have an Euler characteristic of two; it simply cannot be planar. The same logic works for the utilities graph, only it doesn't have enough vertices to form sufficient faces for its Euler characteristic to be 2. They are definitely not planar graphs.

This proves that you cannot add 24 to the planar factor-linking graph, and that the utilities problem cannot be solved. It may seem that it has taken a lot of effort to prove there is no solution to such simple puzzles, but that's how maths works. Mathematicians do it with rigour and for slightly longer than you expect.

It's not all wasted effort, though: these are mathematical techniques with benefits. We saw in Chapter 5 that shapes with holes in them have a different Euler characteristic. So far, we have only managed to prove that you cannot solve the utilities problem on a flat surface or on a polyhedron with no holes. If a shape has a single hole in it, the Euler characteristic is 0, and it turns out that the utilities graph can have *faces* + *vertices* – *edges* = 0 if you draw it on a doughnut. Or any shape with a hole in it. Including a coffee mug. This is why I have the puzzle on a mug: the handle is a built-in torus (doughnut shape, you doughnut). You can draw one edge under the handle and a second edge over the handle to reach the other side without any other edges having to intersect. This is the maths loophole which makes the utilities problem solvable. Worth waiting for, I think you'll agree.

Going Out and Colouring In

We can now use graphs to look at other situations, including going to a party. When you arrive at a party, there is a chance that everyone there already knows everyone else, or it could be the case that no one knows anyone else at all (making for an awkward party). It is more likely to be some combination of people who do know each other and some people who don't. We can use graph theory to say some things for certain about who will know each other at a party. For example, for any gathering of six people, there will definitely be three people who all know each other *or* three people none of whom know each other (for convenience, let's call them mutual strangers).

This seems unlikely, but you can check it the next time you're in a group of six people. One of these two cases will always be true. To confirm that this will always be the case, we can take the complete graph K_6 so that each edge represents two people who could know each other, then make the edges bold where the people are already friends. K_6 has fifteen edges and, as they could all be either bold or not bold, this is $2^{15} = 32{,}768$ possible friendship networks between six people. Which is a lot, but if we could flick through them and check them at the rate of one a second, we will have confirmed that all the networks contain either three mutual friends or three mutual strangers in just over nine hours. Individually checking all the possible situations to prove that something is always true is called proof by exhaustion and, given it takes nine hours of work, you can see why.

Unfortunately, this does not scale well for bigger groups of people. Adding one more person means the time it would take to do all the checking jumps from nine hours to just over twenty-four days. If you wanted to do an exhaustive check of all the possible networks for a party of twelve people, every human alive working together to check 7 billion networks a second would still take over three hundred years. And that's without lunch breaks. While technically possible, it's certainly entering the realms of the unfeasible. This is not what we would call 'in a reasonable amount of time'.

PEOPLE	NETWORKS
1	1
2	2
3	8
4	64
5	1,024
6	32,768
7	2,097,152
8	268,435,456
9	68,719,476,736
10	35,184,372,088,832

Above is a table for the first ten people who may gather in a group. As you can see, the number of possible friendship networks goes up at a truly alarming rate as more people join the party. After the seventh person joins, the number of possible networks is over 2 million, and at ten people it has passed 35 trillion. If we wanted to know something about the human population as a whole, just writing down the number of possible friendship networks for 7 billion people would require over 7 quintillion digits (that's eighteen zeroes). Although these numbers are still finite, they are rapidly becoming unwieldily large. Even if we did have powerful computers to employ, there are still too many cases for exhaustive searches to be an option when investigating large networks.

This is why we have graph theory. The first thing graph theorists can do to help with large networks is to reduce them down to fewer simple cases. Of the 32,768 possible friendship networks of six people, a lot of them have exactly the same structure, so we don't need to check them all.

All seventy-eight ways that six people can know each other. You might want to laminate this and take it to parties.

You are surrounded by five people ready to be amazed by graph theory.

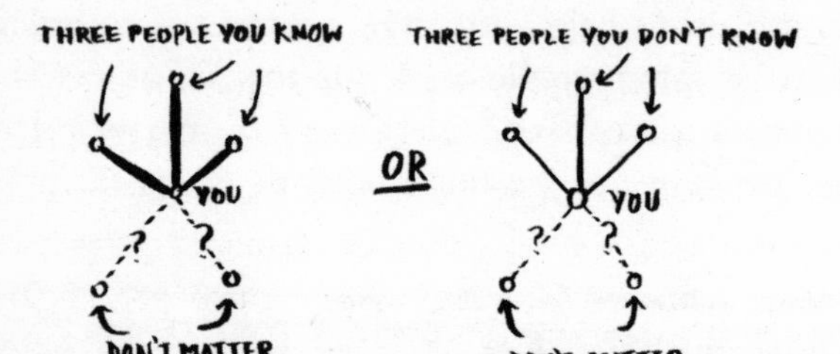

Of those five, your relationship to at least three must be the same.

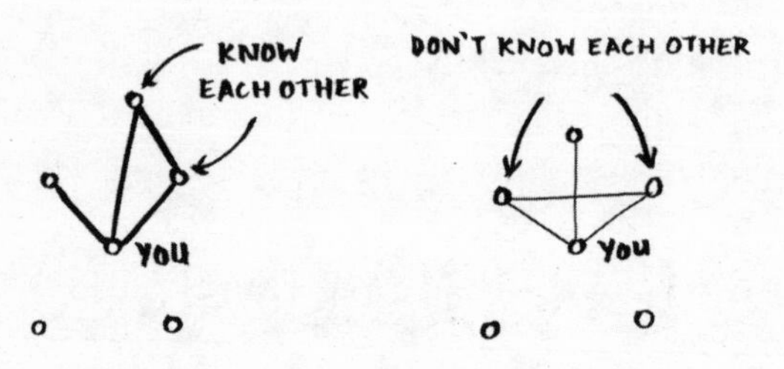

Either two of those people have the same relationship, closing a triangle . . .

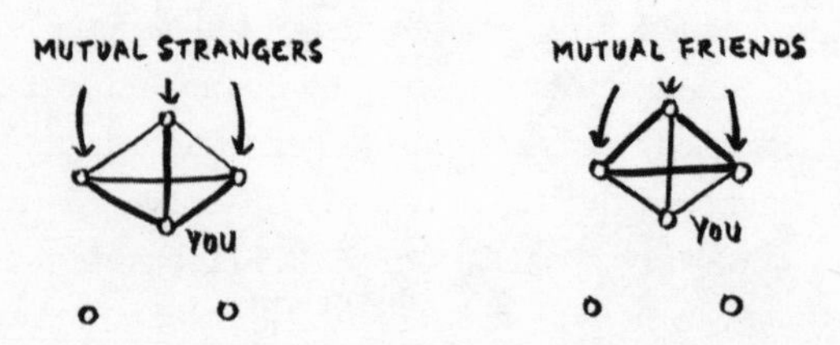

. . . or they all have the opposite relationship, forming their own love triangle (or lack-of-love triangle).

Mathematicians have managed to reduce all these networks to only seventy-eight different situations, which we can then check much more quickly. You can do all seventy-eight of them yourself if you want. This is still proof by exhaustion, as we are manually verifying all the cases, but by being clever and reducing the number of situations, it is much *less* exhausting. Leaving you with some energy left over for actually going to a party.

What graph theorists can sometimes also do – and it is much more helpful if they can – is do away with checking situations altogether. Imagine you are at a party in a group of six people. To break the ice, you can show them why three of you must form a triangle of either three mutual friends or three mutual strangers. Diagrams are optional, depending on the type of party.

This proves that in any network of six (or more) people there must either be one complete triangle of all friends, or one complete triangle of all strangers. And we've proved it without having to check all the cases.

Sadly, graph theorists often take a long time to come up with the non-exhaustive way to prove something, which is a shame. A proof by exhaustion will prove that something is true, but it rarely gives any insight into the mathematical patterns behind why it is true. It's nice to know that we're right, but mathematicians want to know *why* things are the way they are. There are things which we've managed to prove are true in maths, yet people are still working on them, trying to find better proofs. One such thing is how to colour in a map.

Colouring in maps may not seem that difficult to start with. Sure, you need to make sure that no two adjacent sections of a map are the same colour; otherwise, it will be ambiguous where one stops and the next begins, but that can be easily solved by using lots of different colours. However, early map makers noticed that they never needed to use that

Odd map out. Three of these maps can be coloured in with three different colours; one needs four.

many. In fact, any map you are faced with can be coloured easily using no more than five or six – indeed, they realized it could be done with four colours or fewer. You *can* make them more colourful if you want to, but the important point is that you don't have to: you can get away with only four colours and no two contacting sections need be of the same colour.

We could even try to use only three colours. Try colouring the maps drawn above using only three, making sure that, if any two regions contact each other, they are different colours. They range from easy, to hard, to impossible. Yes, one of these maps, you cannot colour in with three colours; a fourth is required. See if you can work out which one it is. It would not be possible to make the same sort of colouring puzzle for four colours; as I mentioned above, *every* map that could ever be drawn can be coloured in with four colours or fewer. There is no way I could throw one into the mix that would require five colours.

For a long time, mathematicians didn't know if there was a theoretical map out there somewhere that would require a fifth colour. There was a chance that, even though all known maps worked with four or fewer, there was some complex monster of a map on a distant horizon needing more colours. On 1 April 1975 the good Martin Gardner published a map in *Scientific American* which he said needed five colours. It was of course a mathematical joke, and while it *could* be coloured with four colours, it was fiendishly difficult to work out how to do it.

The Four-colour Conjecture was formalized in 1852, and by 1880 it seemed to have been proved to be correct two different ways, becoming the four-colour theorem. The 1880 proof was by Tait (who was also a knot theorist with Lord Kelvin), but a preceding proof in 1879 was by an English lawyer named Alfred Kempe. Sadly, they were both wrong, and in 1890 a 28-year-old upstart mathematician in Durham named Percy Heawood published a paper entitled 'Map Colour Theorems' in which he showed that there was 'a defect in the now apparently recognized proof'. So the four-colour theorem was demoted to the Four-colour Conjecture once more, where it would stay for nearly a century.

The 1879 proof deserves a closer look, because it so nearly worked. Alfred Kempe was an Englishman who had a joint love of mathematics and music but then became a lawyer. He certainly never stopped doing mathematics, though, either as a hobby or professionally. He gave a series of lectures at the Royal Institution called 'How to Draw a Straight Line' (later a bestselling book of the same name), was a fellow of the Royal Society from 1881 and its treasurer from 1898 until 1919: just long enough to see Ramanujan become the second-ever Indian fellow in 1918. His 1879 near-proof of the Four-colour Conjecture contained some amazing maths, but the fact that it was slightly flawed was a major source of embarrassment to him. His obituary in 1923 made no mention of it, despite it being perhaps his greatest contribution to mathematics.

The first step of Kempe's near-miss was to turn maps into graphs. Overleaf is a map of Australia with all its states marked. This can be made into a graph by having one node for each state, then, where two states contact each other on the map, they're joined by an edge on the graph. The problem now is to make the vertices different colours so that no two vertices joined by an edge are the same colour.

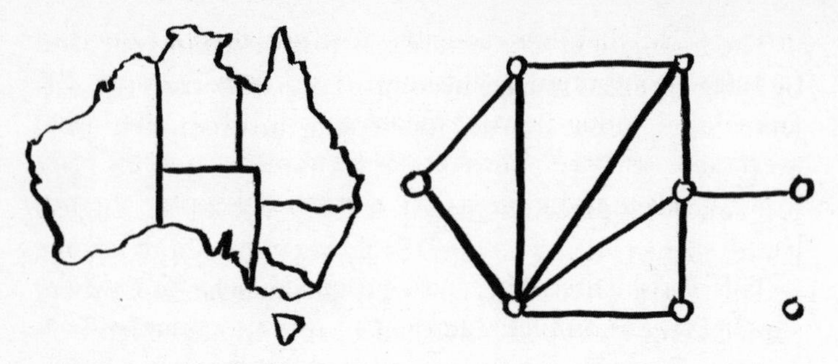

A map of Australia next to a map of Australia.

Kempe's attack was two-pronged: one tine was reducibility; the other unavoidability. The concept of reducibility was that a lot of complicated graphs implied the existence of smaller graphs with the same properties – just like any non-planar graph can definitely be reduced to either K_5 or the utilities graph. Unavoidability means that all the possible graphs in a situation will contain one of a few different subgraphs. The statement that any group of friends must contain either three mutual friends or three mutual strangers is a statement of unavoidability. Kempe showed that even though there are infinitely many possible maps, they unavoidably all contained one of a small list of subgraphs. Once this has been demonstrated, it is then possible to check all those subgraphs exhaustively and prove that they all reduce down to need only four colours. Using this method, he proved the five-colour theorem, but when he then tried four colours, he had too few unavoidable subgraphs. In other words, some graphs slipped past his exhaustive search, which is why counterexamples were later found.

Nearly one hundred years later, the mathematicians Kenneth Appel and Wolfgang Haken realized that for Kempe's method to work you would need to check 1,936 different unavoidable graphs exhaustively. Checking all these by hand/mind and proving that they are reducible was not feasible, so they turned to computers to automate the process. Despite the limited computing power available in the 1970s, they were successful, and the four-colour theorem became the first major theorem to be proved by computer. People were not happy.

This was the first-ever maths proof in which some of the steps had been completed automatically by computer, unseen by human eyes. To this day, not all mathematicians are totally convinced. In 1997, a more straightforward proof was produced (it used only 633 unavoidable cases, as well as simplifying other processes), but it still required computers.* That 1997 proof has been confirmed by computer (much like Hales is doing (update: has done!) with his Kepler conjecture proof), but while that may strengthen the proof, it does nothing to alleviate concerns over using computers.

This is where the four-colour theorem remains today, with most mathematicians convinced it has been proved (despite being deeply unsatisfied by the proof) and a few still considering it the Four-colour Conjecture. There has not been the second breakthrough, as with the six-people friendship network, a proof which builds up by logic, rather than by crunching down through all the possible cases. Yes, map makers can now sleep at night knowing they'll never need to buy a fifth colour, but again, for mathematicians, the quest is not truly over until we know *why* something is the correct answer, not merely that it *is* the correct answer.

* One of the mathematicians behind the 1997 proof has a great breakdown of it on their site and explains why we cannot just double-check the computers by hand: http://people.math.gatech.edu/~thomas/FC/fourcolor.html.

Ten

THE FOURTH DIMENSION

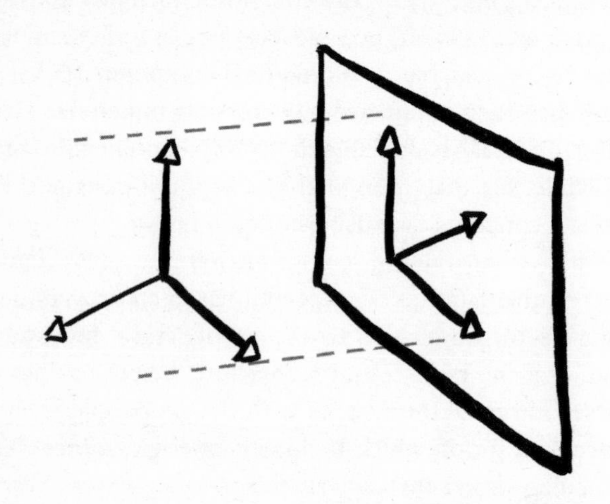

As humans, we live in a three-dimensional world. We are 3D creatures, and as a 3D creature I find no concept as terrifying as the notion of encountering a 4D creature. Such an organism would be god-like to us and, were it the slightest bit malicious, it could torment and destroy us at will. Humans are not equipped physically or mentally to deal with a fourth dimension, so any higher-dimensional being would have the ultimate tactical advantage.

There's a reasonably accurate description of an interdimensional fight in the comic *1963 – Tales of the Uncanny* (published in 1993). The story 'It Came from . . . Higher Space!' features a 4D protagonist attacking a 3D victim. It was written by comic-book legend Alan Moore (author of *Watchmen*, *V for Vendetta*, *The League of Extraordinary Gentlemen*, and many others) and the monster appears as a collection of hovering, disjointed body parts which move and morph around in the air. It starts as a series of wisps in mid-air which balloon out into 3D and are described as being almost like doughnuts made of meat. *Not* appealing.

We can explain the disjointed nature of the 4D monster by looking at what would happen if we, as 3D creatures, attacked a 2D creature. Being 3D means that the space we operate in extends in three different directions: side to side, backwards and forwards, and up and down. A 2D creature can move only in two directions: it's constrained to a flat surface. Let's imagine a hypothetical creature who is completely flat – a hypoflatical, say – living in a completely thin universe,

The 4D monster which came from Higher Space!

so thin that it would appear as a piece of paper does to us. We could loom as close to it in an up or down direction as we want and, because the hypoflatical has no concept of a third dimension, it would have no idea we were there. The third dimension provides perfect cover. Time to mount our own terrifying attack on a lesser-dimensional being – and all it takes is a move in the third direction into its two-dimensional world.

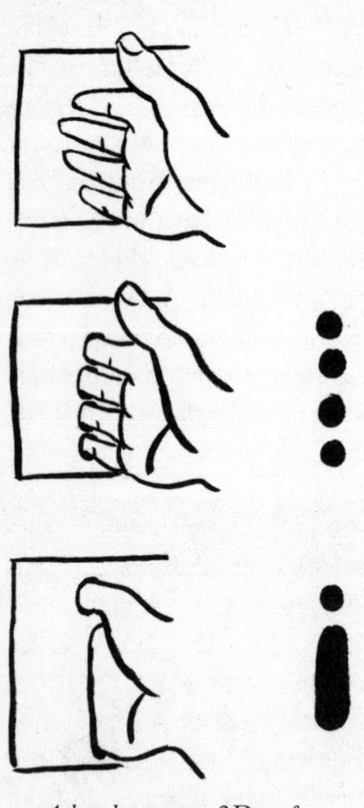

A hand next to a 2D surface would be unnoticed. As it passes through the 2D world, only cross sections would be seen.

Imagine it from the hypoflatical's point of view. As our fingertips pierce its 2D reality, they appear as individual floating circles which grow, move and then merge together as our palm reaches the 2D world. If we pass our fingers through their flat universe, all the hypoflatical sees are shape-shifting 2D cross sections, like flying meat pancakes. Trying to hide won't do it any good either: as 3D creatures, we can see the entire 2D world set out in front of us like a blueprint – which is exactly how the 4D creature in the Alan Moore story describes the 3D world.

So the hypoflatical cannot take cover and lock itself away safely. We can reach into any closed 2D room as easily as we can reach into

the middle of a 2D square. We can also see inside the 2D creature: all its internal body parts are laid bare and available to tinker with. This is what makes the 4D attacker so scary to us 3Ders: it can sit right next to our 3D universe watching everything we do and everything inside us, and needs only to reach in and kill us from the inside. They're ominous and they're gruesome but, thankfully, there's no evidence that 4D creatures exist. But it can't hurt to try to understand a bit more about the fourth dimension, just in case.

Let's say we are benevolent 3D creatures and simply want to show the hypoflatical our 3D world – educate them in what lies beyond their world, say in the spirit of interdimensional good will. We could let them have a look at a 3D cube, for example. If we moved the cube into its 2D world, the hypoflatical would see various cross sections of it. The boring way would be to shove the cube through face first, but this would merely cause a square to appear and disappear in front of the hypoflatical. Slightly more interesting would be to go edge first, causing a rectangle to expand out of nowhere before it shrinks back down to nothing. The best option, though, is vertex first, which causes a triangle to appear, grow bigger, shape-shift, *then* shrink back down to nothing. Far more entertaining for the plane-bound hypoflatical. And, in another of those amazing maths links, if we were to keep track of the total area this moving triangle covers, it would be a perfect regular hexagon, and not only that but exactly the same hexagon cross section of a cube we came across when we were solving Prince Rupert's cube-through-a-cube puzzle.

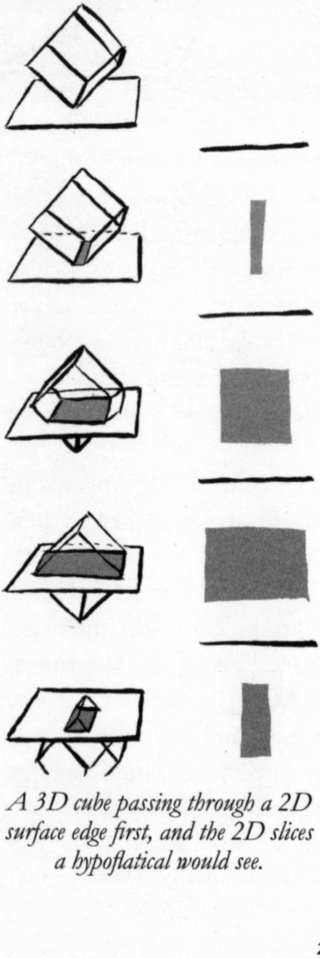

A 3D cube passing through a 2D surface edge first, and the 2D slices a hypoflatical would see.

Now let's take a look at what happens when a 4D hypercube passes through our 3D world. Were a kindly 4D creature to show a 4D cube in the same spirit of cross-dimension relations, we would see a series of 3D cross sections as they passed through our 3D world. The boring face-first option would just look like a normal 3D cube appearing and disappearing. Edge first is far more exciting. That would look like a triangular prism expanding out of nowhere, distorting into a hexagonal prism then contracting back into a triangular prism with a different orientation before disappearing into nothing.

Even better still is what happens if a 4D cube passes through our 3D world vertex first. Now, a tetrahedron appears and expands uniformly out of nowhere, distorts through a strange shape made of hexagons and triangles, is briefly an octahedron and then continues back through those shapes (only orientated differently) before collapsing back down to nothing. Now that's what I call throwing some impressive shapes.

Actually, let's talk shapes for a moment. If you cut a 4D cube exactly in half to get the biggest central 3D cross section, the shape of that cross section is an octahedron – the dual shape of a 3D cube. The total space the 4D cube goes through as it crosses our 3D world is a shape called the rhombic dodecahedron. However, none of this really shows us what the 4D cube looks like in its entirety. For that, we're going to have to move up from talking shapes to making models of shapes.

A 3D cube passing through a 2D surface vertex first, and the 2D slices a hypoflatical would see.

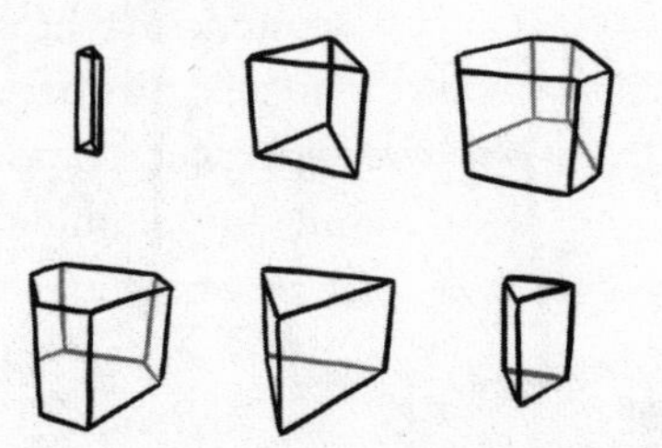

The 3D cross sections of a 4D cube passing through our world edge first. The total space required is a hexagonal prism.

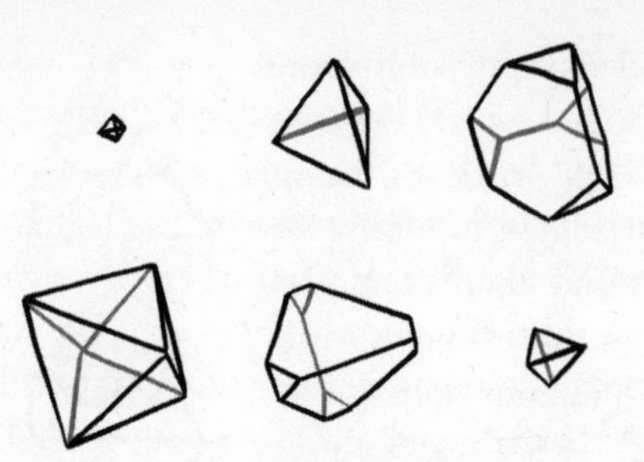

The 3D cross sections of a 4D cube passing through our world vertex first. Fun fact: the total space required is a rhombic dodecahedron.

Build Your Own 4D Cube

We can build a model of a 4D cube out of straws. We'll use coloured straws, as we did for our space-filling Weaire–Phelan model, but now the colours represent different directions. Let's start with fewer dimensions and work our way up. Making a shape in 1D is easy: that's just a single straw. I'm going to use a red straw to represent the one direction available. A 2D square isn't much harder. Use twisted pipe cleaners, as before, to join the corners of two red straws (for the horizontal edges) and two blue straws (for the vertical edges): there are now two directions you can move in. Child's play even for a hypoflatical.

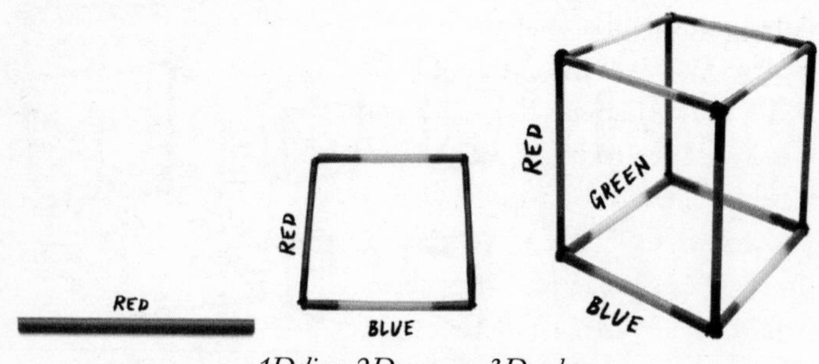

1D line, 2D square, 3D cube.

A good way to look at a square is as a 1D edge which has been duplicated to give two identical lines which are then joined together by new edges that follow a new direction. Likewise, a 3D cube is a 2D square duplicated and the two versions joined together at each of their vertices by edges moving in a third direction. If you make a second square using the same colour scheme as the first and join all the corners together, using green straws, say, you'll have a model of a cube. Out of habit, you'll probably have made this so that it stands upright. However, in the spirit of education, you can squash the whole thing flat and show it to a hypoflatical.

To take your cube into the next dimension, make another 3D cube and join all its vertices to all those of the first cube using a different-coloured straw – say, yellow – to represent the fourth direction. This model is the equivalent of the squashed cube we showed the hypoflatical: the second square *should* be lifted off the flat surface, but instead it lies just off to the side. Our second cube *should* be lifted off our 3D surface into 4D but, instead, it sits next to the first. What you have is a model of a complete 4D cube, only flattened to fit into our measly three dimensions. There's a pleasing pattern to how many edges and vertices there are in each cube as you move from one dimension to the next, but that's in 'The Answers at the Back of the Book'.

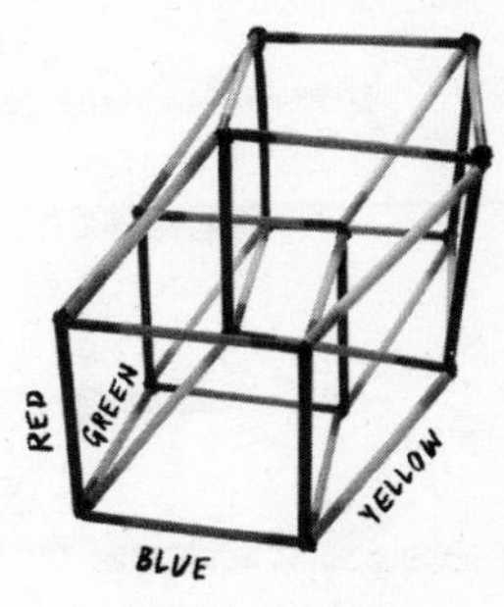

4D hypercube

The flat version of the 3D cube is the same shape as the shadow of a cube. If we took a light and projected a 3D cube on to a flat surface, we would see its 2D shadow, and that's the shape our hypoflatical would see in 3D. Our benevolent 4D creature could, in the same way, project 4D shapes to give 3D shadows: the straw model of a 4D cube you made is the 3D shadow of a 4D cube. Only, there is a second option for casting shadows: you can include some perspective. For the 3D cube, if you had made the second square smaller than the first, you could have suspended it inside the bigger one. This is the planar projection, so none of the edges cross. Likewise, for 4D, you could have made one cube smaller and suspended it inside the bigger one. This is the shadow of a 4D cube, and, similarly, none of the faces intersect. If this is giving you a sense of déjà cube, it's because you have seen this version of a 4D cube before: the cube-shaped bubble. You may have already unknowingly made the 3D shadow of a 4D cube using soap film.

As well as using perspective in this way to examine shapes in a higher dimension, we can also look at the shadows of *rotating* cubes. For the 2D shadow of a rotating 3D cube, it helps to colour in two opposite squares and leave the other four adjoining squares blank. This way, you can track one square as it rotates. We could also use this rotating-shadow system to show a 3D cube to a hypoflatical. Even though the cube is rotating in the 3D space directly next to the hypoflatical's 2D universe, it would be able to tell when a square is drawing closer to them, as its shadow gets bigger. Of course,

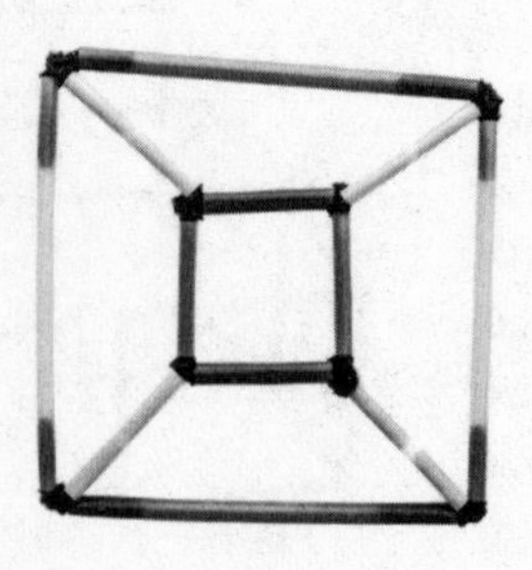
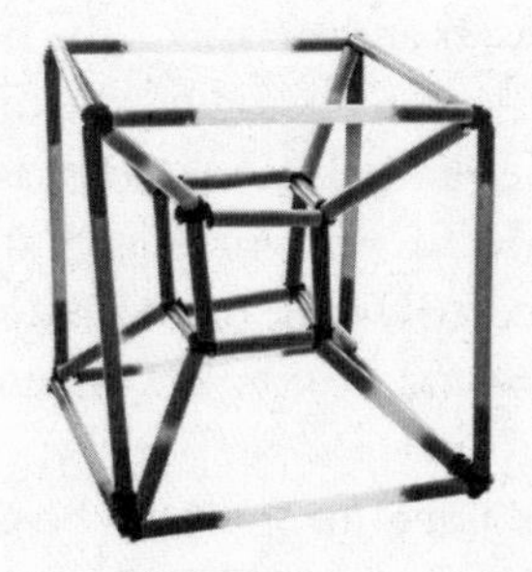

Adding a dash of perspective avoids intersections.

when it's further away, the same shadow becomes smaller. Unfortunately, from its point of view, the squares keep passing through each other, and all the explaining in the world will not help the hypoflatical visualize that they're not going *through* each other but rather in front and behind in a higher dimension.

We can do the same thing by projecting a rotating 4D cube into our 3D world. The two opposite cube faces are coloured in and the other joining cubes are transparent. As the 4D cube is rotating next to our universe, as each cube comes closer to us in 3D it looks bigger, and as it moves away it looks smaller. Again, as with the 3D cube to the hypoflatical, the cubes look to us as if they're going *through* each other as they rotate, when in fact they're going in front of and behind each other in the fourth dimension. And *we* cannot visualize *that*. My favourite moment with the rotating 4D cube projection is when the 3D cells of the 4D cube are exactly edge on with our universe. In the same way that, when one of the faces of the 3D cube is perpendicular to the 2D surface it's being projected on it appears to collapse and disappear for a split second, when a 3D cell is at exact right angles to our reality, it will briefly completely disappear from our point of view. The fourth dimension is weird.

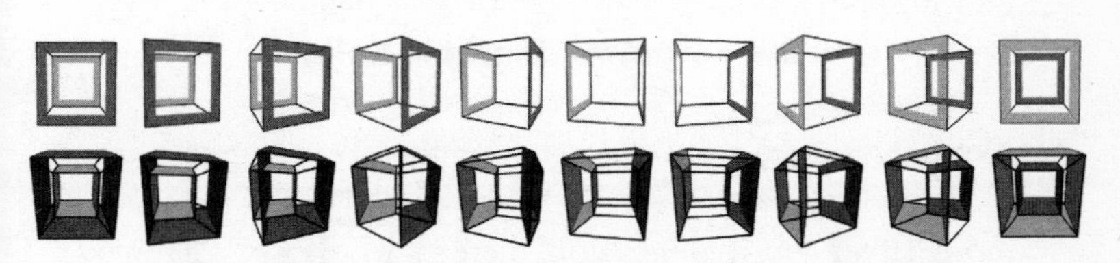

The 2D projection of a rotating 3D cube and the 3D projection of a rotating 4D cube, both completing a half-rotation at the same time.

If you want to have a go at moving a 4D cube around for yourself, I suggest you try to solve a 4D Rubik's Cube. Of course such a thing exists. Unlike a 3D Rubik's Cube, where you need to get all the same-coloured 2D stickers on the same 2D side of the 3D cube, with a 4D cube you need to get all the same-coloured 3D stickers on to the same 3D side of the 4D cube. To do this, you cannot move a 4D cube directly, but you can use the online version where you can drag the 3D projection of the 4D Rubik's Cube around to move it indirectly. It does get very confusing, as the 3D shadow moves in the three directions we're used to, but a 4D cube can also be rotated in a fourth, orthogonal (at right angles), direction (in this case, achieved by holding down the shift key). To make matters worse, because you'll be watching the 3D projection of the 4D Rubik's Cube on a computer screen, what you're actually doing is interacting with the 2D shadow of the 3D shadow of the 4D cube. Good luck.

A 4D Rubik's Cube, ready to solve.

The Forgotten Platonic Solid

Not only is there a 4D equivalent of the 3D cube, there are 4D equivalents of all five 3D Platonic solids. And then one more. In the fourth dimension, there's a whole new Platonic solid called the hyper-diamond which just does not work in 3D. The fourth dimension is not only the home of terrifying hypothetical monsters, it's a whole new world in which we can do mathematics beyond what is possible in a mere three dimensions. That fourth dimension allows more room to

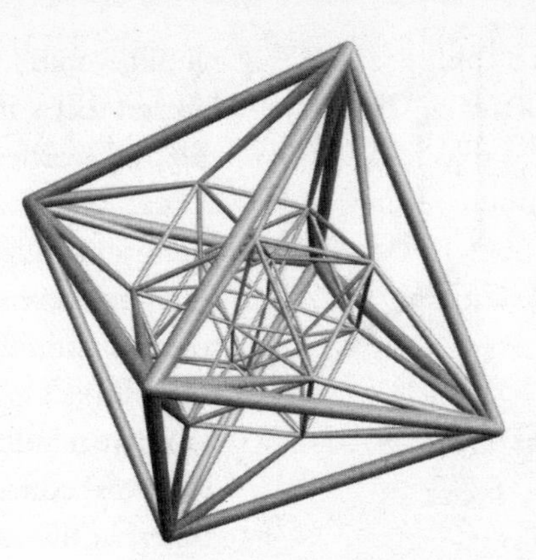

Behold, the hyper-diamond.

move, and *great* shapes. The hyper-diamond (more commonly known as the icositetrahedron or octacube) is my favourite Platonic solid.

First, though, we need to know what we *mean* by the fourth dimension. Famously, Einstein's physics theories require us to live within four-dimensional space-time. He keeps the same three dimensions for the space around us, then uses time as the fourth – which simply means that every single object has not only a position in 3D space but also a position in time. Mathematically, this allowed the same physics equations to apply to spatial positions as well as positions in time. However, this was only a repurposing of 4D mathematics which had been developed to deal with pure four-dimensional space. Mathematicians long before Einstein wanted to explore what would happen if we had four spatial dimensions instead of three.

As we said before, the poor hypoflatical with its paltry two dimensions could never imagine what the third dimension would be like. We could try to explain that it's possible to move in a whole new direction at right angles to the Hypoflaticals' entire universe, but there wouldn't be much point really: it

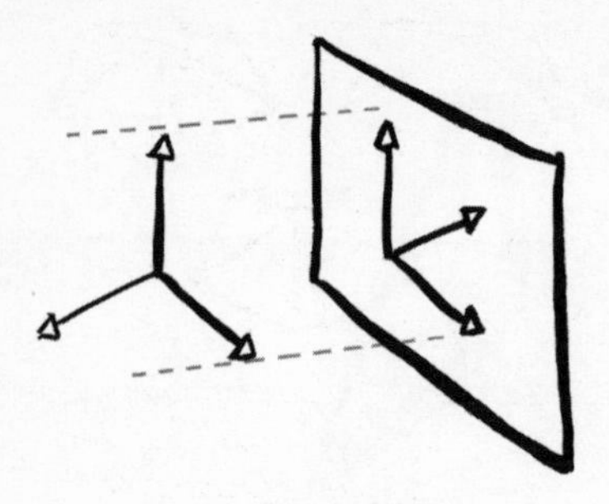

3D axis projected onto a 2D surface.

would, quite literally, be beyond them. If we took the axes of vertical and horizontal and tried to twist them in a third direction – out – this third direction would just look like a diagonal line in 2D to the hypoflatical. As we know, all polyhedrons can be drawn as 2D graphs, so we could persist and show it flat versions of 3D shapes, but it could never know what they really look like in true 3D.

The fourth dimension poses the same problems for us. In terms of maths, it's just a new direction to move in which is at right angles to the three directions we already have. Kind of conceivable. But we cannot imagine what that fourth direction *is*. As the Hypoflaticals are unable to envisage the possibility of moving off their flat surface in a new direction, we cannot picture how to move off our 3D reality in a new direction. Mathematically, though, it's not that complicated.

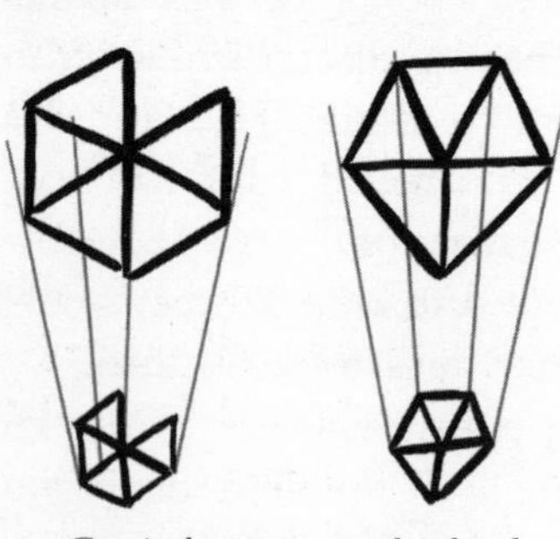

Gap in hexagons can be closed in 3D causing its projection to distort in 2D.

A hypoflatical could certainly do all the maths in Chapter 5 on 3D shapes: they just wouldn't be able to picture the results intuitively. We could show them that the gap left by five triangles could be closed to form a 3D vertex and that this continues to form an icosahedron, and even show them the shadow of that icosahedron, but the hypoflatical could never actually hold one.

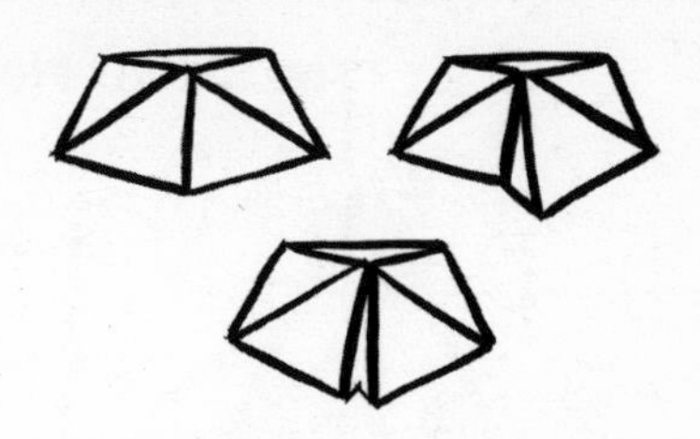

Three, four and five tetrahedrons do not quite meet in 3D, but they can be curved in 4D and closed.

If you take five tetrahedrons in 3D, they do not completely fill all the space around a point: there's a gap. This gap can be closed by lifting the tetrahedrons up a dimension and curving them around in 4D. If this curving around is continued in 4D, six hundred tetrahedrons will all fit together perfectly into the 4D Platonic solid known as the hyper-icosahedron. Likewise, joined four to an edge, sixteen tetrahedrons form the hyper-octahedron, and five tetrahedrons arranged three to an edge is the hyper-tetrahedron. Three cubes to an edge gives you a hypercube and, finally, 120 dodecahedrons can be arranged in 4D with three meeting at each edge to form the hyper-dodecahedron.

In general, if there isn't a name for something in 4D, add the prefix 'hyper-' to the equivalent name in 3D and you won't go far wrong. There is some 4D-specific terminology, though: after 2D polygons and 3D polyhedrons, there are 4D polychorons. When we attach 2D polygons to form a 3D polyhedron, the 2D shapes formed are called faces. When we join 3D polyhedrons to make a 4D polychoron, those 3D shapes are called cells. For this reason, the hyper-icosahedron is sometimes called the 600-cell, and the hyper-dodecahedron the 120-cell. Only two of the 4D Platonic solids get their own special name: the hyper-tetrahedron is called a pentatope and the hypercube a tesseract. If you listen closely in the Marvel live-action films (*The Avengers*, *Captain America*, *Iron Man* and *Thor*), there's often mention of a cube-shaped stone called the Tesseract.

The Six 4D Platonic Solids

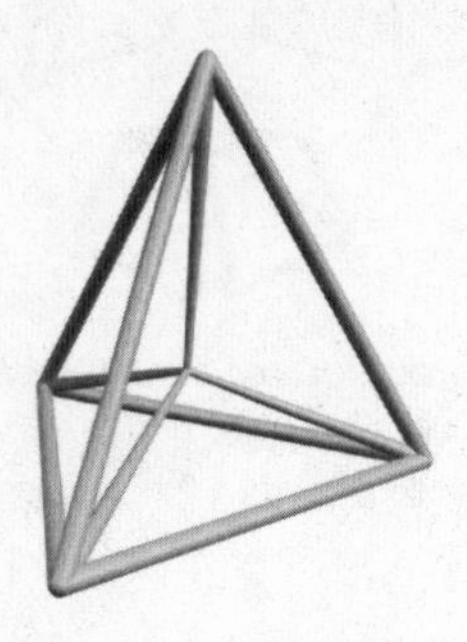
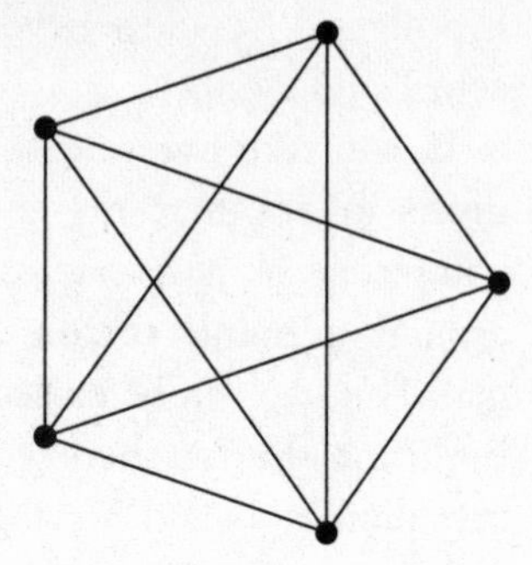

Projection and graph of 5-cell {3,3,3}, aka pentatope, aka hyper-tetrahedron.

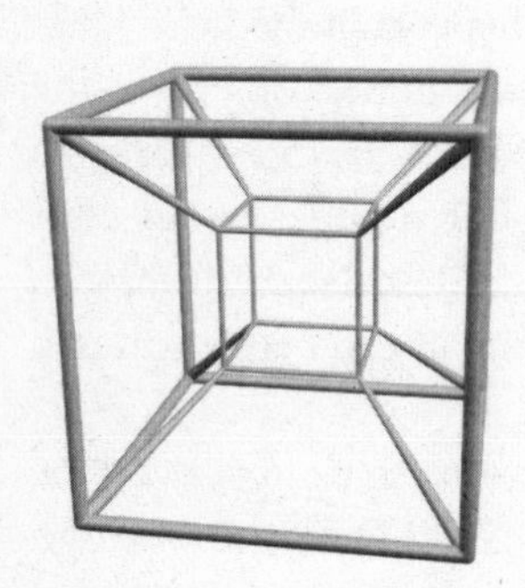
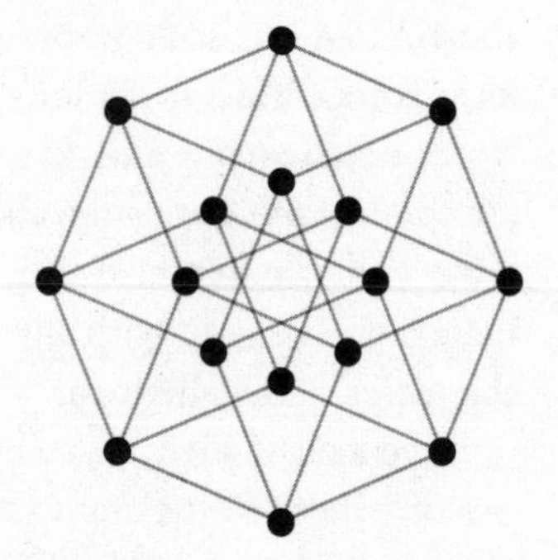

Projection and graph of 8-cell {4,3,3}, aka tesseract, aka hyper-cube.

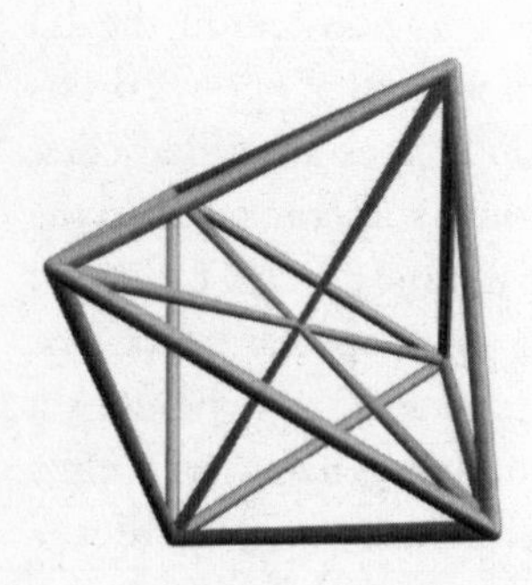
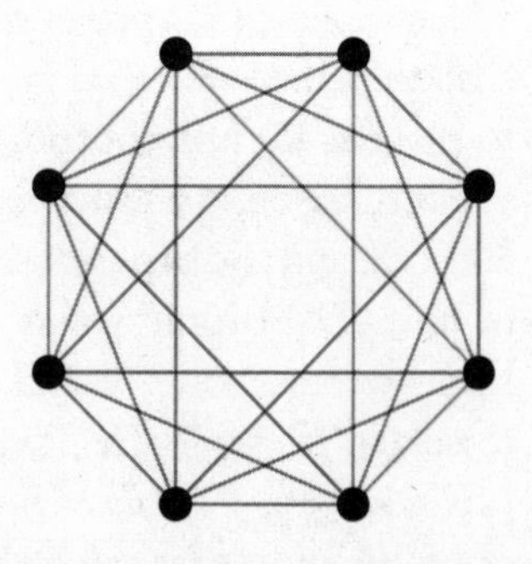

Projection and graph of 16-cell {3,3,4}, aka hyper-octahedron.

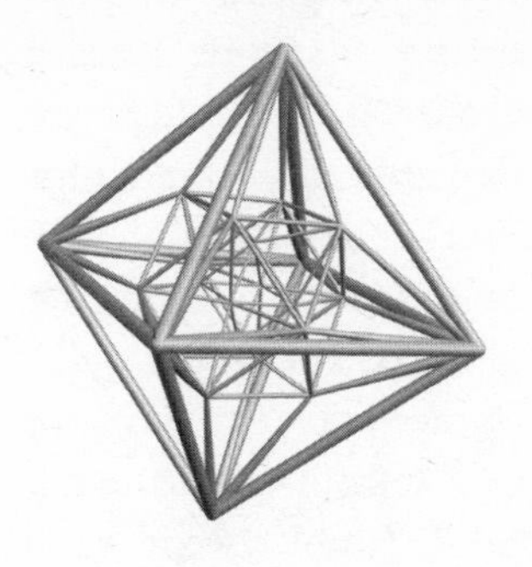
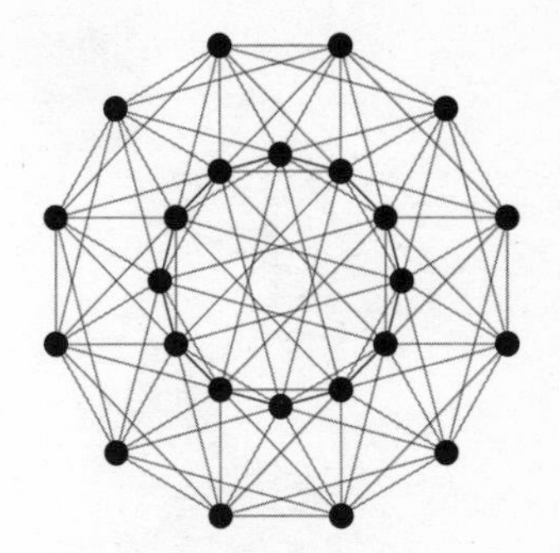

Projection and graph of 24-cell {3,4,3}, aka hyper-diamond.

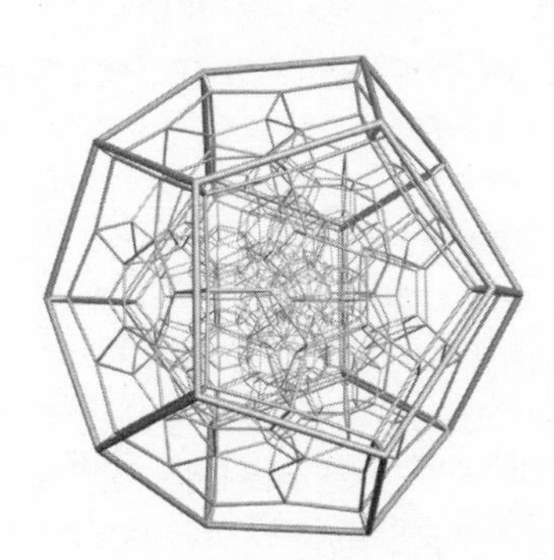

Projection and graph of 120-cell {5,3,3}, aka hyper-dodecahedron.

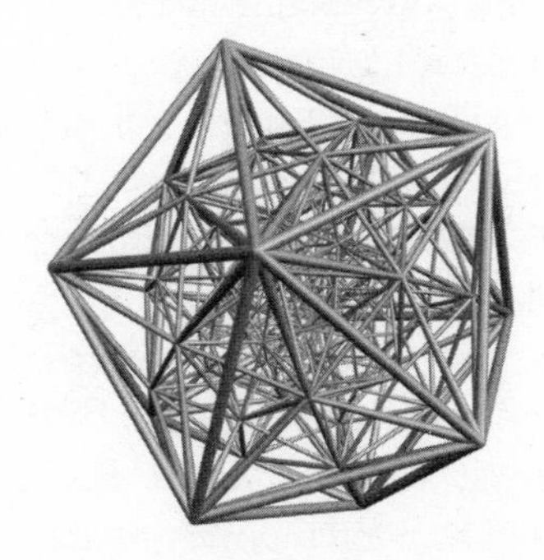

Projection and graph of 600-cell {3,3,5}, aka hyper-icosahedron.

The rhombic dodecahedron is a cube turned inside out.

I nearly dropped my popcorn the first time I heard them say it. These films are based around the search for a 4D cube! Sadly, fellow moviegoers rarely appreciate it when you point this out.

Now, finally, we can have a look at the hyper-diamond. Made from twenty-four octahedrons joined three to a corner, it's also known as the 24-cell. The name 'hyper-diamond' comes from the fact that its closest 3D relative is probably the rhombic dodecahedron, which is a polyhedron with diamond-shaped faces (but in 3D the rhombic dodecagon isn't a Platonic solid). The rhombic dodecahedron is closely linked with cubes (if you turn a cube inside out you get a rhombic dodecahedron) and hypercubes (as we saw before, the cross section of a hypercube is a rhombic dodecahedron). If you turn a hypercube inside out you get the hyper-diamond. The reason I love the hyper-diamond is because I think it's the first true 4D shape. All the other 4D Platonic solids are merely 4D versions of 3D shapes, whereas the hyper-diamond has no direct 3D equivalent. It's a shape that can exist only in four dimensions.

The 4D Platonic solids were discovered by the Swiss mathematician, linguist and schoolteacher Ludwig Schläfli around the year 1850. His work was so ahead of its time, though, that his treatise was rejected by contemporary mathematicians and not published fully until after his death in

1901. It is now highly regarded, and in it Schläfli unearthed and described all the higher-dimensional Platonic solids. In his memory, mathematicians use what is called the Schläfli symbol to describe these hyper-objects. This system denotes normal regular polygons by the number of sides they have, putting them in curly brackets (so a square is {4} and a heptagon {7}). For a 3D shape, you put the number of edges its 2D faces have in curly brackets, add a comma and then how many of those faces meet at a point. At the corner of a cube, three squares meet, so we can write it as {4,3}. An icosahedron has five triangles at each vertex, so it is {3,5} and, likewise, we have the tetrahedron {3,3}, octahedron {3,4} and dodecahedron {5,3}. For 4D shapes – well, we add a third number for how many 3D cells meet in 4D at each of the edges. Get the hyper-point? A tesseract has three cubes meeting at each edge, so we write it as {4,3,3}; the five tetrahedrons at each edge of the hyper-icosahedron means it is {3,3,5}.

The Schläfli symbol is a great way to describe a shape because it also gives an insight into how that shape behaves. Amazingly, if you reverse all the numbers in a shape's Schläfli symbol it gives you the dual shape. In 3D, the dual of the cube {4,3} is the octahedron {3,4}; and in 4D, the dual of the 600-cell {3,3,5} is the 120-cell {5,3,3}. This means that the shapes which are their own dual are the ones with a palindromic Schläfli symbol, such as the tetrahedron {3,3}.* The hyper-diamond is built from octahedrons {3,4} joined three to an edge, so its Schläfli symbol is {3,4,3}. Without

* For 2D regular shapes, all Schläfli symbols are only one digit and so they are all, technically, palindromic and, delightfully, all the infinitely many regular polygons are their own dual.

having to do any complicated working-out in 4D, we now know that the hyper-diamond is its own dual.

Much as Euclid proved that there can only ever be five 3D Platonic solids, over two millennia later Schläfli proved that there are definitely only six Platonic solids in 4D. However, there are loads of other things we can do in three dimensions beyond just making Platonic solids and, of course, there are plenty of things to make and do in the fourth dimension beyond these six shapes. To go any further, though, we need to look at what life would be like in the fourth dimension.

Life of a Hyperthetical

In 'It Came from . . . Higher Space!' the hero eventually defeats the 4D monster by using one of the few advantages a lower-dimensional object has: it is very sharp. As 3D creatures, we know that we can be easily cut by very sharp, almost-2D blades such as razors and knives. Imagine the cutting power of something perfectly 2D attacking us from just the right angle: it could slice through us almost without our offering any resistance. In the same way, we can conceive of 3D objects appearing very sharp to something in 4D, so our 3D hero can slice straight into the 4D monster and infiltrate its brain, thankfully saving Earth in the process. But what would life be like for this 4D monster? Instead of something malicious, what if our friendly hypothetical aliens were 4D creatures? Going back to our prefix for 4D shapes, let's call them hyperthetical aliens.

It took humans a very long time to discover the maths of the fourth dimension. We've known about regular 2D shapes and the Platonic solids for thousands of years, possibly stretching back into prehistory, but it was not until the 1850s that we realized there were more shapes in 4D. Even now, all we have is a sketchy mathematical understanding of how the fourth dimension behaves. Our brains have evolved to deal only with 3D objects. Not so for our Hyperthetticals. In some far-flung universe it could be that 4D is the status quo (the status quattro). They were born in 4D, moulded by it. Life for them would be very different.

For a start, they wouldn't be able to tie knots. Yes, it's impossible to tie knots in 4D. Every way you try to tie one in a piece of string will end up being the unknot in 4D. This is because the crossing-switch move we use to untie knots in 3D happens automatically in 4D. In 3D we have to go to all the effort of cutting the string, moving it to the other side and retying it. In 4D, it can simply slip around in all the extra space provided by the bonus dimension. The problem with having four different directions to move about in is that it becomes much harder to constrain and contain things.* I'm sure there are some nuanced, complicated questions this raises about the Hyperthetticals' biology and what kind of organic chemistry controls whatever long-chained molecules they may use as DNA, but at the very least we know they must keep their shoes on with Velcro.

Stacking and tiling objects also changes completely in 4D. It seems that the way shapes fill space in any given number of dimensions is fairly unique and doesn't tell us much about how they would fill space in another dimension. Conway (and associates), who discovered that you can fill space with one octahedron combined with six smaller tetrahedrons back in

* We have not looked at how objects orbit each other in 3D, but as we know from the predictable year around the sun, our Earth's orbit is pretty stable. All this extra freedom of movement means it is impossible to have stable orbits in 4D. So their solar system must be messed up.

Chapter 6, did try to extend the concept into 4D – but it didn't work. In their words: 'We are not aware of any non-trivial analogs of our tetrahedra-octahedra tilings in other dimensions. These observations imply that tiling problems are generally dimension specific and the results for a particular dimension cannot be simply generalized to other dimensions.'

There are some shapes, however, that do tile in 4D: you can completely fill 4D space with both the hypercube and the hyper-diamond. If you try to stack 4D hyperspheres, it's still very inefficient, but not as bad as in 3D. Just as in 2D there are worse options than the circle (namely, the smoothed octagon), in 4D there are worse options than a hypersphere. Only in 3D is the sphere seemingly the absolute worst option for efficiently filling space. Not only do greengrocers have to stack fruit which is the worst shape we know of, they are doing it in the worst dimension possible as well!

Some of our shape puzzles continue to work nicely in 4D. We can still play around with almost-2D shapes by cutting them out of very thin paper, so Hypertheticals can also make very thin 4D shapes to replicate 3D objects. They can make a hyper-coin fit through a smaller hole cut in hyper-paper. There are other puzzles which generalize completely to 4D. If a Hyper-Prince Rupert stated that it is possible to fit a tesseract through a hole in an identical-sized tesseract, he would still be correct. But it's a bit of a tighter squeeze. A 3D cube has a 2D cross section which can take, at most, a cube with sides 6 per cent bigger, whereas a 4D hypercube has a 3D cross section which can allow another hypercube to fit through with sides only 0.7435 per cent bigger. They may even have 3D flexagons which flex in the fourth dimension, but I fear that would be frighteningly counter-intuitive for us and flex our brains out of shape.

One toy they would have for sure are 4D solids of

constant width. As soon as I learned about the 3D shapes of constant width which can roll between 2D surfaces, I wanted to know if there are 4D shapes of constant width that can roll between two parallel 3D surfaces. I was curious about how such shapes would work. Going from 2D to 3D, there are two options: you can have the solid of revolution of the 2D Reuleaux triangle, or the all-new 3D shape, the Meissner tetrahedron. Going one dimension higher, I figured it might be possible to have the 4D revolution of the 3D Meissner tetrahedron, or maybe you could do something completely new in 4D. It turns out you can do both.

In the maths paper 'Bodies of Constant Width in Arbitrary Dimension', written by Thomas Lachand-Robert and Édouard Oudet, new 4D shapes of constant width are created both ways. There's a rotated 3D Meissner tetrahedron as well as a new shape made from taking a 4D pentatope – {3,3,3}, if that helps – and hyper-rounding off its 3D surfaces. I don't know what the shape is called, but I've been calling it the Robert–Oudet body. If you take a Meissner tetrahedron, all its 2D projections are themselves 2D shapes of constant width. All the 3D projections of the Robert–Oudet body are solids of constant width. The best way to view this shape is as the series of slices it would form if it passed through our 3D world.

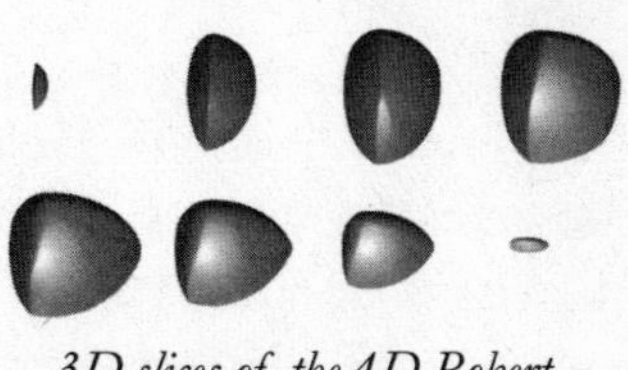

3D slices of the 4D Robert–Oudet body.

Even though all this mathematics of the fourth dimension may seem confusing, it's not beyond our grasp. Sure, we lack the mental capacities to visualize it intuitively, but we can still explore the maths. Beyond that, none of the maths we already know stops working in the fourth dimension. Our arithmetic would make as much sense to 2D and 4D creatures.

If we met our Hyperthetics, their ability to move in 4D will of course make them seem other-worldly and beyond our comprehension, but they add and subtract the same way we do. Provided they're not jerks, we can still communicate via numbers (in whatever base system they use). Even though the context of hyperspace may be entirely alien, the maths is still the same. Should we one day meet aliens from any universe of any number of dimensions, mathematics will still be our common language. We had just better be really, really polite.

Eleven

THE ALGORITHM METHOD

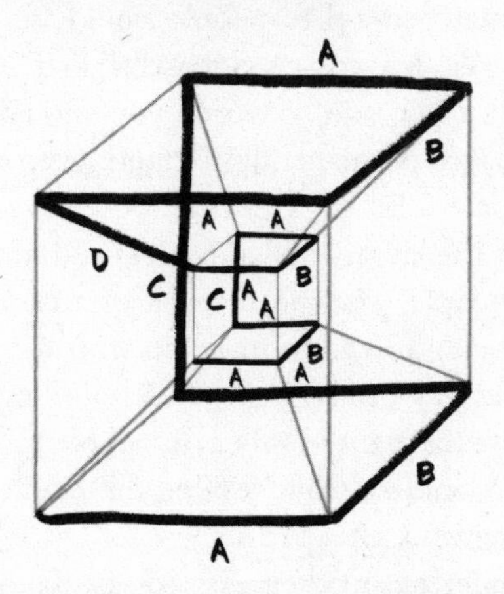

I own ridiculously accurate kitchen scales, because when I'm following a recipe I want to be able to follow it exactly. Life is already full of complicated options and decisions, but at least something like baking a cake comes with complete and detailed instructions so you don't have to worry about making uninformed decisions. I also really enjoy measuring things.

Sadly, there are rarely 'recipe books' for all other aspects of life. But that doesn't mean you can't apply the same style of thinking to them. There are other situations, such as choosing a life partner, where you can still use the same kind of structured decision-making as generally found only in recipe books. A list of pre-determined instructions which guide you through an activity is called an algorithm in mathematics, and it's kind of a cake recipe on steroids.* Instead of telling you exactly what to do, it lets you make decisions and then instructs you exactly what to do next, based on that decision.

Finding a life partner is a delicate balance. When you first start dating people, you don't know, on average, how romantically well matched other people could be to you, and without that base-line you cannot ascertain if someone is an above-average catch and someone you should settle down with. This makes permanently partnering up with the first person you date a bit of a gamble: you should date a few people to get the lay of the land. That said, if you take too long dating people, you run the risk of missing your ideal partner and being forced to make do with whoever is available at the end. It's a tricky one. The ideal thing to do would be to date just the right number of people to gain the best sense of your options while leaving the highest probability of not missing your ideal partner.

Luckily, maths has made it easy for us: that right number of people is the square root of the total number of people you could date in your life. How you estimate the size of your possible dating population is entirely up to your statistical skills and the level of your self-confidence, as is how you then collect your sample. A 'voluntary response sample'

* Do not put actual steroids in a cake recipe.

is generally regarded as socially acceptable, whereas a 'stratified random sample' can land you in gaol.

Putting that aside, here is the recipe for finding optimal love:

Step 1: Estimate how many people you could date in your life, n.
Step 2: Calculate the square root of that number, $\sqrt{n}$.
Step 3: Date and reject the first $\sqrt{n}$ people; the best of them will set your benchmark.
Step 4: Continue dating people and settle down with the first person to exceed the benchmark set by the initial $\sqrt{n}$ dates.

Who knew it could be so easy? Problems like deciding who to settle down with have the mildly disturbing maths name of 'optimal stopping problems'. The original optimal stopping problem was known as the secretary problem, and here it is as originally framed.

You're hiring a new personal assistant and the company's human resources department has put out an advertisement following the guidelines laid down in its official diversity policies. There is now a queue of potential candidates outside your office, spanning a wide range of genders and ethnicities, all ready to be interviewed for the job. Each of the ten candidates will come into your office individually for you to assess their qualifications for the role. After you have interviewed each candidate you need to decide on the spot if you want to hire them for the job or move on to the next interview. If you dismiss someone without a job offer, they'll be snapped up by a rival company: you cannot go back to them later and offer them the job.

You're in an interesting position. Logic suggests that you shouldn't offer the job to the first person you interview, because you have no idea what the general calibre of the candidates is. Nor do you want to wait until the tenth person, because if they're the only one left you're going to be forced to offer them the job regardless of how well suited to it they are. Somewhere in the middle there must be an ideal place to stop interviewing more candidates just to see what they're like, and hurry up and choose a good one. This is the optimal place to stop. It's exactly the same constraint as dating to find a life partner; if you break up with someone you later realize was an ideal candidate, you can rarely go back to re-interview them.

The secretary problem was tossed back and forth between mathematicians in the US during the 1950s, and we don't know who was first to solve it (though people suspect it was the American mathematician Merrill Flood). The first official solution to appear in print was by the British statistician Dennis Lindley, in 1961. Solving this problem involves realizing that all ten candidates could be ranked from best to worst and then shuffled up in some random order. There's a one in ten chance the first candidate through the door is the best one, but the thing is, you just don't know.

By analysing the possible distribution of talent, it was calculated that if you interview the first 37 per cent of any queue then pick the next one who is better than all the people you've interviewed so far, you have a 37 per cent chance of getting the best candidate. So, his algorithm is the same as our dating one, but with $0.37 \times n$ instead of $\sqrt{n}$. The figure of 37 per cent keeps appearing because it is the ratio $^1/_e$, where e is the exponential number 2.718281828... (discovered by Jacob Bernoulli, but more of him later). The short story of why the number e crashes the party is that it is linked to estimating where the best candidate could be in the queue. Using this method gives you a better than one in three chance of choosing the most suitable candidate overall.

However, the original secretary problem assumes that you have an all-or-nothing attitude. Lindley proved mathematically that his 37 per cent method algorithm is the best approach, but that is only if you'll be completely happy with the best person and completely unhappy with anyone else. In reality, getting something that is slightly below the best option will leave you only slightly less happy. A better solution would be one that will give you someone as high up the ranking of candidates as possible, even if they aren't necessarily the best. In 2006, psychologist Neil Bearden calculated the best strategy for selecting the highest-ranking candidate compared to the theoretically best candidate possible, and it was he who discovered the $\sqrt{n}$ method. Out of a choice of ten people, the $\sqrt{n}$ method will, on average, get you someone about 75 per cent perfect; in a queue of a hundred candidates, the figure is around 90 per cent.

You can use this algorithm when faced with any life choice, from choosing life partners to making purchase decisions. I used it when I bought my second-hand car, making sure I saw at least $\sqrt{n}$ of the number of cars I had time to check out before I definitely needed to buy one. I think the biggest advantage is that it makes people wait and not impulse-accept their first option. When examples of real decision-making processes have been analysed, the consistent result is that people choose too soon, and without looking at enough options (with the exception of online dating, where some people become so spoilt for choice they cannot make themselves settle down when there may be better options). An optimal stopping algorithm takes all that indecision away. (See 'The Answers at the Back of the Book'.)

This may all sound very impersonal as a way to find a partner, but maths has been used to locate love. When Kepler (of the Kepler conjecture) lost his first wife to cholera, he decided to make finding his next wife a mathematical process. In a letter written in 1613 he described how he planned to dedicate two years of his life to interviewing and ranking eleven possible candidates and then making a calculated choice. We don't know what his 'wife appropriateness function' to rank them was, but we do know that he felt the constraints of the secretary problem. He tried to go back and propose to the fourth person he interviewed, but she had already moved on, and refused him. He eventually ended up happily married to candidate five of eleven. Now that's some rigorous loving.

What's the Trick?

Leaving sex behind, we can also apply algorithms to numbers. Even better! But performing pre-prescribed step-by-step calculations on numbers may not sound very exciting. Mainly because it isn't. But that's not the point: executing algorithms in itself is not very exciting. Mathematicians don't get excited about algorithms because they enjoy doing repetitive tasks (although many do); that would be like baking a cake just because you really like measuring and mixing ingredients (which I do). Algorithms are enjoyed by mathematicians because of what you can do with them. Mathematicians like to have their calculation and eat it too.

Step 1: Take any number.
Step 2: Add the digits together.
Step 3: If the answer has more than one digit, repeat Step 2.
Step 4: Write the final one-digit answer down.

This algorithm is the recipe to take any number and find its remainder after it has been divided by 9. You can check for any number and the answer will be how much bigger the number is than the first multiple of 9 below it. The output after repeatedly adding digits like this until you get a one-digit answer is called the digital root of a number. The process itself is called casting out nines, because it eliminates any multiples of 9 from a number. Casting out nines is one of the oldest and most important algorithms in mathematics.

Writing around the turn of the previous millennium in what is modern-day Iran, the scientist Ibn Sina (also known by his Latin name, Avicenna) referred to this process of casting out nines as 'the method of the Hindus', implying that it had already been in use for a long time. Casting out nines was used by European accountants long before Fibonacci introduced Hindu-Arabic numbers. The oldest known printed book about financial maths (the *Treviso Arithmetic*, published in 1478) describes the method of making sure the digital root of the answer to some complicated piece of arithmetic is the same as the sum of the digital roots of the numbers being added. They were casting out nines to double-check important financial calculations. We're going to use it to do a magic trick.

Volunteer from audience:

Step 1: Take any number.
Step 2: Multiply it by 9.
Step 3: Read all but one of the digits in the answer out.

Magician:

Step 1: Calculate the digital root of all the digits read out by the volunteer.

Step 2: Know that the missing digit is the difference between the digital root and 9.

Step 3: Announce what the unspoken, missing digit is.

Audience:

Step 1: Have their minds blown.

This is a great magic trick because, whatever number the volunteer chooses, as soon as they multiply it by 9, the digital root of the new number must be 9. By calculating the digital root of the digits they tell you, you know the missing digit is whatever is needed to make the digital root up to nine. Casting out nines to get the digital root is also easy to do mentally, because you never deal with numbers bigger than one digit. It all sounds too simple, but the effect is amazing. I once performed the trick on stage at the Hammersmith Apollo in London in front of over three thousand people. The most nerve-wracking bit was hoping the volunteer didn't make a mistake with the calculator.

The more ambitious version of this trick – which I was too scared to try with such a big crowd – involves camouflaging the process of multiplying by 9. If you ask a volunteer to take a calculator and keep multiplying random digits together until the answer fills the screen, odds are that they will inadvertently either multiply by 9, or at least by two numbers that have 3 as a factor. Stated precisely, those odds give a 96.75 per cent chance that single random digits multiplied to give an eight-digit answer will produce a multiple of 9.*

* To work this out, I wrote a computer program to generate such random numbers 10 billion times, and 9,674,919,018 of the 10 billion were multiples of 9. Writing a computer algorithm to check a magic algorithm pleased me greatly.

However, I wasn't prepared to take that 3.25 per cent chance of looking like an idiot in front of so many people!

There is a whole class of magic tricks called self-working tricks which are really just magic by algorithm. As long as the magician follows the instructions step by step, the magic trick will always work. One such self-working trick is a card trick that almost everyone seems to come across during their life. Commonly called the Three Pile Trick, it's traditionally performed with twenty-one cards, for reasons I cannot fathom: exactly the same trick works with up to twenty-seven cards. The magician boasts (of course) that he will find any card chosen randomly by a volunteer.

Step 1: A volunteer picks one card from a deck of twenty-seven cards, remembers it, and the card is shuffled back in.
Step 2: The magician deals the cards out face up in three piles (consistently dealing one card on each pile at a time, in the same direction).
Step 3: The volunteer indicates which pile their card is in, but not what the card itself is.
Step 4: The magician picks the piles up, putting the indicated pile in the middle.
Step 5: Steps 2, 3 and 4 are repeated twice more.
Step 6: The volunteer's card is now exactly in the middle of the 27-card pack, i.e. it's the fourteenth card from the top.

The magician can now reveal the fourteenth card in a manner determined only by the limits of their creativity. My personal favourite is to start turning the cards over one at a time and claim to be looking for a reaction in the volunteer's face. You go right past the fourteenth card until the seventeenth or so and then claim that the next card you turn over will definitely be theirs. Sometimes they'll even bet you a drink or similar, as they're so sure you're going to get it wrong. Then reach back and turn over the fourteenth card that's already been dealt, technically fulfilling your boast and, if you've had the forethought to make one, winning the bet. Be warned that, if the volunteer has not told an independent party what their card was, or written it down somewhere, they can easily lie their way out of the bet.

This is a nice enough type of trick, but the holy grail of card tricks is a category known as 'any card, any number' tricks. This is where the card the volunteer has selected can be moved to a named position in the deck (rather than being dead in the middle). With a slight tweak, and a bit more thinking, the Three Pile Trick can do exactly this. In my version, while I'm dealing the cards into the piles, I casually ask the volunteer what their favourite number below 27 is. At the end of the trick, their card ends up in exactly that position.

The only difference is that, instead of putting the indicated pile in the middle each time, you sometimes put it in the top or bottom position. To work out where it goes, we call the top pile position 0, the middle position 1 and the bottom position 2. Then simply convert the number of cards you want on top of the volunteer's card into base-3. The first time you recombine the piles, put the indicated pile into position 0, 1 or 2 based on the number of 1s; the second time based on the number of 3s; and the third time the number of 9s. So, if you wanted to put seven cards on top of the volunteer's card, 7 is 021 in base-3 (it has zero 9s; two 3s; and one 1), so you would place the indicated pile in the middle, bottom and top positions in that order (starting at the 1s and working backwards). At the end of the trick you count off the top seven cards, and the selected card is the next one, in the eighth position.

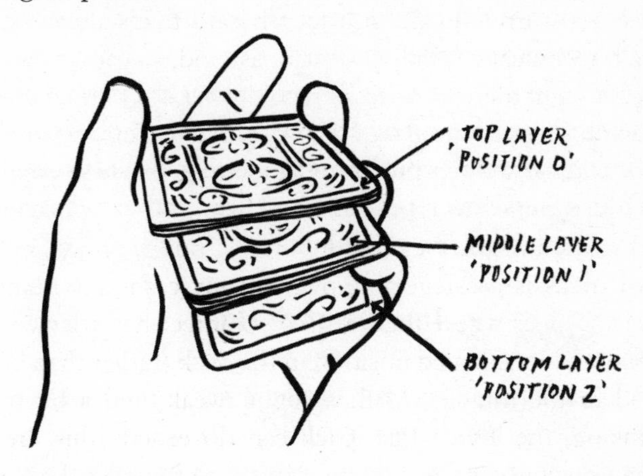

It is with some regret that I explain this trick in my book, as it has served me very well in impressing audiences, as well as professional mathematicians and magicians. I first came across this style of trick in Martin Gardner's 1956 *Mathematics, Magic and Mystery* (his first maths book), where he calls it Gergonne's Pile Problem and implies that it has been well known for the previous century and a half. In the half-century since then it seems to have been forgotten about, and I've had great fun re-exploring the maths behind it and working it up into my favourite performance trick. (See 'The Answers at the Back of the Book'.)

In his book, Martin Gardner dedicates a chapter to variations of the 27-card trick but focuses more on the magic-performance possibilities. He ends the chapter with notes he was sent by a Canadian, Mel Stover, who pointed out the number-base basis of this trick, as well as how it could be adapted to different numbers of cards. He includes instructions on how to scale it up to a hypothetical deck of 10 billion cards. In theory, any chosen card could be moved to any of the 10 billion possible positions by dealing the cards out into ten piles ten times. At a deal per second, that's just under 3,169 years. Mr Stover stresses the importance of dealing all one hundred of these billion-card piles carefully, as a single mistake would ruin everything. In his words: 'this would necessitate repeating the trick, and few spectators would care to see it a second time'.

The Tower of Hanoi

We'll leave magic and try playing a game that doesn't take quite so long. This is a game called the Tower of Hanoi which I first met when I was doing a programming course at university. It's based on a legend of ancient priests in a temple who had to build the tower by moving different-sized discs. Although it may have existed before then, the first mention of it I could find was in the late-1800s book that introduced it to the Western world, *Récréations mathématiques*. It is much like a nineteenth-century version of *Mathematics, Magic and Mystery* and, as mentioned earlier, was written by the nineteenth century's Martin Gardner: none other than Édouard Lucas. Not just a man of cannonballs and primes, he also gave the world several maths games which remain popular to this day, and may have invented the Tower of Hanoi himself.

The game starts with a number of different-sized discs stacked in a tower from biggest to smallest. Traditionally, they have a pole through their centres to keep them in place. There are two other poles: the target pole, which you need to move the tower on to; and a kind of 'holding pole', where you can temporarily put discs to move them out of the way. The only rule is that a bigger disc can never sit on top of a smaller one: any tower on any of the three poles must always go from bigger to smaller.

Whenever I've come across this game, the backstory always seems to involve priests in a temple somewhere who have been tasked via some divine sense of duty to move a huge tower of discs like this from one position to another. So they spend their days shifting discs around from pole to pole. Then, for unexplained reasons, once the tower-moving is complete, the world will come to an end – which does call the priests' motivation into question. But I guess if your entire life consists of monotonously moving discs around, the world ending is not such a bad prospect.

The Tower of Hanoi, or to quote Lucas exactly, 'La Tour d'Hanoï', takes up only three pages in the book, roughly a third of which is dedicated to a large diagram of the game set-up – a large diagram which is no longer in copyright, so I've reproduced it here for you. As I do with all my now-dead maths heroes, I went hunting for some of Lucas's writings so I could get a feel for him as a person. Fortunately, *Récréations mathématiques* is available online as a free ebook which

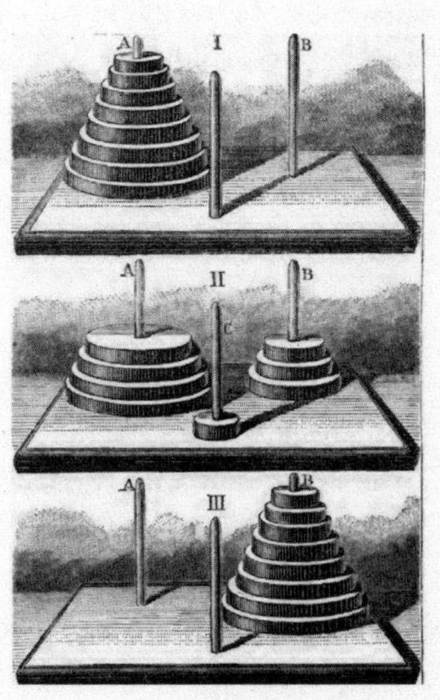

The Tower of Hanoi from Récréations mathématiques.

I could easily download. Unfortunately, as I mentioned before, it is of course entirely in French. So I emailed a copy of the three Hanoi pages to my French-teacher brother-in-law to translate.

When he emailed me his translation, I was surprised to see that half the text isn't about the Tower of Hanoi at all, but rather something called the Wheat and Chessboard Problem. This is a challenge to calculate the total number of grains of wheat on a chessboard if you put one in the first square, two in the second, four in the third, and so on, doubling the amount for each of the sixty-four squares. (hint.subtlety = 'blatant'; there is a link to the Tower of Hanoi.) In the first section of text, Lucas talks about the Tower of Hanoi toy produced around 1883 by a Professor N. Claus (from Siam). Curiously, the original instructions which came with the toy, written by Claus, said that more details about the game were available in the book *Récréations mathématiques* by Lucas. It all seems very circular and incestuous until you realize that 'Claus' is an anagram of 'Lucas', and the full 'N. Claus (de Siam)' is an anagram of 'Lucas d'Amiens' (Lucas was educated in Amiens).* It seems Lucas and his pseudonym are having a great time patting each other's backs.

Back to the game. The two-disc version is nice and straightforward. We'll call the small disc A and the bigger one under it B. Move A on to the holding pole, then shift B to the target pole, before moving A back on top of it. Three easy moves. We could even write them 'A B A' to show the order. If you start with three discs, you can move the whole tower in seven

* Also, while Prof Claus worked at 'Li Sou-Stian' college, Lucas was over at the school of Saint-Louis. He certainly wasn't trying too hard for anonymity.

moves, and a four-disc tower can be relocated in only fifteen disc movements. Find some different-sized discs and have a play around, slowly adding more until you're happy you could relocate a five-disc tower. Most toy versions of the game are sold with around seven to ten discs, which would result in a fairly swift end to the world. Thankfully, the priests of legend have to move a 64-disc tower before the world ends, which buys us all a bit more time.

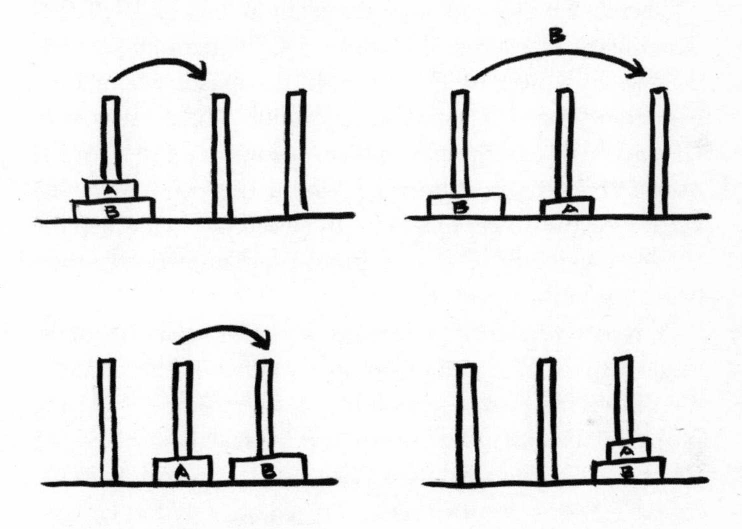

Should we wish to help the priests along, we can describe how to solve the Tower of Hanoi using algorithms. One option would be to write out a long algorithm giving every single move. The algorithm of the two-disc version would be 'Make moves A B A' (assuming they put them on the correct poles). We could scale this up for any number of discs by giving each disc a letter then listing the letters in the order they have to be moved. This is an explicit algorithm. It's like a recipe which micromanages your every move while baking a cake.

1 DISC: A

2 DISCS: ABA

3 DISCS: ABACABA

4 DISCS: ABACABADABACABA

5 DISCS: ABACABADABACABA
EABACABADABACABA

There is a better and more efficient way, though – which is why I first came across the game on a programming course: solving the Tower of Hanoi can be a fantastic example of something called a recursive algorithm. These are a very clever type of algorithm, and form an important tool for the modern creative programmer. My lecturer brought in a small version of the Tower of Hanoi as a teaching aid, to help the lecture theatre of first-year proto-programmers get their heads around recursive algorithms.

A recursive algorithm cheats. It doesn't tell you every single step. It tells you how to do a little bit of the solution, then flakes out with a step 'Use this algorithm to solve the problem.' It's a bit like if I wrote a recipe called 'Matt Parker's Complete Guide to Baking a Cake' and the main step in my recipe is 'Follow the instructions as laid out in "Matt Parker's Complete Guide to Baking a Cake".' You would be rightfully annoyed if you were following those instructions, but that's exactly what a recursive algorithm does – with the slight tweak that the self-referential step in a recursive algorithm applies itself to a slightly smaller task than the problem as originally set out. So, a better comparison would be if my recipe says 'Just bake a small part of the cake first. For the rest of the cake, follow the instructions in "Matt Parker's Complete Guide to Baking a Cake".'

For the Tower of Hanoi, one of the steps in the recursive algorithm is 'Solve the Tower of Hanoi.' (The cheek of it!) Here's how it looks:

Recursive Hanoi Algorithm

Step 1: Count up how many discs you have; call this number *n*.
Step 2: Use the recursive Hanoi algorithm for a tower of $n - 1$ discs to move them out of the way.
Step 3: Move the biggest disc to the target pole.
Step 4: Use the recursive Hanoi algorithm for a tower of $n - 1$ discs to move them back on top.

The amazing thing is that this *works*. Even though that algorithm never explicitly tells you how to solve the Tower of Hanoi, if you follow it step by step, you *will* solve the Tower of Hanoi. For me, that's as close to real magic as an algorithm can get.

How Many Ways to Shuffle a Pack of Cards?

There's a lovely function in mathematics called the factorial function, which involves multiplying the input number by every number smaller than it. For example: factorial(5) = $5 \times 4 \times 3 \times 2 \times 1 = 120$. The values of factorials get alarmingly big so, conveniently, the function is written in shorthand as an exclamation mark. So when a mathematician writes things like $5! = 120$ and $13! = 6{,}227{,}020{,}800$ the exclamation mark represents both factorial and pure excitement. Factorials are mathematically interesting for several reasons, possibly the most common being that they represent the ways objects can be shuffled. If you have thirteen cards to shuffle, then there are thirteen possible cards you could put down first. You then have the remaining twelve cards as options for the second one, eleven for the next, and so on – giving just over 6 billion possibilities for arranging a mere thirteen cards.

For a full deck of fifty-two cards, the number is much bigger. Calculating 52! manually would take a long time, so it's a perfect thing to get a computer to do for us. But in order to ask a computer to do something, you need to state it as an algorithm for the computer to follow. So here's a set of instructions I've written to take an input number and progressively multiply it by every smaller number:

Step 1: Remember the starting value of n as your running total.
Step 2: Subtract 1 from n.

Step 3: Multiply the running total by this new *n*.
Step 4: Repeat steps 2 and 3 until *n* reaches 1.
Step 5: Return the running total.

We can try running this for the first few laps of calculating 52!:

RUNNING TOTAL = 52
52 − 1 = 51
RUNNING TOTAL = 52 × 51 = 2,652
51 − 1 = 50
RUNNING TOTAL = 2,652 × 50 = 132,600
50 − 1 = 49
RUNNING TOTAL = 132,600 × 49 = 6,497,400
49 − 1 = 48
RUNNING TOTAL = 6,497,400 × 48 = 311,875,200
48 − 1 = 47
RUNNING TOTAL = 311,875,200 × 47 = 14,658,134,400
...

We're only six laps in, and the running total has already gone past 14 billion! (No that's not a factorial, I just wanted to emphasize how quickly these numbers grow.) This is definitely the sort of long calculation we want a computer to do for us. The last step is to translate our algorithm steps into a language a computer can understand. So what does an algorithm look like to a computer? Well, much as humans can speak different languages, computers can understand different programming languages. I've selected one called Python, because it has a very simple grammar and is one of the closest computer languages to being readable English. I'll also include some comments to the right of each line to clarify what the code is doing.

Here is the factorial algorithm, coded in Python. If you

are so inclined, you could run this on a computer. Much in the way we were naming functions before, I can give each algorithm a name then write its inputs inside brackets next to it.

```
def factorial(n):                 # I am defining the algorithm called
                                  # 'factorial' and it starts with a number 'n'
   running_total = n              # Remembers n as the starting value of
                                  # the running total
   while n>1:                     # Repeats the next bit until n is no longer
                                  # bigger than 1
         n = n – 1                       # Subtracts 1 from n
         running_total =                     # Multiplies the running total
            running_total * n                   # by the new n
   return running_total                  # Returns the running total
```

I've just double-checked this program to make sure it works. Once loaded up, I can enter 'factorial(13)' into my computer to double-check 13!. And, sure enough, 6,227,020,800 appears back on the screen. So then I let it loose on 52! and here is the exact output from my computer:

```
>>> factorial (52)
806581751709438785716606368564037
66975289505440883277824000000000000
```

That's a number with sixty-eight digits. A truly huge number. To say it the long way, there are 800,000 billion billion billion billion billion billion billion billion billion ways a pack of cards could be shuffled. There are only about 1 million billion billion stars in the observable universe. And the universe is only about 4 hundred million billion seconds old. So this means that if every star in the universe had a billion planets, each with a population of a billion hypothetical aliens, all shuffling a billion packs a second since the dawn of the universe, we would now be only halfway through every possible arrangement. And that's for a pack of only fifty-two cards! Thank goodness we didn't include the jokers.

Anyway, now we know what the answer is, we can try to calculate it in a different way: using a recursive algorithm. We simply need to state the algorithm in terms of itself. Something like: 'The factorial of n is n times the factorial of $n - 1$.' The only extra thing we need to know is that the factorial of 1 is 1. This is the 'get-out clause' in the algorithm which stops the recursion going on for ever.

Step 1: Remember that the factorial of 1 is 1.
Step 2: Multiply n by the factorial of $n - 1$.

Translated into Python:

```
def factorial(n):                # I am calling the algorithm 'factorial'
                                 # and it starts with a number 'n'
    if n == 1: return 1          # The factorial of 1 is 1
    return n * factorial(n − 1)  # Calculate n multiplied by the factorial
                                 # of n − 1
```

Presented with 'factorial(13)', my computer now runs down the chain of recursions until it hits factorial(1), then comes racing back up to give 6,227,020,800 once more. Likewise, putting in 'factorial(52)' gives back the same 68-digit monster. Yet nowhere have I actually told the program how to calculate a factorial, merely how one factorial calculation relates to a smaller one. Once again, recursive algorithms are magic: conjuring answers out of seemingly empty code.

We now have two different computer programs which compute the same answer two different ways. Which leaves us with only one option: make them fight! I just took a quick break to race both factorial programs head to head. I decided that the winner would be the first to compute the factorial of 100. Computing the full 158-digit answer to 100! took the recursive Python program 0.000068 seconds, whereas the normal one came in at only 0.000046 seconds.* A decisive win for the non-recursive program!

However, the recursive algorithm will not always lose to an explicit one. It depends what the algorithm is calculating: some tasks are perfectly suited to recursion. A few chapters back, I was confidently stating different polydivisible numbers in various bases. Finding these took a fair bit of computing grunt, and my program was going to take a very long time to crunch through the base-16 cases – but with some rewritten recursive code I had it done in seconds.

Finally, we can use recursive algorithms to see how long it will be until the disc-moving priests cause the end of the world. Moving each disc requires moving all the discs in the tower above it twice, so for each extra disc in the stack the total number of moves doubles, plus one extra move to shift that bottom disc itself. So moving a tower of n discs takes twice as many moves plus one, compared to the $n - 1$ tower.

```
def Hanoi_moves(n):          # I am calling the algorithm 'Hanoi_moves'
                             # and it starts with number of discs 'n'
  if n == 0: return 0        # If there are no discs, no moves are
                             # required
  return 2*Hanoi_moves(n – 1) + 1    # Double the number of moves
                                     # for n – 1 and then add 1.
```

* 93326215443944152681699238856266700490715968264381621468592963895217599993229915608941463976156518286253697920827223758251185210916864000000000000000000000000, for the record.

```
>>> Hanoi_moves(7)
127
>>> Hanoi_moves(10)
1023
>>> Hanoi_moves(20)
1048575
>>> Hanoi_moves(32)
4294967295
>>> Hanoi_moves(64)
18446744073709551615
```

So if you have a tower of ten discs, you'll need to make 1,023 moves (which you may recognize as $2^{10} - 1$: these are all Mersenne numbers! Any number one below 2^n is called a Mersenne number, even if it's not prime). At one move per second, that is a solid seventeen minutes of disc shifting. Even if you dedicated every moment of your existence to moving discs, a human cannot complete more than a 31-disc version of the Tower of Hanoi in their lifetime, which makes the 32-disc version a perfect gift for the person with too much free time. This 'quick game' can take much longer than the 10-billion-card trick.

For the priests, Hanoi_moves(64) = 18,446,744,073,709, 551,615. Which, at one move a second, is 584 billion years: much longer than the life expectancy of our solar system. Or, as Lucas described it, 'more than 5 billion centuries'. This is also why he mentioned the wheat-on-chessboard problem; the number of grains of wheat for any number of squares is also a Mersenne number. Because a standard chessboard has sixty-four squares, you get the same answer as for sixty-four discs: just over 18 quintillion. Which is a massive number. Even if the priests had started 4.6 billion years ago, as the solar system formed, at fifty moves per second, they would still not be finished 5 billion years from now, when the Sun becomes a giant red star and absorbs the Earth. So I think we're safe.

Moral of the Story

Algorithms may seem cold and calculating, but it's important to remember they're written by humans, and there's no 'correct way' to write one. Despite being the plaything of computers, there's something delightfully human about programming and algorithms in general. Writing good computer code requires a person who is creative as well as mathematical.

Writing efficient algorithms is very much an art, and a highly sought-after one at that. During the summer of 2004 the cryptic text '{first 10-digit prime found in consecutive digits of *e*}.com' appeared on otherwise blank billboards in the US. To find the first 10-digit prime hidden in the never-ending digits of the number *e* would require a very creative bit of programming indeed. But if anyone could write an algorithm to find those digits, and then went to the URL, they would be faced with the website Google was using to recruit programmers!

Epilogue

If the strings of letters describing how to solve the Tower of Hanoi look familiar at all, it's because you have used exactly the same steps to solve a different puzzle already. A B A C A B A not only solves a three-disc Tower of Hanoi, but if A, B and C are directions, they also describe a Hamiltonian path around a cube. So, solving a four-disc Tower of Hanoi is the same as finding your way around a 4D cube, visiting each vertex once.

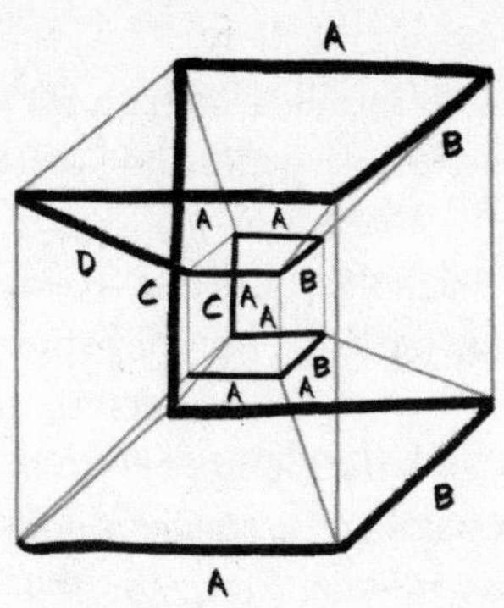

Hamiltonian path around a hypercube

We can do better, though.

The letters A, B and C are only telling you which discs to move but not to which pole, and they're only telling you which edge to follow but not in which direction. We could take the A B A C A B A to go around the cube and add a sign to indicate in which direction. So A means to go forward in one direction and –A means to go backwards. We can use a negative sign to signify moving in the opposite direction to those of B and C too. So the Hamiltonian path going through all the corners of a 3D cube would be A B –A C A –B –A, and a 4D cube would be A B –A C A –B –A D A B –A –C A –B –A. There is, of course, an algorithm for building up these lists of instructions.

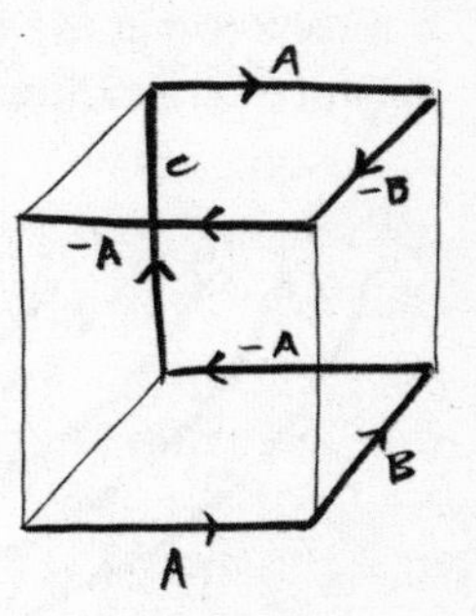

Cube Hamiltonian path with directions

Hamiltonian Path for an n-hypercube

Step 1: List out the Hamiltonian path for $n - 1$ dimensions.

Step 2: Write the new direction.
Step 3: List out the Hamiltonian path for $n - 1$ dimensions in the reverse order and swap signs.

Not only does this solve both the Tower of Hanoi and travelling around a hypercube, but the same pattern solves a puzzle from earlier: how to hang a picture on hooks so that if you remove any hook the picture falls to the ground. Once again, it involves repeating the previous solution twice, with an extra step in the middle. Only, for the hanging-picture problem, we need one bonus move at the end. In this case, A means to put the wire around the first hook in a clockwise direction, and –A is anticlockwise.

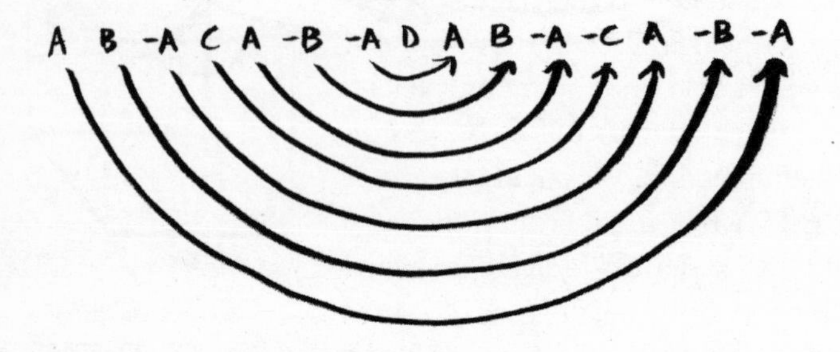

Hanging a Picture on *n* Hooks

Step 1: List how to hang a picture on $n - 1$ hooks.
Step 2: Write the new direction, around the new hook.
Step 3: List how to hang a picture on $n - 1$ hooks in the reverse order and swap signs.
Step 4: Write the negative new direction, around the new hook.

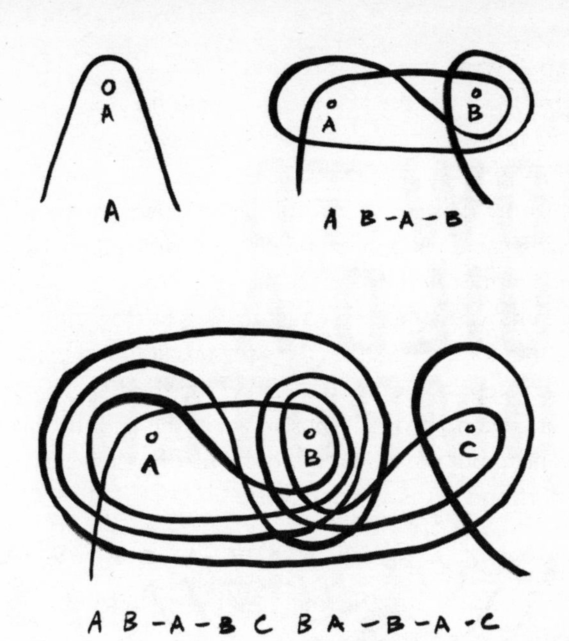

Solving the puzzle of a nineteenth-century French toy is the same as moving around hypercubes is the same as hanging a picture badly. Despite seeming to be completely different, there is the same underlying logic to how these problems are solved. And they can all be solved with algorithms.

Twelve

HOW TO BUILD A COMPUTER

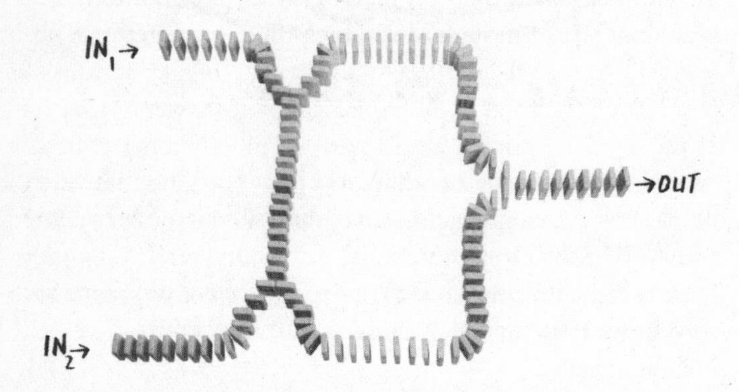

If learning mathematics is like exploring a dense jungle on foot, gradually hacking your way forward, then computers are like having sudden access to a helicopter. Yes, you need to be on the ground to truly understand a terrain, but if you can jump in a helicopter and spend a while soaring above the canopy, you'll quickly get a better sense of the overall landscape, cover more ground than you could on foot and maybe spot something of interest off in the distance. Humans still need to get involved in the nitty-gritty of mathematics – after all, the whole point is for us to learn and understand the underlying logic of our universe – but the whole process can be sped up dramatically with a bit of computational help.

We've already seen how much of a difference this has made to Mersenne primes: computers can check whether a number is prime in a fraction of the time humans need. The ancient Greeks knew of four Mersenne primes, and by the time of Lucas there were still only ten that had been found, the last one in 1883. It took until 1914 to fill in the gaps and find the twelve smallest Mersenne primes. Finding six numbers per millennium is a pretty slow rate of discovery. When a computer was turned on in 1952, five were found in that year alone. Even though Mersenne primes were getting substantially harder to locate as they became bigger and more spaced out, the total number had doubled to twenty-four by 1971. As many primes were found in under twenty years as had previously been found in over two thousand. The total number of Mersenne primes had risen to thirty-six by 1997 and, as of 2014, has quadrupled to forty-eight. The biggest known Mersenne prime in 1914 had thirty-nine digits; a hundred years later, in 2014, the size has exploded to over 17 million digits.

This does not mean that computers have given us more insight into Mersenne primes; we have merely found more of them. The computers involved in GIMPS are still running the same primality test devised by Lucas and Lehmer. The maths hasn't changed, merely the computational firepower. Likewise, with the Riemann Hypothesis so key to understanding the density of prime numbers, computers can check the pattern Riemann spotted in the primes to make sure it holds for bigger and bigger numbers. But this will never prove the hypothesis about why the pattern is there. If we found an exception to his alignment, that would *disprove* the hypothesis and have a massive impact on modern mathematics. But, for now, all we can say is that the Riemann Hypothesis has been checked for over 10 trillion values and everything has lined up perfectly. So far.

Finally, there are areas of maths such as the four colour theorem and the Kepler conjecture about stacking spheres for which we now have proofs done in part by a computer. These computers are not mere supporting calculators: they are doing a huge amount of the checking themselves, because it's impossible for humans to get through it all. And we now have initiatives such as the Flyspeck Project using computers to check the proofs done by computers. Regardless of your thoughts on whether an exhaustive proof crunched by computer counts as 'proper' mathematics, we can all agree that computers are now playing a big role in maths, and it looks like we're going to rely on them more and more.

So, with something so vital to the cutting-edge world of mathematics, we should have a good grasp of how computers work. Yet very few people have any idea. Sure, we may have looked at algorithms and kind of understand how you write a program to tell a computer what you want it to do, but how does it actually carry out those commands? How can it read and understand your instructions? If you ask around, some people will try to explain it by the interaction between the hard drive, memory, processor and other parts of the computer. But the question is not how we happen to build our current computers but rather why we are able to build a computer in the first place. Most people's knowledge of how computers work will hit a cul-de-sac long before you get to how the computer is able to think for itself.

While it is human nature to assign motives to inanimate

items, as if the objects that trip, inconvenience or hide from us are agents thinking for themselves, there are some physical systems for which this *is* true. We're surrounded by devices that are capable of doing some basic thinking and responding to the world around them, from the obvious computers and smartphones to automatic ticket-barriers and washing machines. Inside all of them are circuits which – if we wish to draw the line short of saying they are 'thinking' – are definitely doing some basic computations all on their own.

Inside all these devices are electronic circuits, and they are what are doing the computations. Somehow, a collection of wires and electrical components is able to respond to inputs, compute and put out answers. How a physical object can respond to an input and select the appropriate output baffles most people, but that is the basis of all computing. It need not be mysterious – in fact, you can build your own basic computer without any electronics. You'll just need around ten thousand dominoes. Or, if you can scrape together around a hundred (a tin of double-twelve dominoes will give you ninety-one, close enough), you can at least make a start.

The concept of a machine or object which can automatically think for itself has been around for a very long time; stories of automata reach right back to prehistory. In the 4th century BCE, the mathematician (and member of the school of Pythagoras) Archytas designed and maybe built a mechanical pigeon. Automated creatures reached their surreal peak with the 1739 'Digesting Duck'; a result of the French automaton obsession, the duck could eat grain and excrete waste. It must have been the centre of some pretty wild parties. However, we're not concerned here with inanimate objects that can replicate the movements of life but with ones which can reproduce cognition.

Mechanical calculating devices also go right back into

prehistory. As soon as humans had to do tiresome sums, they found ways to get a machine to do them instead. Some of these ancient devices were concerned with astronomical calculations required to predict events in the sky. Perhaps the most amazing of all of these is something called the Antikythera Mechanism, which was built around 150 to 100BCE. It could do complex calculations to track the position of the sun and moon in the sky, possibly along with some or all of the five planets known at the time, as well as following an intricate calendar. But what may be even more amazing is that we even know it existed.

The only surviving Antikythera Mechanism was found in around 1900, off the Greek island of Antikythera, in a ship that had sunk in about 70BCE. The mechanism looked like one cog, possibly from a later shipwreck, lodged in a large rock. Eventually, several mud-clad fragments of the machine were raised from the ocean floor and found to contain around thirty different cogs, all crafted to an amazing level of detail. It also shows signs of being expertly built to plan, meaning that it wasn't some experimental device but rather one of a number of these contraptions. There are no spare drill-holes or reshaped parts, which would indicate that the machine was gradually refined; rather, it seems to have been custom built to be easy to open and service. Even though all the bronze parts and craftsmanship fit with the technology and skills we know were available at the time, they had been put to use with a level of ingenuity that would not be seen in other clockwork mechanisms for over a thousand years.

It has taken the last century to reverse-engineer the mechanism and work out what it did. Modern imaging has been a huge help, as it means we can peer directly into the surviving fragments and take a close look at the mechanism's inner workings. Using a CT scanner, it's been possible to image the 3D parts of the Antikythera using a series of 2D slices which can then be animated to give a sweeping view right through the machine. As you watch a video of this, different-sized cogs with varying numbers of equilateral-triangle teeth appear and disappear, some intermeshed next to each other and others stacked on top. By reconstructing the mechanisms, we now know these cogs were performing astronomical computations. The mechanism could predict events in the sky long before they happened. There are various theories as to why no other such mechanisms have survived (it seems building them out of a very expensive metal didn't help, and it took a shipwreck to stop the bronze from being salvaged), but my favourite hypothesis is that we have heard so little about them because the Antikythera was a secret military technology of the day. Being able to predict the exact moment an eclipse would occur and the sun would disappear from the sky would be a powerful tactic to control other people.

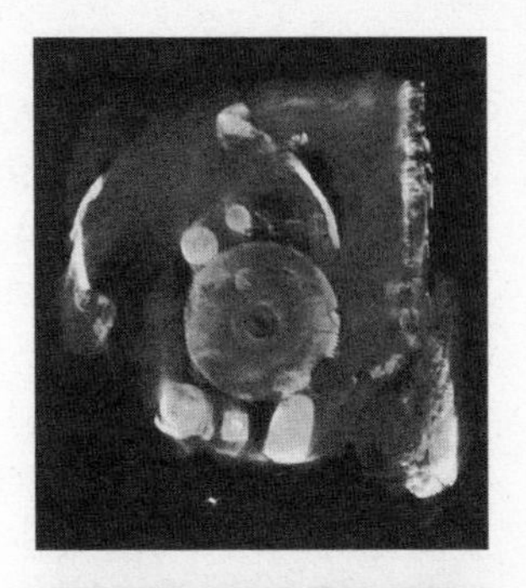

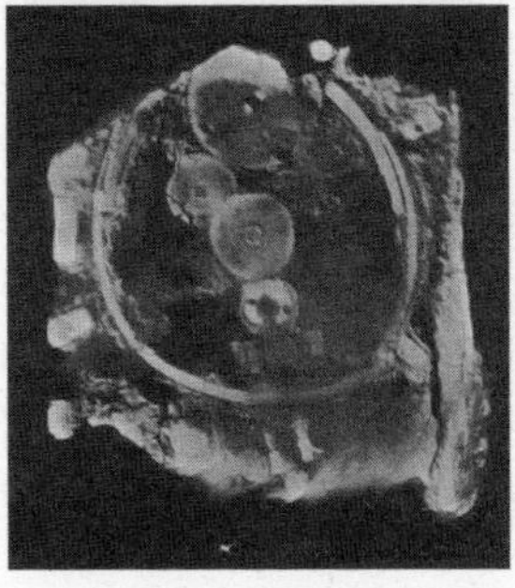

CT scans of the Antikythera Mechanism parts.

It was the number of teeth on the cogs, and how they were all joined together, which allowed the Antikythera to do its computations. Once everything was in place, you merely had to wind the handle forward to start predicting how future skies would look. People have re-created mechanisms using the same cogs as the Antikythera that can perform the same calculations. My favourite is one I saw on display at Google (the physical Googleplex, that is, not the website), which was an Antikythera Mechanism made entirely from Lego. Amazing as it is, though, the Antikythera Mechanism is not really what we would call a computer. You wind the handle and it gives out the same answer every time – like, well, clockwork. The first person who built a computer that could run more exciting algorithms was the nineteenth-century English mathematician Charles Babbage.

By the 1800s, one of the most common aids to calculations was a book filled with tables of numbers. These books were, effectively, giant lists of answers to common calculations which other people had worked out previously, so you could take a difficult calculation, split it into smaller ones and then just look it up. The problem was that producing these books of standard answers was a long, expensive process, and they were far from infallible. Given that they contained

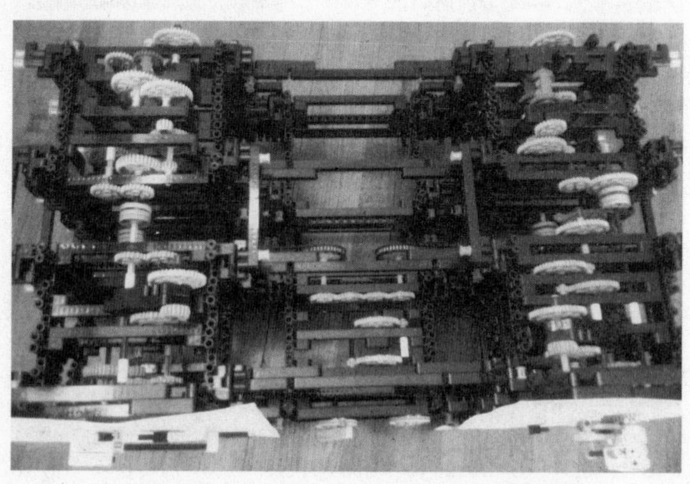

What the Antikythera Mechanism would have looked like, if the ancient Greeks had access to Lego.

thousands of answers, it was extremely likely there would be a few errors in there somewhere. It was hard to know if the answer you had looked up was right or not, and a slight mistake could ruin the rest of your tedious workings out. It was frustration at the mistakes in these tables that inspired the world's first computer.

Babbage credited the genesis of his computation machine to a daydream he had while still an undergraduate at university. He describes in his own words how a fellow mathematics student at Cambridge found him slumped over a book of maths tables. When asked what he was dreaming about, Babbage replied, 'I am thinking that all these tables might be calculated by machinery.' Babbage's epiphany was that it would be quicker, cheaper and more accurate to generate these tables using a machine. He may have been imagining a machine made of cogs and driven by steam power, but it ended up having more in common with a modern smartphone than with the Antikythera Mechanism.

Later, after his university years, Babbage did indeed design what he called a Difference Engine, which could be viewed as a very complicated clockwork system that could be set up to produce values from various equations. It wasn't quite what you could call a computer, but it fixed the problem of mistakes in maths tables. He convinced the UK government to give him some money to build it, but during construction he realized he could design a better machine and never actually finished it. The new machine he came up with he called the Analytical Engine, which was more complicated than the Difference Engine and *was* what you could call a computer.

What made the Analytical Engine such a revolutionary concept was that, instead of being set up in a certain way to solve a specific problem, it could be given different algorithms to run. Babbage borrowed from the other cutting-edge technology of the time – weaving looms – to make his engine programmable. Automatic looms had input slots for punch cards (so called, logically enough, because they were pieces of card that had a pattern of holes punched in them). Hooks in the loom were activated on and off (like rudimentary switches) depending on where the holes were in the card, and the looms wove different patterns into the fabric accordingly. Instead of using the pattern of holes to determine what weaving pattern a loom should follow, Babbage used these cards to tell the Analytical Engine which algorithmic steps to perform.

Because you could enter any program you wanted to run, along with any input data you wanted it to process, this Analytical Engine fulfilled all the requirements to be a computer in its own right as much as any modern computer built today – which is why Babbage is regarded as the 'Father of Computing'. Unfortunately, we do not know if it worked,

Difference Engine built by the Science Museum, London.

because Babbage never built one. Sadly, after his death in 1871, aged seventy-nine, no one else successfully picked up the plans and finished making either of these machines. Well, at least until 1991, that is, when the Science Museum in London spent a few years building a Difference Engine, according to Babbage's plans and using only the nineteenth-century tools and methods that would have been available to him. And it worked. Which bodes well for the Analytical Engine – which works in theory – also working in practice.

You may have noticed that, back in the first chapter, I assigned the title 'Father of Computing' to Alan Turing. This is because, even though Babbage came up with a design for one possible computer, in 1936 Turing described in abstract terms what it would require for something to *be* a computer. Not content with merely explaining such computers theoretically, Turing was at the British code-breaking centre Bletchley Park in the Second World War, where they built the world's first digital, programmable electronic computer, Colossus. Because it was designed and constructed using highly classified wartime technology, no one knew about Colossus until decades later. But immediately after the war, the people at Bletchley Park dispersed to academic institutions around the world and suddenly had amazing success building the world's 'first' computers, as if they somehow already knew it could be done. Turing was involved in a few such projects before settling at the University of Manchester with the world's first electronic computer with digital storage.

The Analytical Engine pre-dated all this, however, and even though it was not digital or electronic, it was programmable. And despite the fact that it was never built or run, algorithms were written for it. When Babbage needed some notes about his machine translated after a series of lectures he gave in Italy, he turned to Ada Byron (better known by her title, Countess of Lovelace), who then expanded the translation with a bonus section entitled 'Notes by the Translator'. In this section, she demonstrated that she was not merely translating the text but that she fully understood how the engine worked and could elaborate further on its potential. Her very first point is the immeasurable increase in potential between the Difference and Analytical Engines. In her words, 'The Analytical Engine, on the contrary, is not merely adapted for tabulating the results of one particular function and of no other, but for developing and tabulating any function whatever.'

But the reason these notes are of huge historical significance today is what happens in the very last section, 'Note G'. I can't tell it better than Lovelace herself: 'We will terminate these Notes by following up in detail the steps through which the engine could compute the Numbers of Bernoulli, this being (in the form in which we shall deduce it) a rather complicated example of its powers.' The Bernoulli numbers form a very complicated sequence and are rather hard to compute. But in this note Lovelace wrote out what is regarded as the first ever computer program, one that would make the Analytical Engine calculate these numbers. Lovelace had worked with Babbage to write an algorithm that could be run on the first ever computer before it was even built. This has earned Ada Lovelace the title of the world's first computer programmer. And all that happened in 1842.

It also earned Lovelace a mention in Lucas's book *Récréations mathématiques*. Published some years after both Babbage and she had died (Lovelace of uterine cancer in 1852, aged only thirty-six), Lucas discussed their work, untainted by the knowledge of what computers would one day become. Much like the book you're holding, Lucas included a chapter on computational devices, but he called his '*Le calcul et les machines à calculer*'. In the middle of it he has a few paragraphs on the Analytical Engine, showing a surprising level of prescience as to how important it would later be recognized to be. Then his prescience made another leap with a section called '*Arithmétique électrique*'.

The other day, I was flicking through *Récréations mathématiques* to read the bits about Babbage and Lovelace when suddenly, even with my trivial knowledge of French, the words '*arithmétique électrique*' – which I instantly translated as 'electric arithmetic' – jumped off the page at me. Lucas was writing at a time when the light bulb had just been invented; when only about half a century had passed since Georg Ohm first analysed electric circuits; and fifty years *before* the first electronic 'switching components' would be devised. Nothing like a modern circuit was possible, or even imaginable, in the late 1800s, but Lucas still saw the potential. He discussed the inventor Henri Genaille's ideas: 'He has just invented a new but as yet incomplete device; however, the principle is still formed: an electronic calculating machine.' He was suggesting, in lilting fashion, and way back when, that one day electronic circuits would be used in computers.

The Electric Brain

Thank goodness Lucas was right and electric machines *can* compute, because we'd be stuck without electric circuits. Even though Babbage had designed a general-purpose computer and Lovelace had written programs for it, it would have been very hard to scale this type of mechanical machine, which was powered by steam or hand-crank, up to anywhere near the kind of computational power we're used to today.

An Analytical Engine, sitting at over 2 metres tall and weighing over 10 tonnes, would have been able to store only around a thousand 40-digit numbers, representing less than 20KB of memory, and perform seven calculations a second. This was a dead-end which could not lead to the computers we take for granted today. There is no way this type of machine could be made to store billions of times the amount of data, while running a billion times faster, while still being small enough to hold in our hand and update Twitter for us.

For modern computers, you need electronic circuits. The fact that they are so incredibly small and fast unfortunately means that most people have no idea whatsoever about what they actually do. A computer processor looks like a small grey rectangle of plastic, when in reality it contains intricate circuits around which electronic signals frantically race. To see what's happening on a human scale, we can build the same types of circuits using dominoes. Get yourself that pack of around a hundred, and we'll start building up some basic computation circuits.

First up are what are called logic gates. These are the building blocks of circuits. Each logic gate can take certain binary inputs and return standard binary outputs. We use the 1s and 0s of binary because a signal is either flowing in a circuit, or it isn't; wire in a computer circuit is either carrying a high voltage, or it isn't. (We're right back to where we started in Chapter 1 . . .) With a chain of dominoes, it's either falling over, or it isn't. In modern electronic circuits, a high voltage is a 1 and the low base voltage is a 0. Copying this, we can have a domino falling over as a 1 and the domino still at rest standing up as a 0. As we know, you can convert any number into base-2, so 1s and 0s are all you need to represent any number and perform any calculation on it.

The two logic gates we're going to build are the AND gate and the XOR gate (which is short for 'exclusive or'). Both have two inputs and one output. The AND gate lets an incoming signal through to the output only if both inputs are triggered together, one *and* the other. An XOR gives a signal through to the output if one input *or* the other is activated (not both). That's the exclusive aspect: it's exclusively one input *or* the other. We can show what each of these are doing with a table, and I've shown how to build them with dominoes as well.

AND gate

IN_1	IN_2	OUT
0	0	0
0	1	0
1	0	0
1	1	1

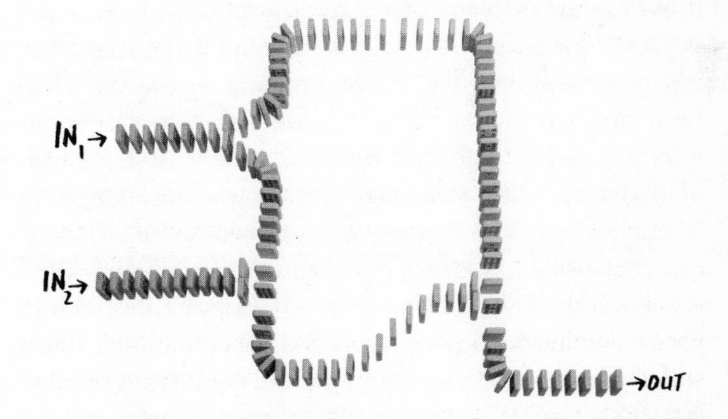

XOR gate

IN_1	IN_2	OUT
0	0	0
0	1	1
1	0	1
1	1	0

We can use these logic gates to do arithmetic. Remember: the XOR gate has an output that indicates if only one input has been activated; the AND gate lets you know when both inputs are active. If we can combine both these gates into one circuit with two outputs, we'll have a network of dominoes which can count how many of its inputs have been activated. The outputs give the number of activated inputs as a binary number. This is known as a half adder, and it only counts as 'half' because we can do better. By combining two of them, we get a full adder. This takes three inputs and gives the binary two-digit sum of how many inputs are activated.

Half adder

IN_1	IN_2	OUT 2s	OUT 1s
0	0	0	0
0	1	0	1
1	0	0	1
1	1	1	0

Full adder

IN_1	IN_2	IN_3	OUT 2s	OUT 1s
0	0	0	0	0
0	0	1	0	1
0	1	0	0	1
0	1	1	1	0
1	0	0	0	1
1	0	1	1	0
1	1	0	1	0
1	1	1	1	1

Full adders are all you need to do binary addition of any size, because binary numbers are really simple to add together. They only ever present a small number of options. While in base-10 you might need to add a 4 to a 7, or a 2 to a 9, in binary you are either adding a 0 to a 1 or a 1 to a 1 (or a 0 to a 0, but that's barely worth mentioning). This limited number of cases makes the whole process very straightforward. To demonstrate this, we can add 11 to 30 in binary. When we convert base systems, 11 becomes 1011 and 30 becomes 11110. As we would in base-10 addition, we write these numbers out one above the other, start with the units on the right and gradually move to the left as we add each column of digits together.

```
   '0 '1 '0 1 1  (11 IN BASE-10)
+   1  1  1 1 0  (30 IN BASE-10)
  -------------
  1 0  1  0 0 1
```

The units column is easy: 1 + 0 = 1, so we can write a 1 under that. The next column is slightly harder: we need to add 1 and 1. The normal answer would be 2, but this is binary and there is no symbol for 'two', so we can't put a 2 as part of the answer. Two, after all, is 10 in binary. So we do the same thing we would do in normal base-10 arithmetic and 'carry' the 1 – which makes nice sense: this column is the 2s column, so we're adding one 2 to another 2, which gives us one 4 to carry over to the next column. The step after next is as hard as it gets: we need to add three 1s. Which simply involves writing a 1 below and carrying a 1 as well. You can see that every single step involves adding up to three 1s and 0s, and then having up to two outputs: writing a number below and possibly carrying a digit. All of which can be done by a full adder. To add binary numbers, you just need a long chain of full adders, one for each column of the sum.

To build your half adder, you're going to need around two or three thousand dominoes. I know, I've tried. I first came across the idea of domino logic gates several years ago when someone sent me a link to a site which described them theoretically. I did a thorough search of the internet and couldn't find anyone who had built even a half adder out of dominoes. There were a few videos on YouTube which were close, but people had cheated by taping dominoes together, and their circuits were fragile; they did not work reliably. Many attempts required chains of dominoes to fall with precise timings, which is not what happens in the real world. I wanted to design a full adder that would work in the real world. Eventually, I cracked it. The circuit design is in 'The Answers at the Back of the Book'. Now I needed enough dominoes to build several of them and chain them together.

So I bought ten thousand dominoes.

For one weekend of the Manchester Science Festival in October 2012, I and a dozen other volunteer 'domino computer builders' balanced thousands of dominoes in intricate circuits which could then perform a calculation. Designing the half-adder had been hard enough. Much like real integrated circuits, which are etched into a surface, the chains of dominoes cannot cross each other. Computer circuits are plane graphs. But the length of each line of dominoes is important. You can see we had to zigzag lines backwards and forwards: these are 'delay lines' to slow one signal down and make sure any preceding calculations are complete before the next step begins. The 10,000-domino circuit presented even more problems, but I had a team of crack mathematicians to help.

At the end of day one, only about an hour behind schedule, the circuit was finished. It could add any two three-digit binary numbers and give the four-digit result. To enter the numbers to be added, we left gaps in the input chains of dominoes

All ten thousand dominoes ready to do a calculation.

which could be left empty to be 0s, and filled in the ones that were going to be 1s. There was one long chain of dominoes which zigzagged along the base of the circuit and branched off to run into all six input chains. We chose the numbers to be added at random, and they ended up being 4 (100) and 6 (110). So, for 4, we left a gap in the 1s and 2s inputs but filled in the

domino chain for the 4s; for 6, the 1s chain was also left with a gap, but the 2s and 4s chains were completed.

Quite a crowd had gathered by this point. We'd been building the computer in the middle of the main hall of the Museum of Science and Industry in Manchester, and people had come by to watch us throughout the day, often waiting in anticipation of someone accidentally knocking some dominoes over. It took the domino computer forty-eight seconds to run the signal through all the various logic gates, and only the 2s and 8s outputs fell over, giving a read-out of 10 (1010). The crowd went suitably wild. Given the six hours of set-up time, this is possibly the most inefficient way ever to add 6 and 4 and get an answer of 10. Or, as some mischievous onlookers pointed out, we managed to take an entire day to show that $6 + 4 = 2 + 8$.

The point remains that what we built is the fundamental element on which all computers rely. If we had a near-infinite supply of dominoes, we could keep building the circuit bigger and bigger and have it perform more and more complicated tasks. Once you can add numbers, you can then expand the domino circuit board to multiply, divide and even find square roots. There are some other technical problems to overcome, such as building memory out of dominoes and a way to automatically re-set them but, in theory, the Skynet is the limit.

As we said before, though, these sorts of physical systems do not scale up well. If we wanted to do a second calculation on our domino computer, it would take another day to stand all the dominoes back up, whereas an electronic circuit is not destroyed in the process of performing a calculation and so, once one computation has cleared all the logic gates, you can send the next one in. A modern processor does this billions of times a second. As I type this on my laptop, the 2.7GHz processor is quietly idling in the background. (2.7 gigahertz is a measurement of how many times its logic gates can be run every second.)* This is slightly faster than the dominoes.

* Actually, this is how many times the processor performs commands in a second, each of which could involve more than one calculation. So this is a low estimate for comparison. A more dedicated me would research how many actual calculations it does per second, aka FLOPS.

At a very rough estimate, the domino computer we built took six hours to set up and run, which means it could do four calculations per day, *if* you had a team of domino-computer builders working around the clock. This is a terrible rate of one calculation every 21,600 seconds, equating to a processor speed of 46.3 microhertz. Which makes my laptop 58 trillion times faster than the domino computer. The calculations that run on my laptop in one second would require a team of domino technicians working around the clock for nearly 2 million years. While I had an amazing team of domino volunteers, this is slightly beyond what I could ask them to do.

We did, however, use day two to redesign the circuit to take two four-digit inputs in order to yield a five-digit sum. Unfortunately, we did not have any extra dominoes or any extra space, so we had to make the circuit both more efficient and more compact, which are also genuine concerns when designing modern integrated circuits for computers. We pushed our luck and, in the final calculation, two things went wrong. We had slimmed down the tolerances in the delay lines to save dominoes, but we shortened one a bit too much and the blocking signal arrived after the other signal had already passed. It was closing the logic gate after the signal had already bolted. Secondly, we ran some of the lines too close and had 'signal bleed' (technical name: 'crosstalk', which sounds much cheerier). A domino on a corner fell over and slid across the ground just far enough to bump one over in the adjacent chain, setting off a phantom signal.

What I loved, though, was that both these problems are issues faced when miniaturizing actual computer circuitry. In addition to all the timing issues you can have in circuits, if you lay two 'wires' too close, the current in one can induce a current in the other. Combined with signal bleed difficulties,

The display resolution is terrible.

both for electrons and dominoes, designing efficient circuits is an art. As you would expect, most modern integrated circuits are themselves designed by computers but, sometimes, they still need a bit of human ingenuity. When Apple released the iPhone5 in 2012, it was soon noticed that the

components which made up the A6 processor inside had been designed and laid out by hand. Circuit design is one of the great applied-maths problems. If you'd like to give it a go yourself, I have ten thousand dominoes you can borrow . . .

Thirteen

NUMBER MASH-UPS

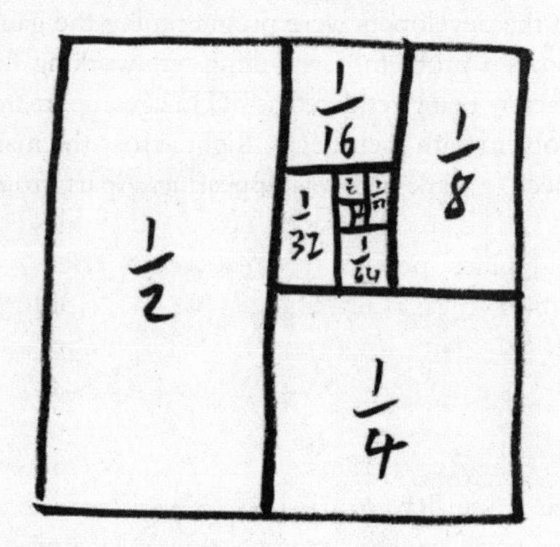

Numbers on their own are great, but when they start joining forces the result is greater than the sum of their parts. Or even the product of their parts, if you'd rather multiply them. Or the square, inverse, or ratio of their parts. However you choose to combine numbers together, they can produce some incredible results – some incredible results without which our modern technology just would not work.

Modern computer graphics in video games and feature films are the result of putting together swarms of numbers behind the scenes and rendering the result as an image. But if those numbers are combined incorrectly, the picture can break. When the video game *inFAMOUS Second Son* came out on PlayStation 4 in 2014, it was a huge hit, selling 1 million copies in the first nine days after its release and causing a jump in the sales of PlayStations in the UK. Yet releasing it depended on fixing a hitch in how those numbers were being put together.

When the developers were programming the game, they came across a problem. Everything was working fine with the characters being rendered as 3D images, apart from one small problem with their necks. Right across the main character's neck, a dark line was appearing. Apart from being ominous and creepy, it just didn't meet the standard of high-quality graphics people would expect. After a lot of investigation, it was realized that some of the programmers working on the game had been combining numbers in slightly different ways, and that this mismatch had come to a head (as it were) at the neck.

Because the face of the main character was more complicated and would have to be manipulated more finely than the other parts of him, it was being calculated and rendered by a different bit of the program to the one that was doing the graphics of his body parts. So there was one subsystem within the software to control the face, and another for the body, and their areas of jurisdiction met at the neck. Each part of the software had to combine all sorts of numbers (the position in 3D space of each pixel, the lighting conditions, the true colour of the object, the angle it's being viewed from, etc.) to give one numerical value for the colour each pixel should appear in the final render. In theory, the images the subsystems were mathematically producing should have melded seamlessly but, for some reason, they were producing slightly different results right on the boundary. I spoke to one of the programmers, and he explained that one part of the program had been rounding its answers down, and the other had been rounding its answers up.

In order to agree on the same way to combine numbers systematically, mathematicians create what is called a function. Even the simple act of rounding a value up or down to a whole number can be represented as several different functions. Always rounding down is called the floor function, and it is represented by strange square brackets that seem to have their top missing, for example, $\lfloor 7.9 \rfloor = 7$. Always rounding up is called the ceiling function, and it's written in strange square brackets that seem to have their bottom missing, so, for example, $\lceil 3.1 \rceil = 4$ and $\lceil -5.6 \rceil = -5$. One part of the computer code for *inFAMOUS Second Son* had been using the floor function when it had to round the values and the other subsystem the ceiling function.

Warning: Graphical Content

At its heart, a function is any systematic method of taking numbers in, processing them then spitting them back out. That said, functions are intimately linked to their graphical representations. Graphing a function is a way of showing the output numbers visually, which most mathematicians would call a plot, having already used the word 'graph' to mean a network. I'll use both interchangeably. Let's get straight down to it: here are my four favourite ways to plot functions graphically.

1. Match the Input with the Output

This is the classic way to plot a function. Take something nice and prosaic – let's say the function $f(x) = x^2$. We have given the function the name f, and the x in the brackets represents the input number. For any input x, we get an output of x squared, aka x^2. For the plot, we look at each value on the horizontal x-axis, and for that input we put a dot above it which lines up with its output on the vertical axis, and all those dots form a continuous line. You can think of the surface of the chart as every possible combination of two numbers; each point has two coordinates. This plot is marking all the coordinates that are a valid (input, output) pair for our function. This works for different functions as well, such as $f(x) = 10x - x^2$.

Not very exciting, I'll agree.

To liven things up a bit, get a tennis ball, a pair of metal tongs, some highly flammable liquid and a video camera. Somehow, you just know an activity is going to be fun if the list of requirements ends in 'some highly flammable liquid and a video camera' – which is why I must insist you do not actually try this activity yourself. But, if you do (which you

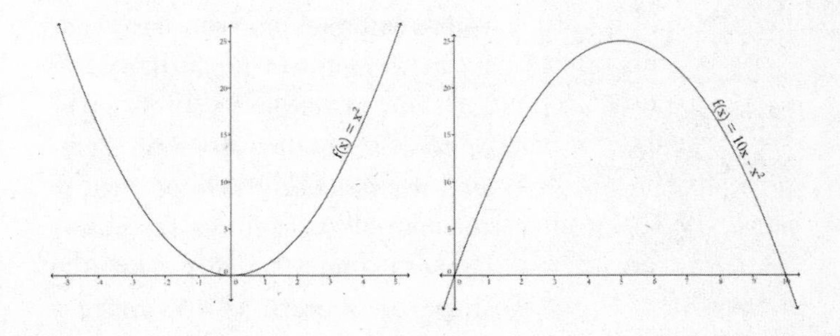

won't), here's what you do (hypothetically): hold the tennis ball in the metal tongs and soak it in the flammable liquid, then have someone keep the camera completely stationary and film you, preferably in the dark, as you set the ball on fire and throw it up in the air. The camera will film the ball's flight and you can then edit the frames together to show its path. What you will end up with is a plot of the ball's height as a function of how long it has been in the air, and it should have the same shape as the $f(x) = 10x - x^2$ function.

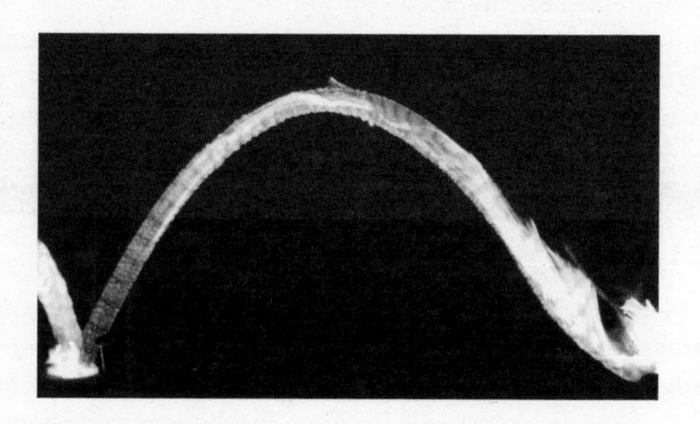

2. One Input, Two Outputs

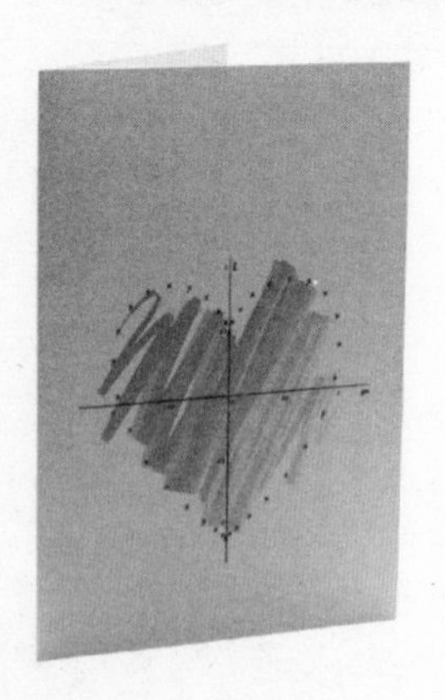

If you have a function which gives you two outputs, you can start feeding input numbers into it then mark each pair of output values as a pair of coordinates on your graph. I created the function below so that, if you plot all the outputs for inputs between 0 and 2π, they form the shape of a heart. If you make a spreadsheet to calculate those values for you and produce a chart, you can email it to the loved one in your life to show them how you feel about them, complete with workings out. Or just send them an 'I Chart You' card. Now you can have your own rigorous loving.

$$\text{COORDINATES} = \left(\sin^3(t),\ \cos(t) - \tfrac{1}{3}\cos(2t) - \tfrac{1}{5}\cos(3t)\right)$$
$$\text{for } t = 0 \text{ to } t = 2\pi$$

3. Points that Pass a Test

There are some functions which check coordinates; only the ones that pass the requirements are plotted as part of the graph. The expression below tells you whether a pair of coordinates is worthy of being part of the graph or not. If you take any pair of (x, y) values, run them through that formula, round the result down to the nearest whole number and plot only those that are greater than ½, you'll end up with a graph that looks like . . . well, like the plot below. Called Tupper's Self-referential Formula, it's a function which, if you plot it, draws a picture of its own equation. Honestly, I didn't believe it myself until I tried it, and it does work.

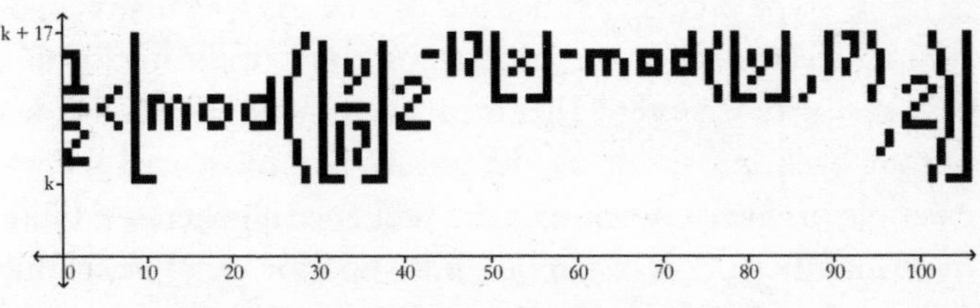

This is both the equation and the plot of Tupper's Self-referential Formula.

4. Two In, One Out: Now in 3D

Some functions take two inputs and return a single output. If you play blackjack, then you know that after being dealt two cards there's a probability that if you ask for a third it will take your total above twenty-one and you'll go bust, losing all your money from that hand. The probability of going bust is a function of the two cards you already have (assuming you have not seen any other cards in the deck), and it's possible to plot that function. We can start with the 2D surface, as before, where the coordinates represent all the possible starting pairs of cards, then plot the height above that in 3D for the probability. You can see a peak in the corner which I call, appropriately enough for some blackjack players, the 'mountain of disappointment'.

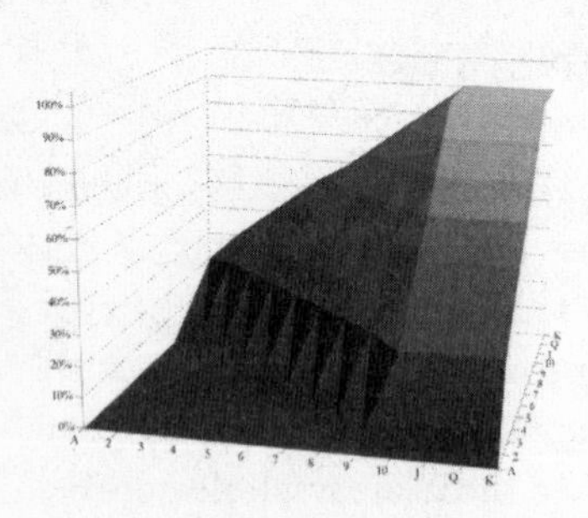

Recursive Functions

We've already met recursive algorithms, so we may as well get to know some recursive functions as well. As a recursive algorithm refers to itself as one of its steps (think of the maddening cake recipe in Chapter 11), a recursive function feeds its own output back into itself as the input. The most well-known example of such a function is the sequence it generates: Fibonacci numbers. A less famous but superior example, to my mind at least, is the sequence of Lucas numbers. Lucas, of cannonball and prime-number fame, is generally unknown these days, yet it was he who named the Fibonacci numbers after the mathematician Leonardo 'Fibonacci' Pisano. With both the Fibonacci and the Lucas sequences, you start with two numbers, then each new output of the function is the sum of the previous two outputs. With Fibonacci, the first two terms of the sequence are 1 and 1; with Lucas, it's 1 and 3.

FIBONACCI NUMBERS:

1; 1; 2; 3; 5; 8; 13; 21; 34; 55; 89; 144;
233; 377; 610; 987; 1,597; 2,584;
4,181; 6,765; 10,946 . . .

LUCAS NUMBERS

1; 3; 4; 7; 11; 18; 29; 47; 76; 123; 199;
322; 521; 843; 1,364; 2,207; 3,571; 5,778;
9,349; 15,127; 24,476 . . .

Once you get the hang of this function going up the number scale and growing bigger and bigger, you can kick it into reverse. If you take any number and subtract the number before it, you come up with the number one before that in the sequence. Try using this to reverse back past the starting point: you should end up with a list of alternating positive and negative numbers. (See 'The Answers at the Back of the Book'.) You can also find functions which link these together. If you add any two Fibonacci numbers which are next to each other but one in the sequence, they will add up to give the Lucas number that falls in the corresponding position in between those two numbers. (So, for example, 3 + 8, in positions four and six in the Fibonacci sequence, gives the answer 11, which sits in position five in the Lucas sequence.)

Fibonacci was active as a mathematician in the early 1200s; he introduced the base-10 number system to Europe and did loads of other interesting things with geometry and numbers. However, his number sequence should, by rights, be a bit of a footnote to his life. He did list the numbers, but didn't do anything else with them. It was only in the 1800s when they were investigated further, by Lucas, that it was realized how mathematically interesting they are. Which is why I think it's a shame that Lucas and his numbers have been relegated to obscurity, while everyone still seems to know all about the Fibonacci numbers.

Infamously, the Fibonacci numbers are associated with the golden ratio. The golden ratio (1.618. . .) is a number represented by the Greek letter phi (φ) and is mathematically significant for being the number with the most made-up things said about it. Don't get me wrong, there are some lovely mathematical things about the golden ratio – it is, for example, the only positive number that is exactly one more than its inverse – but I do find the claims that it appears again and again in art and architecture a bit far-fetched. Neither has anyone ever given me any explanation or

$$\varphi = 1.61803398874989\ldots$$

$$\frac{1}{\varphi} = 0.61803398874989\ldots = \varphi - 1$$

This is impressive.

This is not impressive at all.

evidence for the oft-repeated factoid that a human's height is the golden ratio of the height of their belly-button from the floor.

The Fibonacci numbers' link to the golden ratio is that, as they get bigger and bigger, the ratio between them gets closer and closer to the golden ratio. For some people, the link is so strong that they see the Fibonacci numbers as being synonymous with the golden ratio. Only thing is, it works for any such sequence where you recursively add two whole numbers to get the next one. Instead of starting with 1, 1, I tried starting with random such numbers up to 10 million and, sure enough, within twenty terms in the sequence, the ratio is always 1.618034. Any sequence of integers (whole numbers) where you add two to get the next one will demonstrate the golden ratio. To settle which sequence has the most golden-ratio-ness, start with the golden ratio, raise it up to increasing powers (from 2, 3, 4 . . .) and round each result to the nearest whole number. This will perfectly produce the series of Lucas numbers.

Even more amazingly, the Lucas numbers can be used as a rough primality test. If a number is a prime, then the Lucas number in that position must be one more than a multiple of that prime. For example, the seventh Lucas number is 29, which is one more than a multiple of 7 (i.e. 28). And 7 is a prime. The trouble is that, even though this works for all prime numbers, it also works for some non-prime numbers so, while it can confirm that a number is definitely composite (that is, non-prime) when it *doesn't* match, a match does *not* mean that the number is definitely prime. Numbers

$$\phi^2 = 2.618 \approx 3$$
$$\phi^3 = 4.236 \approx 4$$
$$\phi^4 = 6.854 \approx 7$$
$$\phi^5 = 11.090 \approx 11$$
$$\phi^6 = 17.944 \approx 18$$
$$\phi^7 = 29.034 \approx 29$$

Golden powers give the Lucas numbers.

which pass this test but are actually composite numbers masquerading as primes, are called Lucas pseudo-primes. If we want to make *absolutely* sure that a number is prime, we need to bring in the full force of a stronger test. For Mersenne primes we have the Lucas–Lehmer primality test.

The Lucas–Lehmer primality test also uses a recursive function to produce a sequence of numbers (the Lucas–Lehmer sequence). This sequence starts with 4, and each number after that is the previous number squared, minus 2 (see below). For any Mersenne number (such as $2^3 - 1 = 7$ and $2^8 - 1 = 255$), you take its power of 2 (so, 3 and 8 in our examples) and, if the Lucas–Lehmer number in the position one less than that power (so, second and seventh) is an exact multiple (no remainders allowed) of the Mersenne number, then it is definitely prime.

LUCAS-LEHMER NUMBERS:

4; 14; 194; 37,634; 1,416,317,954;
2,005,956,546,822,746,114; 4,023,
861,667,741,036,022,825,635,656,
102,100,994...

We can test our examples. For example, the Mersenne number 7 is one below the third power of 2. So we look at the second Lucas–Lehmer number, which is 14. Because this is a multiple of 7, we now know 7 is a Mersenne prime. To check the eighth Mersenne number, $2^8 - 1 = 255$, we take the seventh Lucas–Lehmer number (that monstrosity just greater than 4 billion billion billion billion) and see if it is a multiple of 255. It is not.

If the test fails, then it definitely is not a Mersenne prime. This was how Lucas could confirm that the sixty-seventh

Mersenne number was definitely not a prime, despite not knowing its factors. It's also how Lucas proved that the 127th Mersenne number was prime, making it (still) the largest prime ever found without the aid of computers. (More on Mersenne primes in 'The Answers at the Back of the Book'.)

As you can see, the Lucas–Lehmer numbers get ridiculously big ridiculously quickly, so computing them would be a real pain. However, all you need to know at each step is the remainder of that term in the sequence when it is divided by the Mersenne number in question. So, if you wanted to check the Mersenne number $2^{13} - 1 = 8,191$, you calculate each of the values and find the remainder once it has been divided by 8,191. This keeps the values manageable, and the twelfth number in the sequence is indeed 0 – no remainder – proving that 8,191 is prime. Lucas was able to compute the same sequence of numbers for $2^{67} - 1$ and $2^{127} - 1$ by hand, proving them prime and not-prime respectively. If you want a challenge, use a Lucas–Lehmer sequence to prove that $2^{17} - 1 = 131,071$ is a prime.

LUCAS-LEHMER NUMBERS, REMAINDER 8,191:

4; 14; 194; 4,870; 3,953; 5,970;

1,857; 36; 1,294; 3,470; 128; 0

All Together Now

Let's lighten the mood a little with a maths joke.

An infinite number of mathematicians walk into a bar. The first one orders a pint of beer, the next a half-pint, the following a quarter of a pint, the next an eighth of a pint, and so on. The annoyed bartender pours two pints, slams them on the bar and says, 'You mathematicians need to know your limits!'

What if we had a function which took infinitely many inputs and returned a single output? Yes, it is possible to have a function which adds up an infinite list of numbers and returns a single value. It may seem counter-intuitive that an infinite sum will not (automatically) result in an infinite total. If you tried adding all the whole numbers together, you would certainly get that expected infinite total. Because numbers get bigger and bigger, each subsequent thing you add means that the total gets much bigger. The answer explodes and becomes crazy-big. In maths, we call this divergent, because the answer just flies off. There is another option, though, which is if you add an infinite number of things but each one is smaller than the last. In this case, the total will sometimes *con*verge on an actual answer. I'll say it again: it *is* possible for an infinite number of things to add up to a finite answer.

The joke about an infinite number of mathematicians ordering drinks works because each mathematician is ordering a smaller drink than the person before them – specifically, half the size. So, despite there being infinitely many people buying it, the total beer needed is not infinite. It's only two pints. You can see this easily if you use a square to represent a pint. Because each new order is half the size of the previous one, it takes up half of whatever space remains. The total order can never be bigger than those two squares. It would take an infinite queue of people for the total order to be two pints. The answer to an infinite sum like this is called the limit of the sum.

Be warned, though: it's sometimes very hard to tell whether an infinite sum converges on a final total, or if it diverges off and has no neat answer. It's not enough for the numbers in a sequence just to get smaller. The most notorious example of this is the sum of all unit fractions (all the ones with 1s on top). If you add $1 + \frac{1}{2} + \frac{1}{3} + \frac{1}{4} + \frac{1}{5} + \ldots$ each number is smaller than the previous one, yet the sum does not seem to converge to a limit. Called the harmonic series, it was first proven to be divergent by the French mathematician Nicole Oresme in the 1300s.*

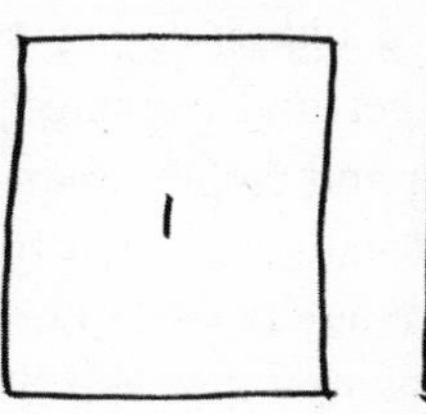

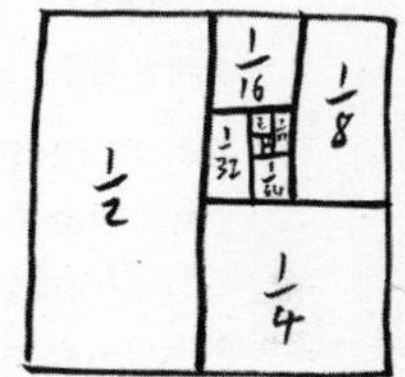

$$1 + \frac{1}{2} = 1.5$$

$$1 + \frac{1}{2} + \frac{1}{4} = 1.75$$

$$1 + \frac{1}{2} + \frac{1}{4} + \frac{1}{8} = 1.875$$

$$1 + \frac{1}{2} + \frac{1}{4} + \frac{1}{8} + \frac{1}{16} = 1.9375$$

$$1 + \frac{1}{2} + \frac{1}{4} + \frac{1}{8} + \frac{1}{16} + \frac{1}{32} = 1.96875$$

$$1 + \frac{1}{2} + \frac{1}{4} + \frac{1}{8} + \frac{1}{16} + \frac{1}{32} + \frac{1}{64} = 1.984375$$

$$1 + \frac{1}{2} + \frac{1}{4} + \frac{1}{8} + \ldots + \frac{1}{1024} = 1.999\ldots$$

Oresme joins Fibonacci in anomaly corner in the history of mathematics. Almost without exception, the Greeks in the distant past did some maths and then absolutely nothing happened in Europe until the sixteenth century or later. Oresme and Fibonacci are the exceptions. The discovery of the harmonic series diverging falls during that gap when, despite other mathematical advances around the rest of the world, not much happened in Europe for about a thousand years. Nicole Oresme not only proved that the harmonic series diverges, he was also a ground-breaker in plotting functions graphically. *And* he used speed/time graphs centuries before Newton.

Oresme's genius was to make a new series which was definitely smaller than the harmonic series. He took the list

* Mathematicians use the word 'series' to mean the 'sum of a sequence of numbers'.

$$\underline{\text{TOTAL}} = 1 + \frac{1}{2} + \frac{1}{3} + \frac{1}{4} + \frac{1}{5} + \frac{1}{6} + \frac{1}{7} + \frac{1}{8} + \frac{1}{9} + \frac{1}{10} + \frac{1}{11} \ldots$$

$$\underline{\text{TOTAL}} > 1 + \frac{1}{2} + \underbrace{\frac{1}{4} + \frac{1}{4}}_{=\frac{1}{2}} + \underbrace{\frac{1}{8} + \frac{1}{8} + \frac{1}{8} + \frac{1}{8}}_{=\frac{1}{2}} + \underbrace{\frac{1}{16} + \frac{1}{16} + \frac{1}{16} \ldots}_{=\frac{1}{2}}$$

$$\underline{\text{TOTAL}} > 1 + \frac{1}{2} + \frac{1}{2} + \frac{1}{2} + \ldots$$

Proof that the total of all unit fractions is massive.

of all unit fractions, and for any of them which did not have a power of two as a denominator, he replaced it with a smaller fraction which did. As all these new fractions were either the same or smaller, the total of this new series had therefore to be smaller than the sum of the harmonic series. But when Oresme grouped these fractions into runs, each of which added up to ½, he was left with a sum of an infinite sequence of ½s, which *definitely* diverges. This meant in turn that the greater harmonic series must also diverge. Oresme had proved that a sequence of ever-decreasing numbers could still be divergent. (His proof was lost for a while, and the same result was independently rediscovered in the 1600s.)

However, even if you know that the sum of an infinite sequence converges, it can be extremely hard to calculate what the total is. Mathematicians knew that the sum of the 'square fractions' (fractions with square-number denominators) converged, but no one could calculate the final total. In 1644, the Italian mathematician Pietro Mengoli challenged mathematicians to find the total, in what is now called the Basel problem, named after the town of Basel, Switzerland, the birthplace of Euler, one of the greatest mathematicians who ever lived and counter of faces, edges and vertices. Solving the (appropriately named) Basel problem is what first put Euler on the mathematical map. In 1735, he revealed that:

$$1 + \frac{1}{4} + \frac{1}{9} + \frac{1}{16} + \ldots = \frac{\pi^2}{6}$$

That fact still messes with my mind. Somehow, adding the square fractions (known as the inverses of the square numbers) give you ⅙ of π squared. There isn't a circle in sight, yet pi suddenly leaps out of nowhere (classic pi). It's another glimpse into the deeper connectivity of mathematics. Finding and proving this result was an amazing feat of mathematics and deservedly made Euler famous. It was not until about a century later that mathematics came up with reliable techniques to deal with these sorts of infinite-input functions.

Meet the Bernoulli Numbers

You may have noticed we've not had need of scissors and paper for a while, or even balloons and straws. This may be causing a queasy sensation. I'm well aware that talking about pushing numbers around sequences and through functions can be a trigger for some people. They have flashbacks to bleak, endless algebra lessons at school. They may even wake up in a cold sweat during the night, screaming, 'SOH CAH TOA!'

But it's OK. You can be brave of chart. I'm not sending you back to school, this is a very different thing (despite a superficial resemblance to the classroom). Frankly, there is no way mathematicians are going to limit themselves only to the wonders of shapes and puzzles you can build when there is a whole area of functions and squares to play with. So we're going to join them. Only now, the fun is slightly more abstract; the things to make and do will exist only in your mind. Don't miss out because you were once bitten by a quadratic equation as a child.

Here we go!

Quick, can you work out:

- the sum of the first four cube numbers?
- the sum of the first ten odd numbers?
- and why these answers are the same?

In this section, I'm going to introduce you to the most powerful function I know. There are others which are more elegant, others which are more useful and, like the function I'll end this chapter with, others which threaten to shake the very foundations of mathematics. But this next one is my favourite, for its sheer power. It's an ugly brute of a function, but it crunches through any problem you throw at it. I'm going to teach you how to use this function, and then I'm going to show you how to hack it.

Like the puzzles I just set, you can get some interesting results if you start adding sequences of numbers. Adding any run of the whole numbers 1, 2, 3, 4, 5 . . . will always give you a triangle number (see p. 49). The sum of the first n whole numbers is the nth triangle number. Slightly more curiously, if you add together the sequence of odd numbers, the total is always a square number. Even more bizarrely, adding the sequence of cube numbers always gives you a total which is the square of a triangle number. Spoiler alert: adding the first ten odd numbers gives you the tenth square number: $10^2 = 100$. And adding the first four cube numbers gives you the square of the fourth triangle number: $10^2 = 100$.

$1 + 2 = 3$ $\quad$ $1 + 3 = 4$

$1 + 2 + 3 = 6$ $\quad$ $1 + 3 + 5 = 9$

$1 + 2 + 3 + 4 = 10$ $\quad$ $1 + 3 + 5 + 7 = 16$

$1 + 2 + 3 + 4 + 5 = 15$ $\quad$ $1 + 3 + 5 + 9 = 25$

Producing the triangle and square numbers.

Adding numbers together, though, can get tedious. Thankfully, there are short cuts.

There's a famous but possibly apocryphal story of the German mathematician Carl Friedrich Gauss adding up a sequence as a child. Apparently, a teacher was handing out busy-work to students and asked a young Gauss to add all the numbers from 1 to 100, to which Gauss almost immediately answered, '5,050'. When questioned, he explained that the numbers 1, 2, 3, 4 . . . 98, 99, 100 can be paired up to give $1 + 100 = 101$, $2 + 99 = 101$, $3 + 98 = 101$, and so on. We can do the same thing ourselves to add numbers up to any number n. The trick is to start with two lots of the sequence and then pair up the values. The answer should look familiar, because it's the same equation for triangle numbers that we had before.

Now for the ultimate trick to get these answers. This function makes mincemeat of these sorts of problems. This is a function which can add together any length of any sequence of powers. If you want the first fifteen cubes, or the first fifty

SEQUENCE FORWARD

$$\text{TWICE TOTAL} = \begin{cases} 1+2+3+4+\ldots+(n-1)+n \\ n+(n-1)+(n-2)+(n-3)+\ldots+2+1 \end{cases}$$

SEQUENCE BACKWARD

← 'n' VALUES →

$$\text{TWICE TOTAL} = \begin{cases} 1 + 2 + 3 + \ldots + (n-1) + n \\ n + (n-1) + (n-2) + \ldots + 2 + 1 \end{cases}$$

$$n+1 \quad n+1 \quad n+1 \quad \quad n+1 \quad n+1$$

$$\text{TWICE TOTAL} = (n+1) \times n$$

$$\text{TOTAL} = \frac{n(n+1)}{2}$$

powers of 4 or the first eighty-seven powers of 13, it's the one function that can compute them all. It's based on the work of Swiss mathematician Jacob Bernoulli. He was born in 1655 (also in Basel), one of a large family of mathematical Bernoullis. He is not the Bernoulli famous for Bernoulli's principle in fluid dynamics (that was his nephew Daniel Bernoulli); nor is he the Bernoulli who discovered l'Hôpital's rule and taught Euler (that was his brother, Johann Bernoulli). Jacob Bernoulli did an amazing amount of his own mathematics, including making significant contributions to the fields of probability and algebra, as well as being the man who brought us the number e. Above all, though, he gave us the numbers which make this function possible:

$$1^m + 2^m + 3^m + 4^m \ldots + n^m = \frac{(B+n+1)^{m+1} - B^{m+1}}{m+1}$$

If you want to find the sum of the first *n* numbers, all raised to a power of *m*, then you put those values into Bernoulli's equation, and it will give you the answer. The letter *B* in the equation is, however, no normal piece of algebra. It's a whole new way of substituting numbers into an equation. You can move *B* around and manipulate it algebraically as you would any other expression but, when you raise it to a power, something different happens. Instead of B^2 and B^3 representing the square and cube of *B* respectively, those little superscript figures tell you to look up the second and third entries (B_2 and B_3, respectively) in a sequence of numbers called the Bernoulli numbers. They can be thought of as a long list of the unusual powers of *B*, and you can see them in the table below.

B_1	B_2	B_3	B_4	B_5	B_6	B_7	B_8	B_9	B_{10}
$-1/2$	$1/6$	0	$-1/30$	0	$1/42$	0	$-1/30$	0	$5/66$

B_{11}	B_{12}	B_{13}	B_{14}	B_{15}	B_{16}	B_{17}	B_{18}	B_{19}	B_n
0	$-691/2730$	0	$-3617/510$	0	$43867/798$	0	$-174611/330$	0	. . .

It's uncertain exactly when Jacob Bernoulli discovered the Bernoulli numbers, but they first appeared in his book *Ars Conjectandi* (*The Art of Conjecturing*), which was published in 1713, eight years after his death. They are not a nice neat list of numbers like the PR-friendly Fibonacci numbers. Except for the first term, every second number is 0. The other numbers lurch between being positive and negative, and are all unwieldy fractions. This is why I described this function as an ugly brute. Definitely something you will never see staring at you from the glossy cover of *Sexy Maths Monthly* (a publication I've just made up). But in the ugliness of the Bernoulli numbers lies a strange type of beauty. To appreciate it, you need to take the function out for a test drive yourself. Use it to calculate the sum of the first four cubes. Be warned that this ride will involve some serious multiplying of brackets and other algebraic gymnastics, which may cause nightmares of being at school.

Actually using this equation does require a bit of work. I forced through $m = 3$ (setting the power to cubes) and $n = 4$ to calculate the sum of the first four cube numbers and, sure enough, the Bernoulli numbers gave me an answer of 100. Yes, it would have been much quicker to add $1 + 8 + 27 + 64$ to get 100, but that's beside the point. This is a way to add any number of powers. It's a generalized method which encompasses every other technique we've used before. If you put in $m = 1$ to add the straightforward (non-squared, non-cubed, just linear) numbers $1 + 2 + 3 \ldots$ and leave the n where it is, the equation rearranges to become the equation $n(n + 1)/2$ for the triangle numbers we know so well (more on these in 'The Answers at the Back of the Book'). This is the grand prize of mathematics: one concise theory that links together loads of other methods. This is why Ada Lovelace wrote the first ever computer program to calculate the 'Numbers of Bernoulli'. It's a beautiful thing to behold.

Now we see if we can break it.

The first hack that mathematicians found was to make the Bernoulli numbers yield the answer for the sums of infinite

$$1 + \frac{1}{2^m} + \frac{1}{3^m} + \frac{1}{4^m} \ldots = \frac{2^{m+1}|B^m|\pi^m}{m!}$$

sequences of numbers. This new function gives you the sum of the reciprocals (fractions) of even-powered numbers. All you need to do is put an even value of m in for the power, and you automatically get the total of the infinite sum. The standard $m = 2$ will give you the same value of $\pi^2/6$ which Euler found, but you can increase it to get the sum of the reciprocals of fourth powers, sixth powers, and so on. The straight-line symbols in $|B^m|$ represent the absolute value function and are there so that the negative sign on any Bernoulli number is ignored and it's simply flipped to become a positive number.

The next hack gets a bit silly. Mathematicians found a way to adapt the Bernoulli numbers to give the sum of an infinite sequence of whole numbers each raised to a power. We would expect this answer to be infinite, but it turns out there is a way to sneak up on an answer. Mathematicians used the function for the sum of the reciprocals of powers and changed it to work for a negative value of m. But the reciprocal of a negative power is simply that normal power. So this would effectively add an infinite list of numbers raised to powers.

$$\frac{-B^{m+1}}{m+1} = \frac{1}{1^{-m}} + \frac{1}{2^{-m}} + \frac{1}{3^{-m}} + \frac{1}{4^{-m}} + \dots$$

$$= 1^m + 2^m + 3^m + 4^m + \dots$$

The first thing to do is put in a value of $m = 1$, which should return the total, if you add together every last positive whole number. Obviously, this answer should be infinity, but that is not the answer which the Bernoulli numbers give. They claim that the answer is $-1/12$. That is to say, $1 + 2 + 3 + 4 + \dots = -1/12$.

One option here is to dismiss this answer as meaningless. We must have broken the Bernoulli numbers, because only an insane person would think that the sum of all numbers is $-1/12$. Or we could risk contemplating the possibility that this is an unexpected insight into the nature of the sum of all numbers . . .

If you recall, in Ramanujan's letters to English mathematicians, he claimed that $1 + 2 + 3 + \ldots = -1/12$. He was so surprised when Hardy took him seriously that he replied on 27 February 1913 in the following words:

> I was expecting a reply from you similar to the one which a Mathematics Professor at London wrote asking me to study carefully Infinite Series and not fall into the pitfalls of divergent series. If I had given you my methods of proof I am sure you will follow the London Professor. I told him that the sum of an infinite number of terms of the series: $1 + 2 + 3 + 4 + \ldots = -1/12$ under my theory. If I tell you this, you will at once point out to me the lunatic asylum as my goal.

It turns out that not only had Ramanujan independently rediscovered the Bernoulli numbers, but he may have found more than one way to prove that $1 + 2 + 3 + 4 \ldots = -1/12$. This is now called Ramanujan summation and gives us an insight into the ways in which the sum of a sequence can be divergent. Of course, the sum of all the positive whole numbers is infinite, but if you can somehow peel that infinity back out of the way and look at what else is going on, there's a $-1/12$ in there.

The Zeta Function

'With the assistance of these methods, the number of prime numbers that are smaller than x can now be determined.' So said Bernhard Riemann . . . which leads us nicely on to one of the most famous functions in mathematics: the Riemann zeta function. In the chapter on prime numbers, I mentioned Bernhard Riemann's 1859 paper 'On the Number of Primes Less than a Given Magnitude'. In it he found a method of calculating how many primes there are below any given number. This would give mathematicians an amazing insight into the distribution and nature of prime numbers. The only problem was that he couldn't prove that this method definitely worked. He did, however, prove that if an apparent alignment in the zeta function was real, then the prime counting method was real. Then he failed to prove that too.

At the time of his paper, the situation was that one of the most amazing insights into prime numbers possible was available to mathematicians . . . if only they could confirm an alignment in a function. Sounds fair enough. Riemann's paper was only ten pages long; I guess it's OK that he left some of the work for other people to do. If anything, he got too excited about the possibilities if it *were* all true – too excited to prove that it was true himself. So, sure as eggs = eggs, other mathematicians got to work – but they couldn't prove it either. What has since been called the Riemann Hypothesis remains unproved to this day.

So, what is the Riemann zeta function?* Well, we met it

* Since his paper introducing this zeta function was published, a whole family of other, less famous zeta functions have sprung up. But when anyone, including myself, writes 'zeta function', they are normally referring to Riemann's original function.

before. It's based on the Basel problem that Euler solved. The zeta function is the sum of an infinite sequence of inverse powers. Using the Greek letter zeta (ζ) as a shorthand for the Riemann zeta function, we can write that:

$$\zeta(m) = 1 + \frac{1}{2^m} + \frac{1}{3^m} + \frac{1}{4^m} + \ldots$$

Which is great, because not only did Euler crack some of the values of this function, the Bernoulli numbers now give us loads of extra values and, using them, we can easily calculate $\zeta(m)$ for all even values of m. But, to see the complete zeta function, we need it to work for any value of m. In fact, not only did Riemann extend this 'Basel function' to all the non-whole-numbered values of m, but he adapted it to take two inputs at the same time.

But we're getting ahead of ourselves. Discussions about the Riemann Hypothesis and the zeta function are often guilty of a bait-and-switch. You're promised a deep insight into how many prime numbers there are and their distribution, and suddenly you're adding reciprocals of powers. Where did the prime numbers go? There is, however, a bizarre link between the two. And, once more, it comes down to Euler.

Along with working on the Basel problem, Euler realized that adding an infinite sequence of reciprocal powers for all whole numbers will give you the same answer as multiplying together an infinite sequence of fractions which use only the prime numbers. So the zeta function can be written as two different equations, one of which relies only on the prime numbers. The one which uses all the whole numbers gives the same result as the prime fractions, but it's easier to work with. We know what all the whole numbers are, but we don't know what all the primes are. So we can substitute one for the other. This link to the primes was the springboard for Riemann's paper. Investigating the sum of reciprocal powers gives insights into the product of all the primes (in fraction form).

The versatility of the zeta function is what makes it so useful, but it also makes it difficult to work with. It is, without doubt, a slippery and evasive function. It turns out that

$$1 + \frac{1}{2^m} + \frac{1}{3^m} + \ldots = 1 \times \frac{1}{1-\frac{1}{2^m}} \times \frac{1}{1-\frac{1}{3^m}} \times \frac{1}{1-\frac{1}{5^m}} \times \ldots$$

On the left, all whole numbers. On the right, all the primes only.

there are many, many different equations which represent different bits of the zeta function. I find the best way to think of the zeta function is to imagine it as a mysterious unknown function which humans are only discovering piece by piece. Every so often we get a little glimpse into how it works, but only a small bit at a time. We saw already that, using the Bernoulli numbers, there's an equation which gives the values for even-number inputs. It was Ramanujan who first found an equation for the positive odd-valued inputs, working in isolation; he had rediscovered the zeta function entirely separately from Riemann, but even he could only crack a bit of it.

$$\zeta(2) = \frac{1}{1^2} + \frac{1}{2^2} + \frac{1}{3^2} + \cdots = \frac{\pi^2}{6}$$

$$\zeta(4) = \frac{1}{1^4} + \frac{1}{2^4} + \frac{1}{3^4} + \cdots = \frac{\pi^4}{90}$$

$$\zeta(6) = \frac{1}{1^6} + \frac{1}{2^6} + \frac{1}{3^6} + \cdots = \frac{\pi^6}{945}$$

$$\zeta(8) = \frac{1}{1^8} + \frac{1}{2^8} + \frac{1}{3^8} + \cdots = \frac{\pi^8}{9450}$$

It would be handy to have a plot of the zeta function for all inputs so we could see all the function at once. But finding the exact values for just the 'easy' parts took the efforts of Euler, Ramanujan and some of the greatest mathematicians who ever lived. Getting the harder values was looking to be impossible. But the prime numbers are not going to give up their secrets if we come begging with only part of the zeta function. We need to fully understand the zeta function for *all* values. And given how primes underpin our modern data security, it's an exploration worth pressing on with.

Thankfully, there is a way to get halfway by cheating. Instead of getting exact values, it's possible to calculate ones which are 'good enough'. Imagine we didn't know Euler's result that $1/1^2 + 1/2^2 + 1/3^2 \ldots = \pi^2/6$ but we wanted to cheat and find that answer anyway. Instead of getting the final answer for the infinite sum, we could just add together enough of the fractions to get close enough. After just the first three, the value is 1.361111111, whereas $\pi^2/6 = 1.644934\ldots$ – not very close. The first ten fractions do better, getting to within 5.8 per cent of the correct answer. I got bored and wrote a computer program to add the first 1 billion fractions together, and the answer of 1.64493405783457 is *definitely* close enough.

$$\frac{\pi^2}{6} = 1.6449340668482264 3647\ldots$$

FRACTIONS	VALUE	ERROR
10	1.549767731	5.8%
100	1.634983900	0.6%
1,000	1.643934567	0.061%
1,000,000	1.644933067	0.000061%
1,000,000,000	1.64493405783457 5	0.00000055%

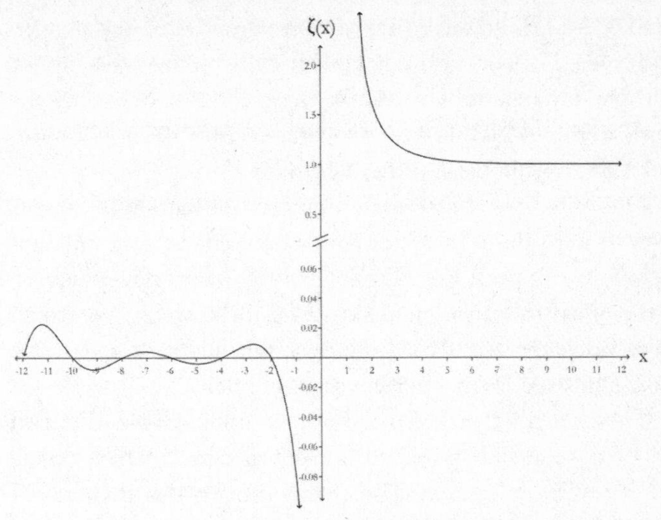

Values of zeta function for values between −10 and 10.

This does still involve a fair amount of cunning. Part of Ramanujan's genius was to find a way to get values out of the zeta function for negative values. As we saw before, these sums diverge and go racing off to infinity, but Ramanujan was able to extract the bit of the answer which explodes and leave the important bit behind. Using Bernoulli numbers, he could produce values for the negative half of the zeta function – which, eventually, gives us a complete plot. This is the zeta function in graphical form.

From this plot, you can see that, for positive inputs, the output drops down from infinity and then slowly tapers out just above a value of 1. Nothing that exciting is happening on the right-hand side. Over on the negative side, though, the line is weaving backwards and forwards. It regularly crosses the horizontal axis, each crossing point representing a value of zero. These points are called, unsurprisingly, zeroes. And these zeroes are no surprise either. They are trivial zeroes. They occur predictably for all even negative inputs, and we understand why they're there. These are exactly the same zeroes which appear as every second Bernoulli number. Once again, we saw them coming.

One thing we didn't see coming is the family of zeroes over to the side. When Riemann extended the zeta function to take two inputs, the 3D plot of it shows a new patch of zeroes next to our original axis. And these are zeroes which were not expected. Not only that, but all these zeroes we have glimpsed form a perfect straight line.

These are not trivial zeroes, they have not obviously dropped out of the equation: these are 'non-trivial zeroes', which we still do not fully understand. The creepy thing is that there could have been zeroes in other places on the plot, but they have all stood to attention in one perfectly straight line. And we don't definitely know why. The line they're on is at right angles to the original number line and passes exactly through the value

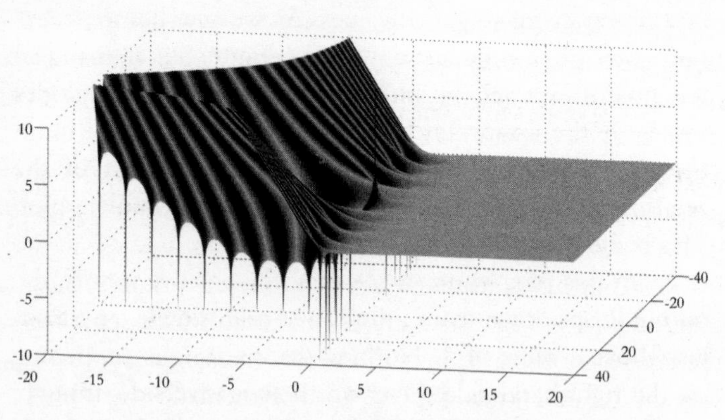

3D view of the zeta function. You can see the trivial zeroes running across the middle, as well as a new line of unexpected zeroes. This is a 'log plot' of the zeta function to accentuate the locations of the zeroes.

of ½. They are haphazardly distributed on the line, but they are all, eerily, exactly on it. The Riemann Hypothesis states that all the non-trivial zeroes of the zeta function are on this line. If we can prove the Riemann Hypothesis is true, then we'll also have proved the method for counting the prime numbers. In some weird, twisted act of mathematical logic, at a fundamental level the alignment of these zeroes stems from the same logic as the density of the primes. It doesn't seem to make sense. But if we can understand this mysterious alignment, we understand where the numbers are hiding their primes.

There has been some progress on proving the Riemann Hypothesis, but it remains unsolved. In 1914, Hardy managed to prove that there are infinitely many zeroes on that line, but he couldn't prove that there aren't any extra zeroes *off* the line. We currently know that 40 per cent of the non-trivial zeroes are definitely on that line, but we need to know it's true for 100 per cent. So much as a single zero somewhere else, and the Riemann Hypothesis would be disproved, causing our apparent understanding to come crashing down. But it hasn't. Everything has uncannily indicated that we are on the right track, but we can't yet prove it for sure.

When the German mathematician David Hilbert compiled a list in 1900 of the most important maths problems to work on during the next century, the Riemann Hypothesis was on it. Because, without the Riemann Hypothesis, we will have lost our only major lead on understanding the primes. However, when the Clay Mathematics Institute was assembling the equivalent list a century later, the Riemann Hypothesis was still on it. To this day, the Clay Mathematics Institute's bounty of $1 million for anyone who can prove that all the non-trivial zeroes of the zeta function are on that line has gone unclaimed.

The problem is that many mathematicians have done the same thing Riemann himself did: surged ahead using the prime counting method, assuming that someone would prove it later on. It looks like a safe bet: as we know, using computers, the first 10 trillion zeroes have been checked, and all of them are on that line. That said, mathematical theories have been disproved with numbers bigger than that, so there could be a zero off the line that we've simply not reached yet. Proving or disproving the Riemann Hypothesis will make a lot of people happy and/or sad, as well as winning one lucky person the million-dollar prize if they prove it correct (and I'm sure a consolation prize if they prove it incorrect and ruin the fun for everyone).

Mind you, Hilbert himself remained sceptical, suggesting that, even after a thousand years, the hypothesis might still remain unproved: 'If I were to awaken after having slept for a thousand years, my first question would be: has the Riemann Hypothesis been proven?'

Fourteen

RIDICULOUS SHAPES

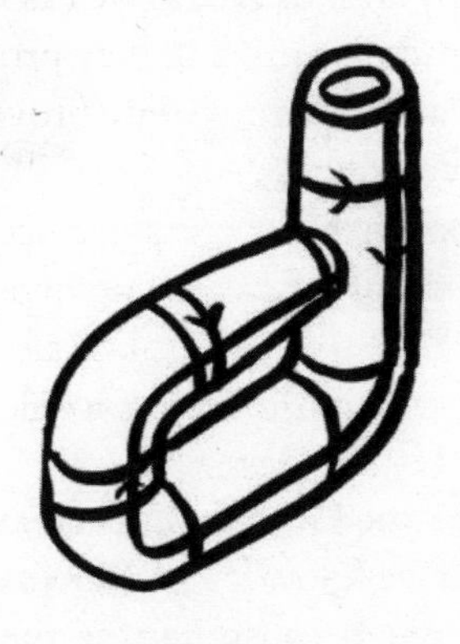

Imagine a shape which cannot be cut in half. Not some hypothetical maths shape – a real one you could build out of paper. A shape which, if you took a pair of scissors to it and cut all the way through it, would remain in one piece. It is possible for such a shape to exist, and you can make one right now. Start with two long strips of paper and join the first one end to end so that it forms a loop, then join the other one up the same way, but with one end turned over so there's a twist in the loop. If you cut along the centre line of the non-twisted loop, you'll go all the way around and cut it completely in half, leaving two thinner loops. Not so for the twisted loop. It is a shape which cannot be cut in half.

You can still cut the whole way along the centre line of this twisted loop, back to where you started cutting, so you'd think it would by now be in two pieces. But, after you have completed the circle, the paper will still unfold as one unbroken loop, only twice as long as the original loop. The untwisted loop is typically called a cylinder, because that, in a way, is what it is: a very thin slice of a cylinder. The twisted

Go on, try and cut a Möbius loop in half. Just try it.

loop is named after the German mathematician August Ferdinand Möbius, so we know it as a Möbius loop or strip. It's named after him not because he was the first to discover such a shape; he was beaten to that by another German mathematician, Johann Benedict Listing. They were both independently investigating the properties of a twisted loop in 1858, and Listing came up with the goods first, albeit by a narrow margin. History has decided, though, and 'Möbius loop' has stuck. Possibly because it sounds cooler than 'Listing loop' (and has more umlauts).

Surely we can cut a Möbius loop in thirds . . .

The Möbius loop is well loved in the world of mathematics because of the counter-intuitive ways in which it behaves. Not being able to be cut in half is only the start. All its strange properties come down to the twist: that's the only difference between a Möbius loop and a cylinder. One important property of the Möbius loop is that it only has one side. Yes, if you try to colour in one side of a Möbius loop (or, for the less artistic, draw a line down the middle), you will find that you can't. When you start with a pen on one side, you'll soon find yourself drawing on the other side – except that it's not really the other side, it's the same side. If you draw a continuous line from one point to another on a Möbius loop, then both of those points will be on the same side. If you start drawing a line on one side of a cylinder, you'll never get over to the other of its two sides.

Make yourself a new Möbius loop: we're going to try something else. Instead of cutting right down the middle of the strip of paper, come a third of the way in from one edge then cut all the way around, staying the same distance from the edge. After the first lap, you'll find yourself cutting a third of the way in from the other edge – or, rather, the same edge. As well as having only one side, a Möbius loop also has only one edge. If you stay true and keep cutting in a straight line, you'll complete a second lap before arriving back where you started. This time, you'll end up with two pieces!

In fact, you'll have two different-length loops which go through each other. The smaller one is a Möbius loop, and the longer one is the same loop produced by cutting a Möbius loop in half. If you take a closer look at the single loop you made when you cut down the centre of the Möbius loop, you can see it has more than one twist. By cutting a loop with a single twist in half, you now have a loop with four twists! Logically enough, we'll call this a four-twist loop. Loops with

Three-twist loop: It's like cutting through a recycling logo.

different numbers of twists behave differently.

Making a two-twist loop and cutting it in half gives you two identical loops that go through each other; a two-twist loop *can* be cut in half. My favourite, however, is the three-twist loop. If you have a loop of paper with three twists in it and try to cut it in half, not only will you end up with only a single loop, but that loop will have a knot in it. By cutting an untangled loop in half, this knot will appear from nowhere. From here, you can try as many twists as you like: even numbers of twists will always give you two pieces; odd numbers of twists, one piece. The more twists, the more intertwined or knotted the results will be.

Now for some Möbius-on-Möbius action. Make two normal one-twist Möbius loops and stick them together. Importantly, though, the Möbius loops need to be subtly different. When you twist a piece of paper, you are faced with a choice: you can twist the end either clockwise or anticlockwise. Turning the paper different ways gives you different Möbius loops (it's like having right-handed and

Two Möbius loops and two cylinders attached at right angles.

left-handed options, different chiralities), which are mirror images of each other, and you need to make one of each. Glue them together so they cross at right angles. Cut them both in half, one after the other, and you'll get two hearts that loop through each other. If you make the two glued Möbius loops out of romantically coloured paper, you've got one of the nerdiest Valentine gifts imaginable.

As we get caught up in all this one-sided romance, it's easy to forget about the zero-twist loop, the cylinder. Make yourself two new cylinder loops, then stick them together at right angles the same way we just attached the Möbius loops. Cut both of these cylinders in half, and you'll get not two hearts but . . . I'll leave it to you to find out. Do give it a try. The answer will not be in 'The Answers at the Back of the Book'.

Starting to Surface

If any map on a plane surface can be coloured with only four colours, how many would we need for any map drawn on a Möbius loop? It turns out we need six. The four-colour theorem applies only to flat surfaces: if you start working on the surface of a Möbius loop, everything changes. The Möbius surface doesn't just behave oddly when you cut up its paper incarnation, it also produces a wide range of weird bits of maths.

To understand why the Möbius loop cannot be cut in half, try it again and watch the scissors really closely as they cut around it. When they get back to where they started, something very important has happened: the scissors are now upside down. This doesn't happen with a cylinder – when you cut around and back to where you began, the scissors are still exactly the same way up, and you get two pieces. The journey the scissors have gone on around the Möbius surface is called an orientation-reversing path. By following this path, the orientation of the scissors has changed.

I find the fact that a Möbius loop has an orientation-reversing path even more bizarre than it having only one side. But it gets stranger still. We're building our models out of paper, but that can be a bit misleading. A piece of paper has a front and a back and there is some thickness to keep them separate. Mathematicians consider things like the Möbius loop to be made from an infinitely thin surface: there's no thickness at all between the front and the back. A surface in maths is a 2D world with no thickness; it would make for a very strange universe to live in.

This surface is infinitely thin.

Earlier, we met our Hypoflaticals, who lived on a completely flat 2D surface. But a surface can be 2D without being flat as such. We've already used a balloon to represent a 2D surface which loops around to join itself. A hypoflatical living in a spherical 2D world could start walking in one direction and eventually find itself back where it started! But what if we took a hypoflatical's 2D world and turned it into a Möbius surface? They could start walking in a straight line and go all the way around the loop and back to where they started, only this time they would have flipped upside down. Left and right, up and down – none of these have any permanent meaning on a Möbius surface, because if you follow an orientation-reversing path, they all get swapped around. A Möbius surface is said to be a

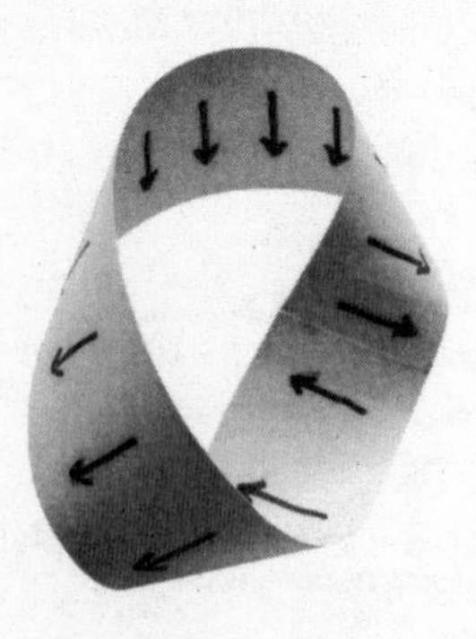

Just by moving the arrow, it swaps from down to up.

non-orientable surface. Living on a Möbius surface would be a very strange experience indeed.

This is why a map drawn on a Möbius surface can need up to six colours. The lack of thickness means that if you colour one part of the surface, both sides are the same colour, and one edge of the map is connected to the opposite edge, but upside down. Cut out the uncurled Möbius strip below to make a Möbius map which requires six colours. No section can be the same colour as its neighbour, so colour each of the six sections differently, making sure to colour the back in the same way. Once you've joined it with a twist, you can check that all six sections touch all five other areas. If any two colours don't touch, it means they could be the same colour. But if each of the six touch all other five colours, this map cannot be completed with fewer colours.

What if we took our Hypoflaticals and made them live on a toroidal (doughnut-shaped) surface? Well, there are no orientation-reversing paths on a torus, so they can have left and right, which is nice. A torus is an orientable surface. But, oddly, any maps drawn on one might need up to seven colours. The utilities problem can be solved on a torus, as we saw on my coffee mug. If you wanted to colour in a pattern on a mug, it could take up to seven colours to make sure that no two the same touch. Get a white mug and a whiteboard marker and see if you can do it.

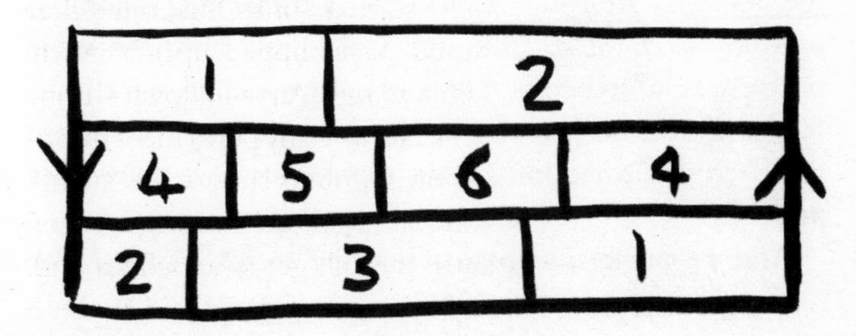

The interesting thing is that these properties (how many colours a map might need, whether the utilities graph can be drawn without any crossings) do not depend at all on the actual shape of these surfaces. A coffee mug, for example, is a completely different shape to a doughnut, yet they are both at heart a torus. Unlike geometry, where the actual shape of things is very important, this new area of maths, where we investigate the properties of a surface, is much more flexible.

This study of surfaces started in the early 1700s, and Euler wanted to call it *geometria situs*, meaning the 'geometry of position'. This name caught on for a while, but when this area of maths came into its own in the 1800s, it was the forgotten discoverer of the Möbius loop Johann Benedict Listing's suggestion of 'topology' which stuck. Topology has since grown to be a huge area of mathematics, covering the investigation of surfaces, regardless of what shape they've been stretched into.

The idea that the same mathematical thing can be distorted into new shapes should be familiar enough to us by now – we've already done the same thing with knots and graphs. And, as we call a class of knots an isotopy class, in topology shapes that can be deformed into each other are called homeomorphic and they live in the same homeomorphism class (the name being pieced together from 'homeo' for staying the same and 'morph' for form). So, a coffee mug is homeomorphic to a torus. Although topologists have the same problem as knot theorists: how to quickly identify which surface is which, no matter what position the knot or surface has been deformed into. Only, topologists have solved the problem.

And we've already come across the secret: it's the Euler

characteristic! In Chapter 5 we saw that any shape on a closed surface with a single hole through it can always be expressed as *vertices* – *edges* + *faces* = 0, which means that all toroidal surfaces have an Euler characteristic of 0. To work out the Euler characteristic, you can use the formula 2 – (2 × *number of holes*), so it's the number of holes that topologists use to categorize surfaces, only they call the number of holes the genus (*g*) of the surface (another word stolen from biologists). It was Riemann himself who in 1851 came up with the idea of defining the genus of a surface by its Euler characteristic, but he didn't prove that it always works in identifying any surface. In 1863, Möbius proved that any two orientable surfaces with the same genus must be the same surface, albeit distorted in different ways. Then, in 1882, fellow German mathematician Felix Klein proved the subtly different result that not only do orientable surfaces with the same genus always have the same surface, but that it is impossible for two surfaces to be the same without their genus matching.

The genius of genus, though, is that, even though it started out being based on the number of holes in an orientable surface, it was adapted beyond that. For an orientable surface, you can calculate the genus from the Euler characteristic using *Euler characteristic* = 2 – (2 × *genus*). The genus of a closed non-orientable surface is also linked to the Euler characteristic, but via the relationship *Euler characteristic* = 2 – *genus*. The concept of a genus now cuts across several areas of maths. You can assign a genus to anything from polyhedrons, to graphs, to surfaces, and it will tell you how these shapes are related. The pentatope hyper-tetrahedron is linked to the K_5 graph, is linked to the torus surface. Genus is an incredibly powerful concept which ties huge sections of maths together.

We can also use the genus of a surface to check how many colours maps may require. This formula will tell you the maximum number of colours needed to cover any map on an orientable surface without any two touching:

$$\text{MAX NUMBER OF COLOURS REQUIRED} = \left\lfloor \frac{7 + \sqrt{48g + 1}}{2} \right\rfloor$$

You can check it by putting in $g = 1$ to give the seven colours required for a torus. Were you to need to colour a coffee mug armed with two handles, this formula shows you that you'll need eight colours. The rounding-down part of the formula means that sometimes different numbers of holes require the same number of colours. Twelve different colours are required for a six-handled mug, but will also cover you for seven handles. Once you have a hundred colours, that's good enough for mugs with 776 handles through to 792. I'm not sure which is harder to imagine, how a 792-handle mug can be divided into a hundred sections, all of which contact all ninety-nine others, or how to drink tea out of it.

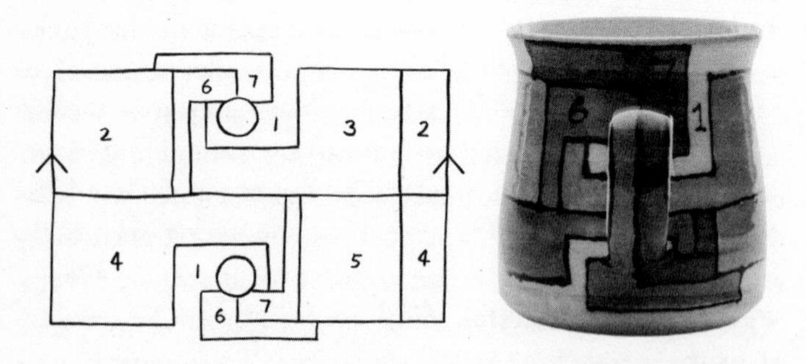

Seven colours on a mug, every one contacting all other colours.

The Genus of Knots

When you cut the three-twist loop in half, the knot should have looked vaguely familiar. Yes, it's a trefoil knot. There is of course a link between knots and surfaces. For every odd-twist loop you cut in half, a different knot will be produced. The reason for this is that the outside edge of the surface is itself a knot! If you colour in the edge of a three-twist loop (which has only one edge), you can see that it's a trefoil. But this isn't the only surface which has a trefoil as its edge. To avoid any confusion between the edge of a surface and the edges of graphs or polygons you may draw or project on to them, the outer edge of a surface is officially called its boundary.

We can take a knot and turn it into the boundary of a surface using an old friend of ours: soap film. Make a knot out of sufficiently rigid wire and dip it in some bubble solution. The soap film will cling to the wire frame and produce a surface. It will give you the smallest surface for any arrangement of the knotted wire, but by moving the wire into different shapes you can produce all sorts of different surfaces. The challenge is to find the orientable surface with the smallest genus possible.* This is then the official genus of the knot. Much like finding the minimum crossing number for a knot, it can get quite fiddly.

An orientable surface made by a knot is called a Seifert surface, named after the German topologist and knot theorist Herbert Seifert. Germany was the centre for amazing advances in mathematics at the turn of the last century, until Hitler's rise to power in January 1933. Caught on the wrong side of the

* Finding the genus of a surface with a boundary is slightly more complicated than what we've looked at so far with closed surfaces, but it certainly can be done. You effectively glue the boundary shut.

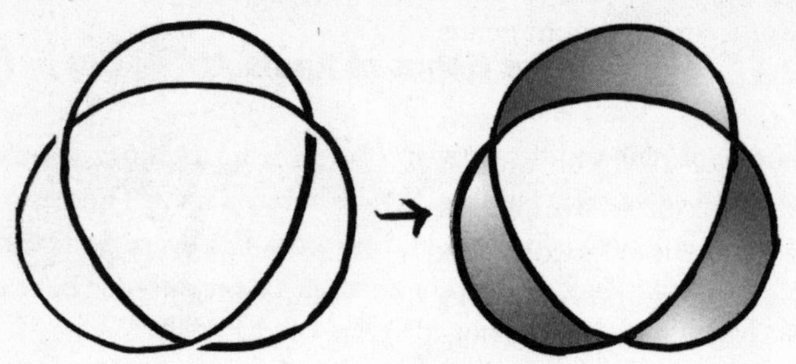

Trefoil knot becomes the boundary of a surface.

Nazis, in 1935 Seifert was moved into a position at Heidelberg University, which had been vacated when Hitler forced the retirement of all non-Aryan-descent professors. Certainly not a Nazi sympathizer, Seifert slipped political statements into the epigraphs of maths papers during the Second World War and, post-war, he was one of the few professors trusted by the Allies.

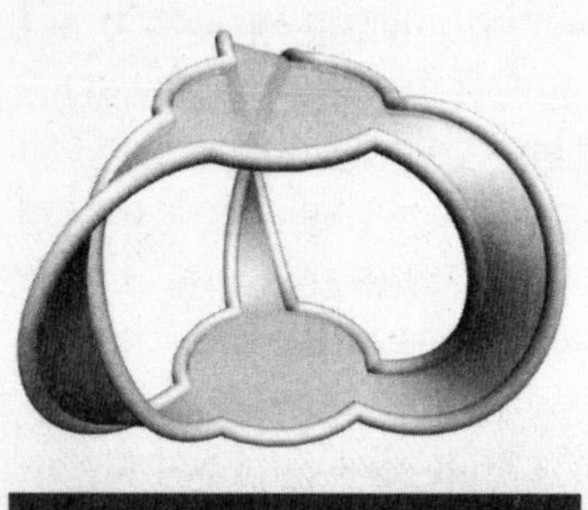

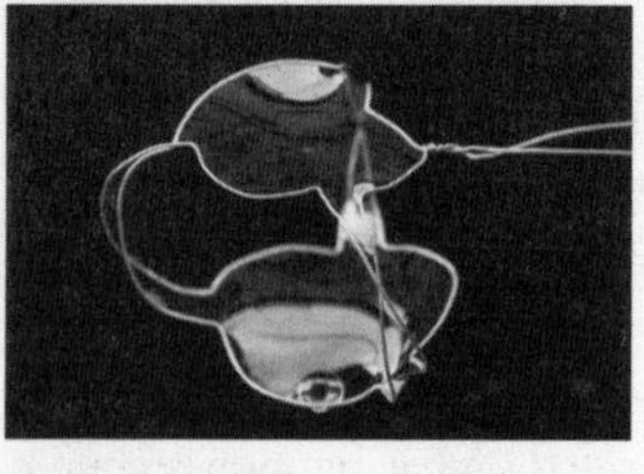

A trefoil knot as the boundary of a genus-one surface.

Seifert came up with his surfaces in 1934, and it looked like a very powerful tool for knots. If it is possible to match any knot to a surface and genus, then the mathematics developed for other objects with the same genus can be used in knot theory. As always with knots, however, everything was harder than it looked. Finding a Seifert surface by trial and error was not good enough. But then Seifert came up with the Seifert algorithm, for matching any knot to its minimum genus Seifert surface. Probably. Annoyingly, it doesn't always work. Even though knot theorists have proved that it definitely works for whole categories of knots, in 1986 it was proved that, for some knots, the Seifert algorithm will not produce the Seifert surface with the smallest possible genus. As always, there is much knot research still to be done.

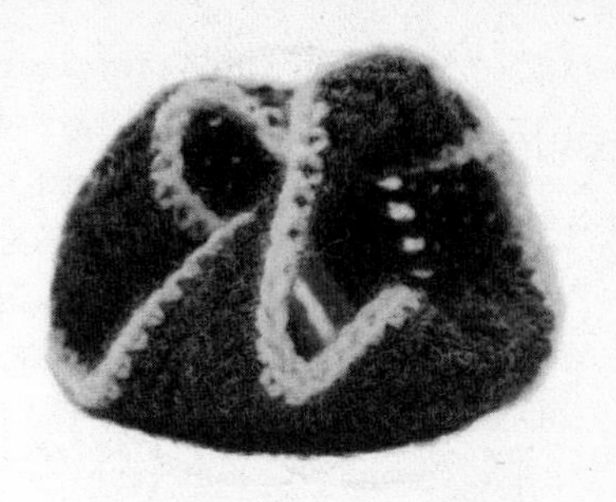

A trefoil Seifert surface, as knitted by knot theorist Julia Collins.

Despite sometimes still requiring a lot of work to calculate, the genus of a knot can be very useful. The unknot is the only knot with a genus of zero and, under knot addition, the genus of a summed knot is equal to the genera of the two knots added together. At this point, we've left holes long behind. They were one way we found to give surfaces a genus, but now we can consider them an innate property of knots as well, no holes required. Using *g*(*knot*) as the generic function for finding a knot's genus, we can write that *g*(*knot* A # *knot* B) = *g*(*knot* A) + *g*(*knot* B). The kickback to this is, because the genus of a surface is always a positive number, adding two knots together can never give a resulting genus of zero (unless they're two unknots already), which proves that knots cannot cancel each other when being added. We now know you can never add a knot to untangle another knot.

The link between knots and twisted surfaces is also the reason why DNA becomes so knotted. As we know, cutting along the mid-line of a twisted loop will always result in knots and tangles – which is exactly what happens when a circular piece of DNA divides. The double-helix shape of a

DNA molecule can be thought of as a long, twisted piece of paper. Some cells then do go the extra step and join the two ends of the DNA together to form a loop; many kinds of bacterial DNA are circular. For DNA to reproduce, it needs to unzip itself down the middle, and the topology of the DNA chain dictates which knots are formed afterwards. This is the situation knot theorists are studying as part of modern microbiology research.

Meet the Whole Family of Surfaces

Let's take a simple game and make it a bit more challenging. The problem with a game of Noughts and Crosses is that, most of the time, no one wins. If both players have even the slightest knowledge of the strategy of the game, every match will end in stalemate. Wouldn't it be great if there was a way to make sure that every game produced a winner? Well, have I got the surface for you. Play Noughts and Crosses on a cylinder. On a cylinder, someone is guaranteed to win every time.

Or play on a Möbius surface. Or on a torus. You can play Noughts and Crosses on any surface. If you want to play on a cylinder, you can of course make an actual cylinder out of paper and draw the board on it. Or, if you're lazy, just draw a

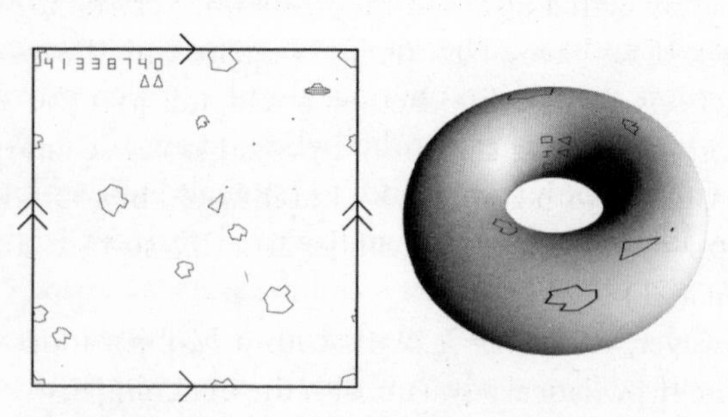

The flat Asteroids *screen is actually a toroidal universe.*

normal board and remember that one side wraps around to the other. In maths, this is indicated by drawing two arrows, and the rule is that the board should be twisted so that the arrows match up. These matching instructions can be extended to using two pairs of arrows, making your square board into a true torus. You may think you've never played a game on a torus before, but if you've ever tried the 1979 video game *Asteroids*, you have. Going off one edge of the screen maps you back on to the other side as if you were flying in a toroidal universe.

As mathematicians are always striving for completeness, these matching instructions raise an interesting question: what sorts of shapes will other arrangements of arrows give? There are actually only two other options for matching the opposite sides of a square, and neither of them produces shapes which can be made. At least not in 3D. The final two surfaces work only in 4D. All these are 2D surfaces, but the number of dimensions they need to join together varies across 2D, 3D and 4D.

Not having any matching instructions at all leaves you with a blank square, deceptively called a disc (don't forget: this is topology, and the exact shape doesn't matter: a disc is the same as a square). This is a flat 2D surface, and it doesn't need more than two dimensions. A Hypoflatical would be very happy with a disc, and they'd also be very happy with a cylinder. If we gave a hypoflatical a square with the matching instructions for a cylinder, they could still join the arrows together in their flat 2D world. It would look like an annulus (that is, a disc with a hole in it), but it would be topologically identical to a cylinder. A cylinder is a 2D surface that can exist in a 2D world.

However, this doesn't work with a Möbius loop. If we gave the hypoflatical a square with the matching instructions for a Möbius loop, they'd be stuck. It's impossible to join

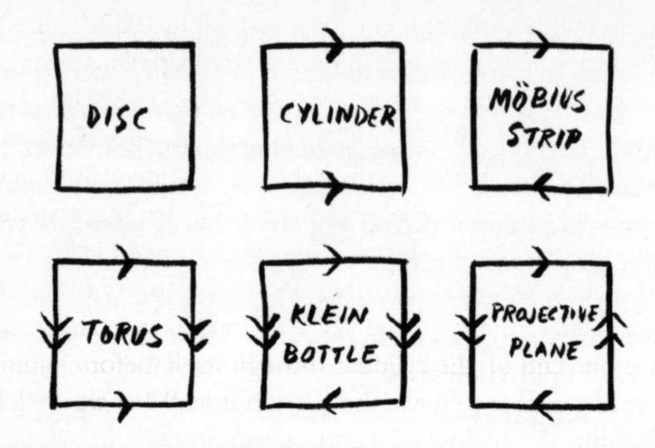

those two arrows without lifting one end up and turning it over. Because of the twist, a Möbius loop is a 2D surface which requires 3D space to join it together. Likewise with a torus. Its surface may still be infinitely thin, but it can join together only in 3D. The final two surfaces are called the Klein bottle and the projective plane, and they are both 2D surfaces that need 4D space to join together properly.

My favourite is the Klein bottle, named after Felix Klein,

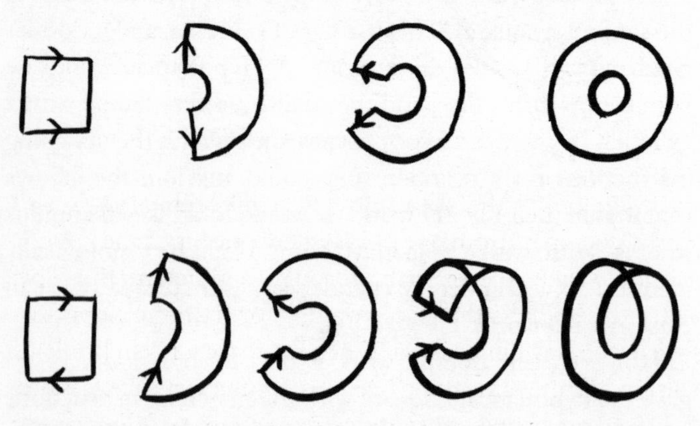

Making a cylinder and a Möbius loop in 2D. The Möbius intersects itself.

who we met before, when he discovered it in 1882. This is a torus with one pair of matching arrows swapped. It can be roughly thought of as a torus that has been twisted, in the same way that a Möbius loop is a cylinder that has been twisted. Except it's not a twist in the way we normally understand it: the Klein bottle is a torus made from a cylinder where one end of the cylinder has been turned over in 4D before being joined to the other. If we were to try to follow the matching instructions in 3D, we would have to cheat and shove one end of the cylinder through itself before joining the ends together to make the Klein bottle. What we are left with is the 3D shadow of the 4D Klein bottle.

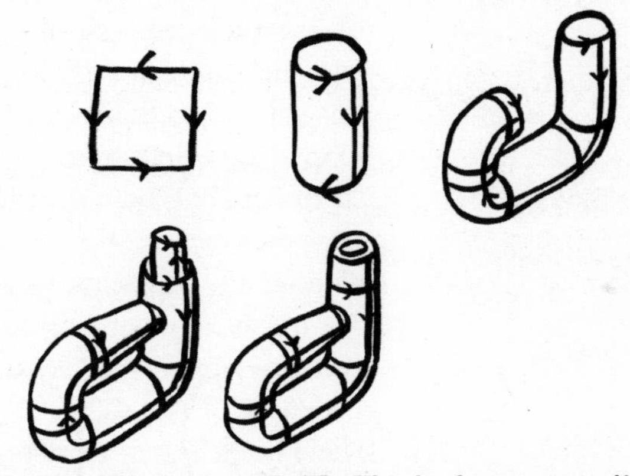

Making a Klein bottle in 3D. The Klein bottle intersects itself.

If you look at the 2D shadow of a 3D Möbius loop, it all works fine, apart from one point where one bit of the boundary looks like it goes right through a different bit. This self-intersection is the price you pay for trying to put a surface in too few dimensions. When we force a Klein bottle to exist in 3D, the same thing happens: it intersects itself. When a surface can happily exist in a certain number of dimensions without any self-intersections, we say it can be embedded in that space. A cylinder can be embedded in 2D space, a Möbius loop in 3D. When there are not enough dimensions for the surface to fit in properly, it has to start crossing through itself. Sure, we can still squash a Möbius loop down and cram it into 2D space, but it doesn't belong there. We can immerse a Klein bottle in our 3D space, but it will never be embedded.

The remaining surface, the projective plane, is what happens when both the matching arrows on a torus are switched to twists. Both it and the Klein bottle are non-orientable surfaces, but I find the projective plane the hardest to imagine. Most of the projections in 3D are not very helpful either. One way to imagine it is to think of a Möbius loop in 4D being stretched around so that its one boundary forms a circle – which in 4D can happen without any problems, whereas in 3D the surface gets in the way – and then zipping that circle closed. Didn't help? Oh well, at least the Klein bottle is easy to picture. The 3D versions can even be made out of glass to look like a bottle. Or they can be knitted out of wool to form a hat.

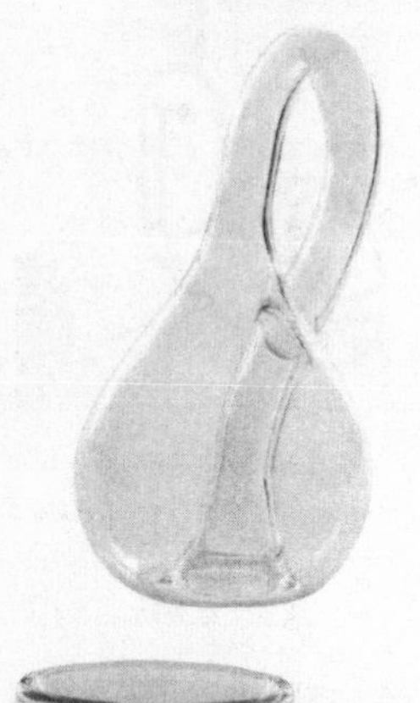

A standard Klein bottle, and one distorted to be a Klein stein.

I had seen a few knitted Klein bottles online, so I asked my mother if she could knit me the 3D immersion of the 4D Klein bottle so I could wear it as a hat.

She just looked at me. We then had a long conversation where I was talking Maths and my mum was speaking fluent Knitting, but because she is quite the dedicated knitter, she eventually got it to work. The first one she knitted for me (the prototype, as I called it; or as she called it, a perfectly good gift) worked well, but it was all the same colour. I asked if she could knit me one that was stripy. Apparently, this is easy enough to do when knitting – you just change colours after a few rows – so I gave her a long list of numbers for the thickness of each stripe in the hat. The photo here is me wearing my Klein hat. The stripes are the digits of pi. Should anyone have a nerdier hat than this, I'd like to hear about it.

If we ever meet Hypertheticals, we should wear these 4D hats to greet them.

I think a 4D hat is a good place to finish our journey through topology. If you want to knit your own, you can use the same striped pattern as mine, or, if you want a challenge, this is the ultimate. Any map on a Klein bottle can be coloured with six or fewer colours, so try to knit it as six different-coloured sections, each of which contacts all five others. And what if you accidentally cut your Klein bottle in half? Is it impossible to cut in half, like a Möbius loop? Well, if you cut it in half just the right way, you'll be left with . . . a single Möbius loop.

Fifteen

HIGHER DIMENSIONS

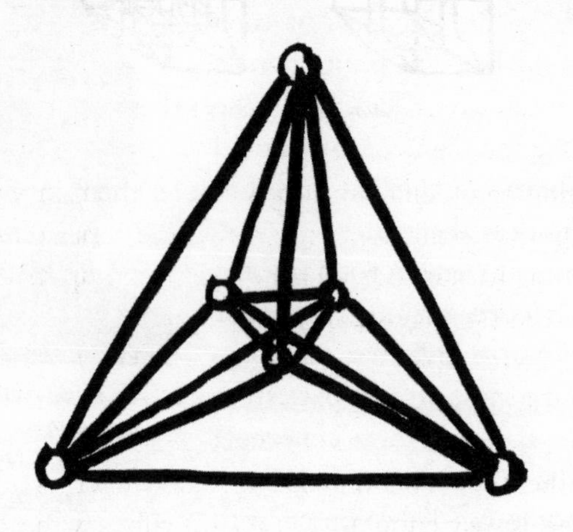

Although I dedicated an entire chapter solely to the fourth dimension, all the others above it will have to share just the one. This isn't because they become easier to visualize – far from it, high-dimensional objects will shank you in the visual cortex without a second thought – but because, once you get over the conceptual hurdle from 3D to 4D, it's easy to keep extending patterns further, into 5D and beyond. Accepting dimensions higher than what you can perceive takes some

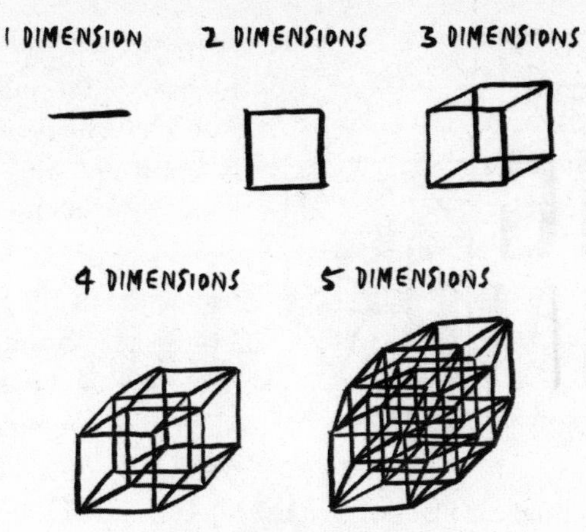

Cubes from 1D to 5D.

convincing arguments, but once you let them into your life it's a slippery uphill slope into more and more dimensions. We're going to start our journey in 5D, end up in 196,883D and see some strange sights along the way.

We'll start simple. Can you find your way around a 5D cube? Well, for the fifth dimension, we take the first four dimensions and add another new direction at right angles to all the other four. Thinking of a 5D cube in terms of vertices and edges is easy enough: it's two 4D cubes with new edges linking their matching vertices together. There will be thirty-two vertices all up, joined by eighty edges. If you want to visit all those vertices, then a Hamiltonian path around the 5D cube is just the same as solving a five-disc Tower of Hanoi. Likewise, everything continues in the same way for 6D cubes, 7D cubes, and so on.

We mentioned the 4D Rubik's Cube online earlier. Well, there's a 5D Rubik's one too. These higher-dimensional cubes still *project* down into lower dimensions, so we can see their shadows, only now we are looking at shadows of

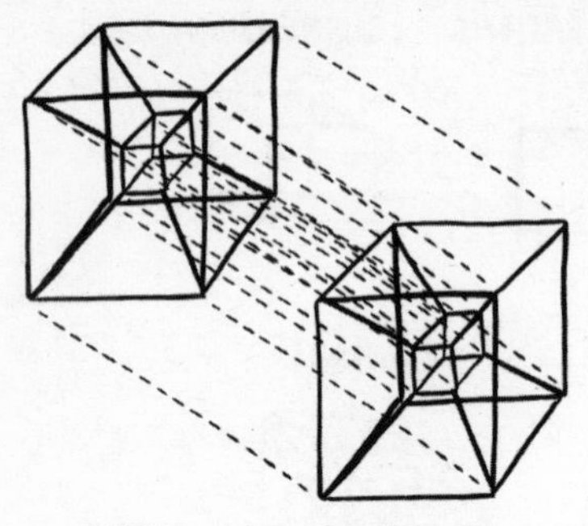

Different view of a 5D cube.

shadows. Should you try to solve the online 5D Rubik's Cube, your screen is showing you the 2D projection of the 3D projection of the 4D projection of the 5D cube. Confused? Good. I find that anything past about 5D requires so many shadows-of-shadows that the resulting image is just not that useful. Beyond 5D, we need to rely much more on the maths and less on trying to see how the shape looks projected down into 3D or 2D.

However, many higher-dimensional shapes refuse to behave like the normal 3D shapes we're used to, so our intuition is of very little use. Some of them act in incredibly strange ways, which, initially, seem impossible. A cube may be easy enough to grasp, but spheres in higher dimensions are just downright bizarre. For example, what happens when we try to trap a hypersphere in a box? It breaks itself out.

To demonstrate this, first we need to build a box – any handy hypercube will do. Then we need a way to measure

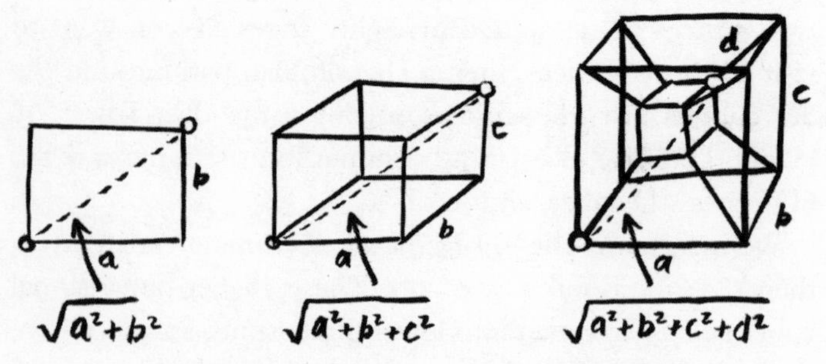

Space diagonals. Pronounced 'spaaaace diagonals'.

things in higher dimensions, for which we can use Pythagorean theorem. We've already seen how this can be used to calculate the diagonal of a rectangle, but this continues all the way up through the dimensions. Pythagoras works the same in 3D to calculate the space diagonal between opposite vertices ('space diagonal' has to rank as one of the best names in mathematics), and consistently works to measure the hyperspace diagonal in any number of dimensions.

Interestingly, even though in 2D it is possible to find integer-length rectangles which have an integer-length diagonal (such as 3 × 4 with a diagonal of 5; 5 × 12 with a diagonal of 13; and infinitely many more), no one has ever found a 3D integer-length cuboid which has integer-length face diagonals as well as an integer-length *space* diagonal. This isn't to say that these perfect cuboids don't exist – no one has proved that one cannot be found. The best mathematicians can do for now, however, is an integer-length cuboid with all integer-length face diagonals (so not the space diagonal), called an Euler brick. The smallest Euler brick has sides of 44, 117 and 240, and there are infinitely many more of them. The search for the perfect cuboid continues.

OK, so what is a higher-dimension sphere? On the surface, it might seem simple: a sphere is all the points which are a certain distance – the radius – from a fixed centre point. A circle is a 2D sphere. A sphere is a 3D sphere. (OK, that one's obvious.) A 4D sphere is all the points which are the same distance from a centre, and so on. Unfortunately, these spheres are much more slippery than you would expect. Which is why, for safety reasons, we need to keep them locked in hypercubes so they cannot get away. We'll pack the rest of the hypercube full of padding spheres so that the sphere at the focus of our attention cannot move around inside its box.

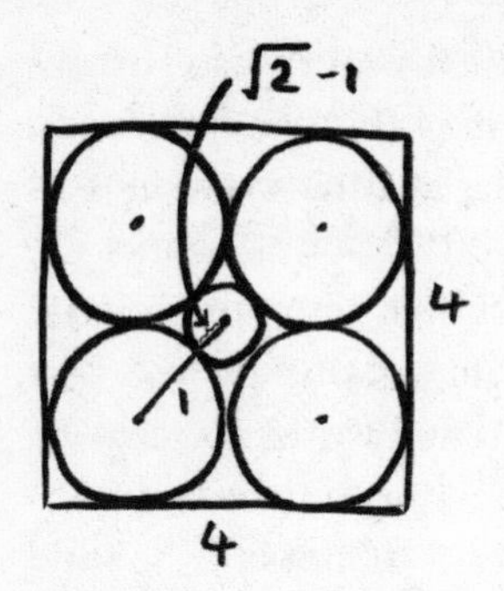

The trapped centre circle has a radius of 0.414; that's as big as it can get.

All these precautions to keep our hypersphere under control may seem disproportionate. Well, let's find out if they are or not. We'll start in 2D and work our way up. Our 2D cube is a square, and I'm going to use one with sides which are 4 units long. This is because it neatly fits four circles of radius 1 unit into it. The centre of each of these circles will sit a quarter of the way in from two of the edges. We're now going to carefully put our specimen circle (a 2D sphere) in the very middle of this box full of other circles. Unlike the padding circles, which remain a fixed radius of 1, we'll let the middle circle expand to be as big as the space in the middle of the box allows.

We can now calculate exactly how big this middle circle can be. The box-length of 4 makes sense now, as it means that the distance from the middle of each padding circle to the centre of the box can easily be calculated using Pythagorean theorem. For our 2D square, this diagonal length is $\sqrt{2} \approx 1.414$; the first 1 of this is used up by the padding circle, so our middle circle can only have a radius of up to 0.414 before it touches the padding. It's not a very big circle, but at least we know exactly how big it is, where it is and what's containing it.

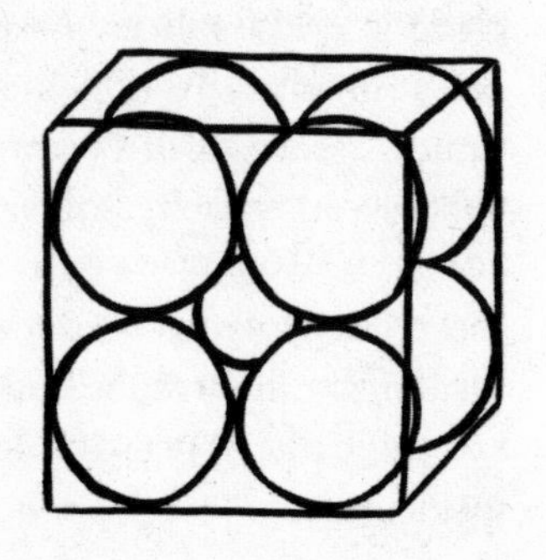

We can fit eight padding spheres into our $4 \times 4 \times 4$ 3D box. The number of unit spheres is going to double as we go up each dimension, and they'll always be centred one unit from all their nearby edges so that they're just kissing them. The diagonal from the centre of each padding sphere to the centre of the cube is $\sqrt{3} \approx 1.732$, which allows our specimen sphere to expand to a radius of 0.732: not a big leap. It makes sense that, as this sphere has more directions to move in, it can grow slightly bigger. However, as we go up each dimension, we're adding more and more padding spheres to keep the specimen sphere/hypersphere/and so on, trapped, so surely it will not be able to escape. Its growth should tail off.

However, as you may have guessed, the centre sphere is not going to behave itself: somehow, it will escape, even though, by definition, it will remain centred in the middle of the cube and be surrounded by other spheres holding it in place.

Checking the numbers for 4D shows us that everything continues to line up for a $4 \times 4 \times 4 \times 4$ tesseract, with sixteen padding unit spheres. The distance from the centre of each padding 4-sphere to the centre of the 4D box is a very exact and tidy 2 (i.e. $\sqrt{4}$), giving our centre specimen sphere a radius of 1. It seems that four dimensions have given it enough freedom to expand out to be the same size as the padding spheres around it.

Onwards to five dimensions, and things start to get a bit strange: our specimen sphere has continued growing and now has a radius of 1.236, bigger than the padding spheres around it (from now on, I'll just call it a sphere, regardless of how many dimensions it's in). What started as a small gap in the middle when we were in 2D is now a large chasm in 5D. Originally, we expected the middle specimen sphere to get a bit bigger as it gains more dimensions but that this expansion would eventually, and rapidly, tail off. It doesn't. Our centre specimen sphere continues to get bigger and bigger at an alarming pace as more dimensions are added.

The big surprise is when the box enters ten dimensions and the inner sphere's radius hits 2.162, which means that it is actually reaching outside the box. Not only has this sphere managed to expand despite the padding spheres around it, it has now escaped the box entirely. In nine dimensions, it manages just to touch the box, with a radius of 2, and then, for any dimensional space beyond that, it sticks out through the sides of the hyper-box (or hypercube). From twenty-six dimensions onwards, the sphere is more than twice as big as the box it's inside. And it shows no sign of slowing down. The size of this centre sphere does not converge as the dimensions get higher: it continues to diverge off into the distance.

The numbers do not lie, but we still need to explain how the sphere reaches out of the box. The box doesn't change shape – it's always 4 units long in all directions. Importantly, the other spheres are a fixed radius of 1. We do not let them expand as much as they could; we just arrange them all so they touch the outside wall of the box and the other spheres next to them. As the number of dimensions goes up, the gaps between these packing spheres get bigger and the centre sphere somehow grows spikes which can reach through those gaps and out of the box. It was mathematician Colin Wright who first gave me this puzzle and, in his words, it's best to think of higher-dimensional spheres as being spiky. It seems these spheres are covered in multi-dimensional bristles. Now *that* I didn't see coming. (See 'The Answers at the Back of the Book'.)

The Hunt for More Platonic Solids

Moving up from 3D to 4D gave us a whole new Platonic solid, the hyper-diamond, which was very exciting, I'm sure you'll agree. We can now keep moving up through more dimensions to see what shapes are regular enough to be called Platonic. Surely new shapes are lurking there, just waiting to be discovered.

First, we'll get the predictable (as such!) ones out of the way. As we already know, the cube works perfectly in 5D, and all dimensions above that. In 5D, you have the 5-cube with Schläfli symbol {4,3,3,3}; and all higher dimensions have their own hypercube with Schläfli symbols {4,3,3,3,. . .,3}. These Schläfli symbols mean that you start with four-sided squares, join three of those per vertex to get a cube, join three 3D cubes to an edge to get a 4D cube, join three 4D cubes . . . and so on, always combining hypercubes in bunches of three to go up a dimension. Likewise, the 2D triangle, the 3D tetrahedron and the 4D pentatope family carry on for as many dimensions as you can muster, with Schläfli symbols {3,3,3,3. . .,3}. Instead of being called hypertriangles, the word 'simplex' is used, the simplex in the nth dimension being the complete graph K_{n+1}.

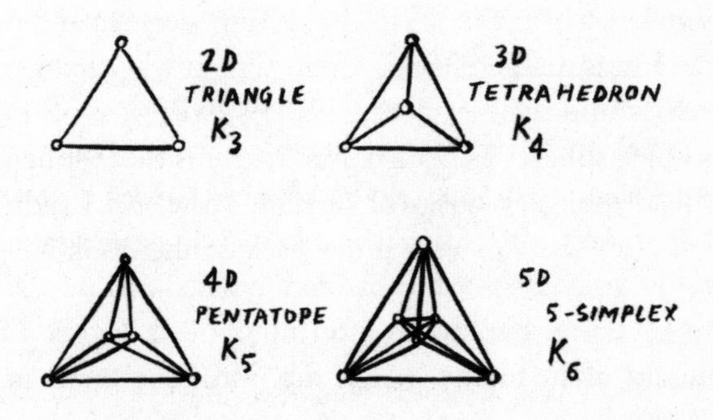

Slightly off topic I know, but I find it amazing that the formula for the highest-dimensional measurement for the non-regular forms of these shapes forms a pattern. As in, 2D shapes have area, 3D shapes have volume and (we can say) 4D shapes have content. As we saw many chapters ago (those were simpler times), any 2D triangle has an area of ($width \times height$) $\div$ 2. Width and height are the extent the triangle reaches in the available first two dimensions, so we could rewrite this $(d_1 \times d_2) \div 2$. The volume of any tetrahedron, however irregular, is $(d_1 \times d_2 \times d_3) \div 6$; the content of a pentatope is $(d_1 \times d_2 \times d_3 \times d_4) \div 24$; the hypercontent of a 5-simplex is $(d_1 \times d_2 \times d_3 \times d_4 \times d_5) \div 120$; and right up to the *n*-hypercontent of any n-simplex being $(d_1 \times d_2 \times d_3 \times \ldots \times d_n) \div n!$.

The final 5D Platonic solid is the next member of the octahedron/hyper-octahedron (aka 16-cell) family. In general, they're called the cross polytopes. The 5D version has Schläfli symbol {3,3,3,4} and it also works in all higher dimensions with Schläfli symbols {3,3,3, . . .,3,4}. The Schläfli symbol shows that the hyper-octahedron family starts with a three-sided triangle and these are bunched in groups of three in all dimensions apart from the actual dimension the Platonic shape is in, where it is closed off with a bunch of four. This means that, in any dimension, the cross polytope is always a dual shape with the hypercube, while the simplex continues to be its own dual.

So far, so predictable (for some definitions of 'predictable'); now, on to the *un*predictable. Sadly, there are no 5D equivalents to the dodecahedron/hyper-dodecagon (120-cell) family, or to the icosahedron/hyper-icosahedron (600-cell) family. Both of those family lines end in 4D. Nor is there anything that comes after the 24-cell. It was a freak 4D mutant that briefly flared up into existence in the fourth dimension and then disappeared without leaving any descendants.

That's OK. I'm sure the fifth dimension has new shapes we can be friends with instead . . . But it doesn't. There are only three Platonic solids in 5D. No problem; we'll check 6D. In six dimensions, there are still exactly the same regular shapes: the hypercube, simplex and cross polytope. Now, that wasn't very exciting. We'll go on to 7D. Same three regular shapes again. Nothing changes in 8D either. Dimension after dimension, only those three Platonic-solid shapes seem to appear. The search is futile: for every dimension from 5D onwards there are always the same three regular shapes. If you drop by the nth dimension, I can guarantee you'll find the hypercube {4,3,3,3, . . . 3} and the cross polytope {3,3,3, . . . 3,4} being duals of each other, and the simplex {3,3,3, . . ., 3} off dualling itself in the corner. It was Schläfli himself who proved this bleak Platonic-solid landscape for all dimensions, back in 1852.

I find it fascinating that regular shapes can go from there being infinitely many in 2D, to five and six options in 3D and 4D respectively, and then suddenly only three for all other dimensions. The problem with 2D is that there aren't enough options to place any significant constraints on regular shapes: they're allowed to run wild. 2D is too limited to have any nuanced behaviour. 3D and 4D are the sweet-spot for enough freedom to build something interesting but there not being too many options to ruin everything. From 5D onwards, there's too much freedom to be able to lock anything down. For me, this explains why we live in a universe with around three to four dimensions: it's enough to make the options interesting, but not too many that any attempt at anything too complex falls apart. Throw in the lack of knots and orbits in 4D and it tips the scales to why we experience a 3D reality.

All right, brace yourself now. Schläfli did come bearing some good news: despite unearthing a desolate landscape of only three Platonic hyper-shapes, he also discovered a small nugget of maths wonder. Only it's buried fairly deep. Somehow Schläfli managed to generalize the incredibly useful Euler characteristic so that it works in any dimension. We've already seen how useful the Euler characteristic can be and so to generalize it pan-dimensionally is enough to give anyone a nerdgasm. It just takes a bit of concentration to follow his logic. Totally worth it.

Imagine a line-up of the various components of a 3D polyhedron in increasing order of the number of dimensions each part has. Like some kind of shape-crime has occurred. At one end are 0D vertices, followed by 1D edges and then 2D faces. For a 4D polychoron, the line-up would be: 0D vertices, 1D edges, 2D faces and 3D cells. For a 3D polyhedron, the Euler characteristic relationship is: *vertices* – *edges* + *faces* = 2. First, a subtraction and then an addition, and if you carry on up through the dimensions, this alternating pattern continues. For polychorons, for example, the Euler characteristic is *vertices* – *edges* + *faces* – *cells* = 0. So, same pattern, except now it equals 0 instead of 2. Hold on to your caps: that switch from 2 to 0 is part of yet another alternating pattern.

1D: $V = 2$

2D: $V - E = 0$

3D: $V - E + F = 2$

4D: $V - E + F - C = 0$

5D: $V - E + F - C + H = 2$

For generic parts P with different dimensions:

$$nD: P_0 - P_1 + P_2 - P_3 + \ldots \pm P_{n-1} = \begin{cases} 0 \text{ WHEN } n \text{ IS EVEN} \\ 2 \text{ WHEN } n \text{ IS ODD} \end{cases}$$

For the fifth dimension, I've used H to represent a hypercell. There is no agreed terminology for what to call what comes after vertices, edges and faces in higher dimensions. For the general case, we can use P to represent a general polytope part, where polytope is a generic name for a shape in any number of dimensions. So you can see I've put P_2 for a face, because faces are 2D shapes.

The 2 or 0 thing can even be fixed by including an extra term on each series: the full polytope itself. Technically, a polygon has one face and a polyhedron has one cell. This corrects all of them to give an answer of 1. And there you have it. A single pattern which encompasses shapes in any number of dimensions. This amazing pattern was the brainchild of Ludwig Schläfli. And it means there will always be some things we can be sure of in higher dimensions, no matter how far we go.

$$1D: V - E = 1$$

$$2D: V - E + F = 1$$

$$3D: V - E + F - C = 1$$

$$4D: V - E + F - C + H = 1$$

$$nD: P_0 - P_1 + P_2 - P_3 + \ldots \pm P_n = 1$$

A polyhedral formula for every dimension.

A Lot of Space to Fill

Get yourself some oranges: it's time to do some more packing. Regular 3D oranges are fine, or any spherical fruit of choice. This time, instead of stacking them on top of each other, we're going to try to wrap them. The challenge is to encase them in cling film in the most efficient way possible (or in wrapping paper, should you wish to give them as a gift to someone). If you have only two oranges, it's trivial, because there's only one option: put the oranges next to each other and start wrapping. If you have three oranges, though, you're faced with a choice. You could put them in a triangle shape before wrapping them, or you could position them in a straight line. Try doing it both ways and see which uses the least cling film. There's always going to be some wasted when it overlaps to form a seal but, if we ignore that, what we're looking for is the minimum convex surface area required to completely enclose three spheres.

For four oranges, it gets even more complicated, because of all the different ways to position them before wrapping. But have a go. Feel free to try as many options as you have patience and/or oranges for. If you are one of the vast majority of people who are not going to try this, then have a think about fifty oranges and decide if they would be more efficiently wrapped as a long, sausage-shaped line or all clumped together. Most people guess that packing the spheres tightly would be more efficient when wrapping them, but a lot of mathematicians opted for the sausage option. This is the basis of what is known as the sausage conjecture.

So, who was right? It turns out both camps were kind of correct. For three and four oranges, the sausage option is indeed more efficient, as are all the options with more oranges

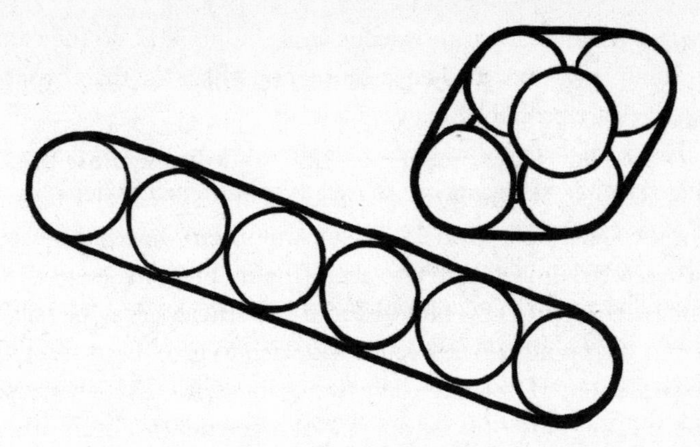

Sausage vs haggis. Who will win? They both have a lot of balls.

than that, including fifty, all the way up to fifty-six oranges. Then, completely unexpectedly, the moment you add a fifty-seventh sphere, the whole sausage collapses and it's more efficient to wrap the oranges clumped up together. This can be entertainingly referred to as a sausage catastrophe, and the resulting clump of spheres is known as a haggis. It is more efficient to wrap spheres as a haggis rather than a sausage for all groups of fifty-seven and bigger. At least, it is in 3D.

But what if we have some hyper-fruit to wrap? As we already know, the more dimensions you're eating fruit in, the spikier it gets. For 4D spheres, a 4D hyper-sausage is definitely the best option, all the way up to 50,000 spheres. We don't know exactly when, but somewhere between 50,000 and 100,000, there's another sausage catastrophe and, by 100,000 spheres, you'll definitely be better off with a 4D haggis. We have no idea what happens in five dimensions, or in any number of dimensions up to and including forty-one. Interestingly enough, once you hit forty-two dimensions, the best way of wrapping spheres is always a hyper-sausage, and the same is true in all dimensions higher than that. It is obviously infuriating mathematicians that there is a gap in their knowledge from 5D to 41D, so they are working away on it. The current consensus is that hyper-sausages are probably the way forward.

Much of our knowledge of higher-dimensional spheres is guesswork at the moment. As well as spheres acting differently in higher dimensions, as we already heard from Conway (on p. 221), the space-filling patterns we have discovered in 3D do not generalize easily to higher dimensions. We do know that, from 4D up to 8D (and probably beyond), the hypersphere is not the worst shape to try to stack; it looks like it's only the worst case in 3D. But when you do stack hyperspheres, strange things happen. An important part of Hales's proof of the Kepler conjecture is based on the fact that, if you're packing spheres in 3D, then a nice neat lattice is guaranteed to be the most efficient method. For higher dimensions, it's conjectured that the best arrangements of spheres will be ad-hoc, non-lattice clumps. Hales himself points out that a non-lattice arrangement is currently our best packing for dimensions ten, eleven, thirteen, eighteen, twenty and twenty-two.

It took until the late 1800s to prove Newton's suggestion that, in 3D, a maximum of twelve other spheres can touch and kiss a central sphere simultaneously, but, for higher dimensions, it took much longer. Mathematicians had no idea what happens in any higher dimensions until 1979, when it was proved that the kissing number is 240 in eight dimensions and 196,560 in twenty-four dimensions. Then, in 2003, we discovered that the 4D kissing number is 24. But that's it: we do not have an exact answer for any other dimension. We're narrowing in on some of them: the 5D kissing number is somewhere between 40 and 44; and 17D is between 5,346 and 11,072. Amazingly, we do know how to get a lower limit for any dimension and, completely unexpectedly, we find this by using the Riemann zeta function! In *n* dimensions, the kissing number must be greater than or equal to $\zeta(n)/2^{(n-1)}$. Mathematicians still have much to learn about kissing.

But why should we care about how spheres can be wrapped, stacked and kissed in higher dimensions? Well, it turns out that Colin set me the original hypersphere-in-a-box challenge for quite a practical reason. He was talking to me about spiky spheres because his day-job involves making numerical processing run more efficiently on a computer for use in nautical radar systems. There may not seem to be a link, but when you're trying to solve a problem with *n* variables, then you're doing the same maths as navigating an *n*-dimensional landscape. Much as the order in which you remove *n* objects from a table will give you the graph of an *n*-dimensional cube because there are *n* degrees of freedom, a function which takes *n* different variables can be accurately plotted as an *n*-dimensional landscape.

This same limitation came up in Hales's work on the Kepler conjecture. As a simple example, think of a non-regular tetrahedron in 3D. It has six edges and can be defined exactly by listing their six lengths. If I give you the set of numbers {8,10,12,9,10,11}, and we agree where each edge goes, they specify an exact tetrahedron which you could build. They could also be coordinates which specify an exact point in 6D space. So, each 3D tetrahedron is equivalent to a point in 6D space; exploring different distorted tetrahedrons is like moving around in 6D. Hales was looking at things far more complicated than a tetrahedron, but his computer was unable to handle the computations of anything above 6D space. In his words:

> My computer was generally able to prove statements about a single tetrahedron, but failed to prove anything about more complicated geometrical objects. In other words, my computer could tell me about the six-dimensional space

> parametrizing the edge lengths of a tetrahedron, but was too slow to handle seven dimensions. Given that the Kepler conjecture is an optimization problem in about seventy variables, I found this limitation to be a frustrating one. The challenge of the problem was to come to a thoroughly six-dimensional understanding of a seventy-dimensional space.

Likewise, when Colin is trying to optimize computational problems with many variables, he is effectively navigating a high-dimensional shape. At this point, intuition is of no use. But if he can build up a few tools and reference points – such as to expect spheres to be spiky – then that at least gives him a fighting chance to find his way around. So, the next time you're on a ship that is not crashing into another ship, you can be grateful that people like Colin can feel their way around spiky balls.

The Monster

To finish this chapter on higher dimensions, we should of course see just how far we can go. At the moment, our record is the seventy dimensions Hales mentioned as part of his solution to the Kepler conjecture. Seventy dimensions is well beyond anything we, or I believe any hyperthetical creature, could possibly process visually. But the maths holds up. Hales's problems with seventy variables relate exactly to points in 70D in a very real sense, and their relationships correspond to 70D shapes. But how far can we go?

There is one last shape I'd like to show you, but it's a long way up. Mathematicians had caught glimpses of what might be a shape possible only in very, very high dimensions. It was named 'The Monster', and it wasn't caught on paper until 1982, by American mathematician Robert Griess. This shape was hunted down by looking at its symmetries. A humble cube in 3D is symmetrical in a variety of ways; you can spin it around, reflect it in different ways, and it will still look the same as how it started. If you group together all the different symmetries of a cube, they have a collective name: C_3. Because its dual shape has the same underlying structure, it also has the same symmetry group C_3 (whereas the dodecahedron and icosahedron have symmetry group H_3 and the tetrahedron is by itself with group A_3).

Looking at the symmetries of shapes is an area of maths called group theory, and it extends well beyond the simple symmetries we're used to in 3D.* Each group of ways in which something can be symmetrical is considered to be a mathematical object in its own right. Much as we have come to accept things such as shapes, graphs, surfaces and knots to be mathematical objects, we can now add groups as new collections of objects. And, in the same way that a polyhedron has a corresponding graph and a knot has a corresponding surface, these groups we're going to look at match up with the symmetries of different shapes.

The letters in the names of these groups aren't important; what is important is how they form families which match up with families of shapes. We can look at the families of different regular polytopes as they go up through different dimensions, and which family of symmetry groups they correspond to. In 2D, things are not very exciting, because two dimensions, as we said before, do not give enough room to move, so the symmetries are extremely limited. Each regular *n*-sided 2D shape has symmetry group

* Group theory is amazing and an area of maths I myself need to learn more about. There are also different naming systems within it; I'm going to use Coxeter notation throughout, as I find it the easiest to follow.

D_n. Things get a bit more interesting from the third dimension onwards.

TABLE OF SHAPES WITH THEIR SYMMETRY GROUPS (WHERE THEY EXIST)

SHAPE	3D	4D	5D	nD
CUBE	C_3	C_4	C_5	C_n
OCTAHEDRON	C_3	C_4	C_5	C_n
DODECAHEDRON	H_3	H_4	–	–
ICOSAHEDRON	H_3	H_4	–	–
SIMPLEX	A_3	A_4	A_5	A_n
DIAMOND	–	F_4	–	–

You can see that the infinite families of the C and A groups carry on for any number of dimensions, always matching up with the same regular polytopes. More interesting is the hyper-diamond. It appears as a regular shape only in 4D, and its group, F_4, is not a member of a big family either. It's called an exceptional group, because it's specific to only one situation and does not generalize up to higher and higher dimensional spaces. There are other exceptional groups which also relate only to the symmetries of a shape in a specific dimension.

Up until now, we have been hunting only for Platonic solids in higher dimensions, but there are of course other, less regular shapes floating around. Groups can correspond to the symmetries of other semi-regular uniform shapes, and the exceptional group E_6 matches up with two shapes in 6D. There are other special shapes which appear

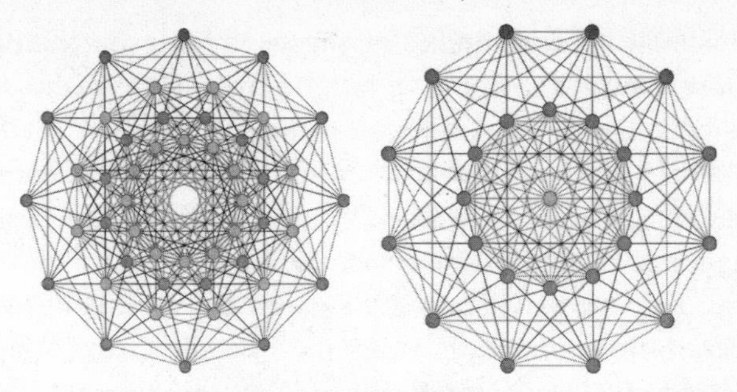

The 1_{22} and 2_{21} shapes.

in higher dimensions! In this case, E_6 has a 6D shape with seventy-two vertices called the pentacontatetrapeton or 1_{22}, as well as the 27-vertex icosiheptaheptacontidipeton or 2_{21}.

Much as I consider the hyper-diamond to be the first truly 4D shape because it's not merely a version of a lower-dimension shape, I consider the 1_{22} and 2_{21} to be truly 6D shapes. Sure, there are also 6D hypercubes and simplexes, but those exist in every dimension. They're nothing special. Likewise, we can look to other sporadic groups to find more interesting structures in higher dimensions. To pick a few more arbitrarily: there are three groups found by Conway – Co_1, Co_2 and Co_3 – which are based on symmetries of a particular 24D lattice; the Mathieu group M_{24} is based on symmetries of the 4D Klein quartic (a different shape to the Klein bottle). Our question is: of these sporadic structures, which one needs the most dimensions?

There are only a limited number of sporadic groups, and one of them does indeed have the geometric interpretation with the highest number of dimensions. It's the Monster Group, and the shape it corresponds to can exist only in 196,883 dimensions. This boggles my mind. As you travel up past hundreds of thousands of dimensions, with only a few predictable infinite families of shapes to keep you company, suddenly, out of the blurred monotony, a shape flashes into existence for a single dimensional space. It wasn't there in 196,882D and has gone again by 196,884D. In that one tiny window, a shape beyond any human comprehension exists. It is a real mathematical object, as much as a triangle or a cube. The title of Griess's 1982 paper gives the Monster its other, more affectionate name: the Friendly Giant. We will never be able to picture the Friendly Giant, but we know it exists.

We've gone from shapes in 5D all the way up to the Friendly Giant in 196,883D and, along the way, we've met some unexpectedly spiky balls. All this mathematics takes place in a world beyond what our 3D brain can visualize. But, amazingly, maths still gives us a tool to explore these strange landscapes. The fact that we even know that the Friendly Giant exists and can study some of its properties amazes me. Mathematics allows humans to explore reality far beyond our rightful place within it. And if we can discover the existence of a Friendly Giant almost 200,000 dimensions beyond ours, who knows what intrepid mathematicians of the future may come across.

Sixteen

GOOD DATA DIE HARD

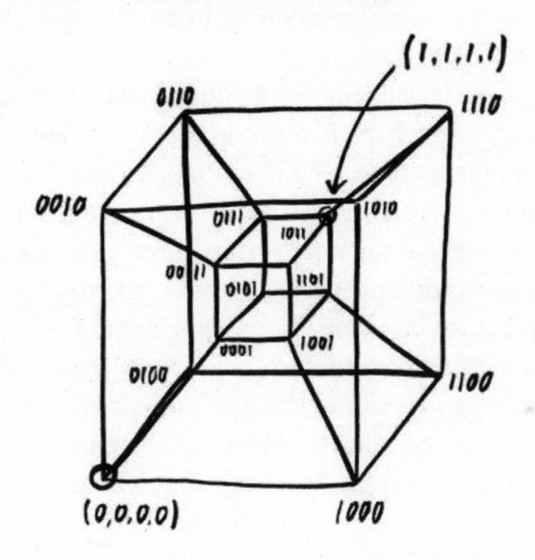

Take out a bank card and find the sixteen-digit number which appears on the front. Now email that number to me, along with your date of birth and mother's maiden name. Or, if you're slightly more security conscious, just write all sixteen digits out on a piece of paper where only you can see them. Now cross out the first digit, then every alternate number and, above each, double the number and write the new digit above. If doubling a digit

3 3 3 3 9

~~12~~ 6 ~~12~~ ~~12~~ 6 6 ~~12~~ ~~18~~

~~6~~+4+~~3~~+5 + ~~6~~+1+~~6~~+3 + ~~3~~+3+~~3~~+9 + ~~6~~+9+~~9~~+7

TOTAL = 80

Every second digit is replaced by the digital root of its double.

gives a two-digit answer, add those two digits together (the digital root from earlier) and write the total above. If you now add all the numbers together, I can guarantee that your total is a multiple of 10.

This pattern is deliberately put into all bank card numbers so that it's easy to check if it is correct or not. This is how a website can instantly tell you if you made a mistake entering your number; it hasn't had to check with a bank somewhere to see if it's right; it's just performed that calculation. If the answer is not a multiple of 10, it knows you haven't entered a valid bank card number. This is error detection.

The pattern in our modern bank cards traces back to a 1960 patent for a 'Computer for Verifying Numbers' by American inventor Hans P. Luhn. He invented a machine which, if you entered any number into it, would tell you what extra digit you needed to add on the end to complete the pattern: the check digit. Pre-dating portable electronic computers, this was a hand-held mechanical device which could calculate the check digit using cogs (more Antikythera Mechanism than Apple MacBook) and had a built-in stamp to print the number. He imagined it would be used in complicated manufacturing processes, each part number being entered into his calculator, which would then stamp that number and its check digit on to a package. Or, a number could be entered back into the computer and, if it was correct, it would stamp a tick on to the package. It so happened his patent came out right when bank cards were first being developed, and his check digit pattern was adopted for them.

In the patent, he explains why he added the extra step of needing to double every second digit. When numbers are copied out by hand, 'it often happens that an error occurs by transposing two of the digits'. If a check sum just added the digits as they were, it would be the same regardless of what

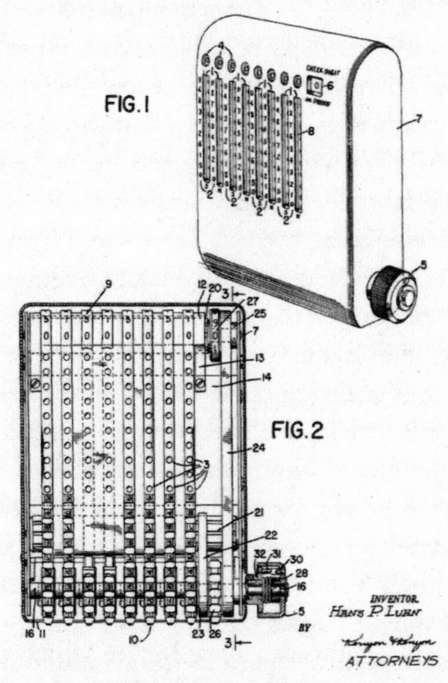

Fig. 1 and Fig. 2 from Computer for Verifying Numbers (US patent 2950048A).

order those digits were in. But, in Luhn's words, 'if every other digit is a substitute digit in accordance with the system herein set forth, such an error will be detected'. Moving the digits around can change the total. His mechanical 'computer' and imagined factory use may seem antiquated to us, but to this day Luhn's method is still spotting transposition errors when people type their bank cards into websites manually.

Very similar patterns now appear in almost every number which needs to be frequently copied and is susceptible to mistakes. Exactly the same pattern appears in barcodes, except you start with the second digit (not the first), then multiply every second digit by 3 (but don't take the digital root) to get a multiple of 10. This is so that when barcodes are scanned at the checkout in a shop, it's possible to make sure the number has been read correctly. It would be very expensive to build lasers and electronics to ensure that barcodes are always scanned correctly; now they only need to be 'good enough'. As long as the laser does enough scans sufficiently quickly, one of them is bound to be correct. Thanks to the check digit, the checkout machine can discard all the incorrect numbers, waiting for the one number that matches the barcode pattern.

Much like the 'multiply by 9 digital root trick' we came across earlier, it is also possible to use these check digit patterns in a magic trick. If someone were to read all but one digit of a number to you, you can mentally calculate this missing digit. I taught myself how to do the calculation for barcodes (because very few audience members are prepared to call out their bank card number). The pattern is easy enough: it's only a matter of how much of your free time you're prepared to dedicate to learning how to do this mentally. When I was learning, I made a spreadsheet to generate barcodes then programmed my computer to read them out with one digit missing. Time well spent.

Most people live their lives unaware of these check digits going on in barcodes and bank cards, but that's only because they don't need to know about them. They could learn them if they wanted to. Beyond these, however, are other furtive, secret check digits which are deliberately hidden so that no one knows they are there. The UK tax office hid a top-secret check system into VAT numbers, for instance. The pattern is: multiply the first digit by 8, the second by 7, the third by 6, all the way down to multiplying the seventh digit by 2; then take the sum, add 55 to help obfuscate the process, add the final two digits as a two-digit number – and the grand total is always a multiple of 97. The idea was that no one would notice this pattern, but Her Majesty's Revenue and Customs (HMRC) could use it to quickly identify anyone who included fake receipts in their tax return. However, it was leaked and is now on Wikipedia. There are undoubtedly other secret check digits around us which no one has noticed yet.

Companies also use check digits to protect themselves and their employees. If you had blindly followed my instructions and emailed me your bank card number, and you did it while at work, there's a chance that the email would never have left the building. In a large company, it's almost unavoidable that every now and then someone will fall for a phishing email at work and send off their or the company's credit card number to criminals who will waste little time putting that number to good use. So some companies scan every series of sixteen digits in all outgoing email traffic and look for Luhn's pattern. If any match, they are flagged up and sent off to be double-checked

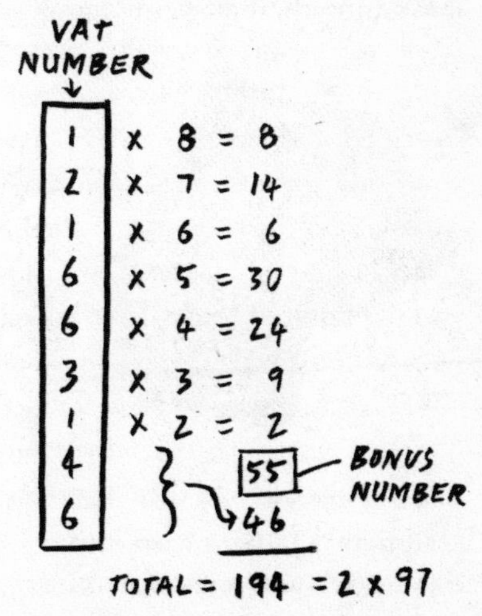

The hidden VAT check system known as Modulus 9755, and now no longer a secret.

to see if it was a random match or if someone really was emailing a bank card number when they shouldn't be. All because of Luhn's one weird trick. Nigerian princes hate him.

Everything Can be Numbers

On Christmas morning 2009 I was at my parents' house unwrapping my presents, just like many Christmases before. I opened the gift from my mother and was faced with what may be one of the greatest gifts of all time: it was a hand-knitted scarf consisting of green 1s and 0s on a black background. It was a binary scarf. I was obviously very excited – I'm a big fan of binary numbers – but then I realized that it was not just made up of random digits. These were the 1s and 0s behind computer messages! So I quickly got a pen and used the back of the wrapping paper to work out what my scarf said. Yes, it was a lovely Christmas morning, sitting there with my family, decoding my present.

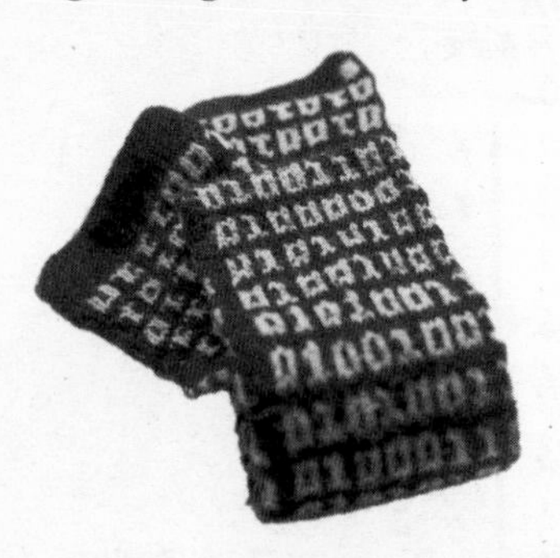

The sort of computer code which made up my scarf is commonly called ASCII, out of force of habit (ASCII was the first mainstream method developed to turn messages into computer code; it stands for American Standard Code for Information Interchange), but we now use something called Unicode. As the heading of this section indicates, everything can be turned into numbers, and there are two reasons why you would want to turn messages into numbers: it means it's possible to add error-detecting patterns on to them, as with barcodes; and it means they can be stored, processed and transmitted by computers. Ever since the first computers were invented, mathematicians have been finding ways to convert things such as writing, photos and music into numbers. Computers deal only with numbers, so if we want them to be more than fancy calculators, we need a way to turn other things into digits.

Turning letters into numbers is easy enough: you can just use their position in the alphabet. The very top line of my scarf reads 010 01101, which is binary for the numbers 2 and 13, meaning it is referring to the thirteenth character of the second alphabet. By 'second alphabet', it means upper case; the thirteenth letter of the alphabet is M. The first five lines on the scarf are 01001101 01000001 01010100 01001000 01010011, which spells MATHS. For each of them, the first three digits give you the alphabet, the next five the position (there is no helpful space to split them apart).

LETTER	POSITION		BINARY
M	13	=	010 01101
A	1	=	010 00001
T	20	=	010 10100
H	8	=	010 01000
S	19	=	010 10011

When ASCII was developed in the United States in 1963, a list of four alphabets (numbered 0 to 3), each with thirty-two characters (numbered 0 to 31) was agreed upon. The zeroth alphabet was actually a list of various computer commands, including strange things such as Carriage Return and End of Transmit Block (which you'll never need unless you're trying to talk directly to a printer in its native tongue), along with a few familiar faces such as Horizontal Tab and Escape. The first alphabet is punctuation marks, starting with the character for a blank space (deservedly in zeroth position) and followed by things such as the comma, ampersand & exclamation mark!

The second and third alphabets are lower and upper case respectively, with some punctuation marks thrown in to make up the twenty-six letters to the full thirty-two spots available. The reason for four 32-space alphabets is that the original ASCII was based on seven-digit binary numbers. The first two digits were for the alphabet (00, 01, 10, 11) and the next five for the position (00000 up to 11111). The absolute last character is Delete, and its placing at 1111111 is no accident. At the time, data was commonly stored on a paper tape with holes punched out for 1s and left intact for 0s. Taking a row of tape and punching out all seven of the holes was a perfect way to obliterate any previous data. Whenever you hit the delete key on a modern keyboard, it is still sending a signal meaning to punch out all seven holes on a piece of paper.

The original four ASCII alphabets

	Alphabet			
Position	0	1	2	3
0	Null char	space	@	`
1	Start of Heading	!	A	a
2	Start of Text	"	B	b
3	End of Text	#	C	c
4	End of Transmission	$	D	d
5	Enquiry	%	E	e
6	Acknowledgement	&	F	f
7	Bell	‘	G	g
8	Back Space	(	H	h

	Alphabet			
Position	0	1	2	3
9	Horizontal Tab	)	I	i
10	Line Feed	*	J	j
11	Vertical Tab	+	K	k
12	Form Feed	,	L	l
13	Carriage Return	-	M	m
14	Shift Out / X-On	.	N	n
15	Shift In / X-Off	/	O	o
16	Data Line Escape	0	P	p
17	Device Control 1	1	Q	q
18	Device Control 2	2	R	r
19	Device Control 3	3	S	s
20	Device Control 4	4	T	t
21	Negative Acknowledgement	5	U	u
22	Synchronous Idle	6	V	v
23	End of Transmission Block	7	W	w
24	Cancel	8	X	x
25	End of Medium	9	Y	y
26	Substitute	:	Z	z
27	Escape	;	[	{
28	File Separator	<	\	|
29	Group Separator	=	]	}
30	Record Separator	>	^	~
31	Unit Separator	?	_	Delete

Since 1963, ASCII has grown a lot. The main expansion pack was in 1985, when the European Computer Manufacturers Association devised a system with the catchy name

ISO-8859 which expanded ASCII to eight-digit binary numbers. This not only included most variations on Latin characters (such as the letter a wearing a variety of different hats: à, á, â, ã, änd å), but there was a bounty of new characters for mathematicians: the fractions ¼, ½ and ¾ got their own symbol, squared² and cubed³ superscripts were included, as well as ×, the multiplication symbol. At last, mathematicians did not have to make do with the letter x when multiplying, everyone in Europe could put the accents back on their letters, the English could now use £ and the Spanish ¿. It was ¡*excelente*!

When the 256 characters encoded by ISO-8859's eight digits started to feel a bit claustrophobic, it was taken over by the even-bigger Unicode, which does what it says on the label: IT ENCODES EVERYTHING. Using binary numbers of sixteen, thirty-two or even more digits, Unicode can cover over a million different letters and so provides a binary code for any symbol you care to think of. Because of Unicode, it's possible to email in Egyptian hieroglyphs (if you care to do so). Humans now have an agreed systematic way to turn any sentence, written in any language, into a string of digits.

You can of course learn the agreed alphabets and convert to and from binary code by hand (I'm now fluent in ASCII, but I've not learned Unicode), or you can use an automatic converter. When my mother was knitting my binary scarf, she asked my brother Steve to calculate the 1s and 0s for her (because he's also a massive nerd) and he just put the message through an online converter. He is used to doing this, because my brother and I email each other in binary, which I cannot over-recommend. The next time you need to send an email, convert it to binary and paste the 1s and 0s into the email instead. That'll sort out the nerds from the geeks. And/or lead to a substantial decrease in traffic to your inbox.

It's not just messages – we can turn any picture into numbers as well. By dividing an image up into a grid, each section can be assigned a number to describe what colour it is. The way this is done is to split a colour apart into what combination of red, green and blue it would take to make that colour. When you see something described as being RGB colour, that's what it means. The values for red, green and blue are normally stored as an eight-digit binary number, which means they range from 0 (00000000) to 255 (11111111).

It may seem to take away from that unique special moment, but a digital photo file is in fact a long list of numbers from 0 to 255. I thought I'd start with those numbers and try to turn them back into a photo of myself, and I thought I'd do it in the most inefficient way possible: I did it using a spreadsheet. I took all the numbers from a digital photo, pasted them into an Excel spreadsheet then coloured the backgrounds of the rows red, green and blue, but so the background of each cell was brighter if the number was bigger, and darker if the number was smaller (I used the conditional formatting feature, in case you want to do this at home). Up close, it looks like a lot of red, green and blue cells but, as you zoom out, it becomes a photo of me!

Do have a go yourself. One way to make a spreadsheet picture is to go laboriously through and enter all the numbers by hand then set up all the formatting yourself. The easy way is to go to my website makeanddo4D.com, where I've set up an automatic converter. Upload any photo you want, and download it as a spreadsheet. If for a second you are not convinced that this is how digital images and displays work,

Zooming out on my spreadsheet of me. I've really Excel-ed myself.

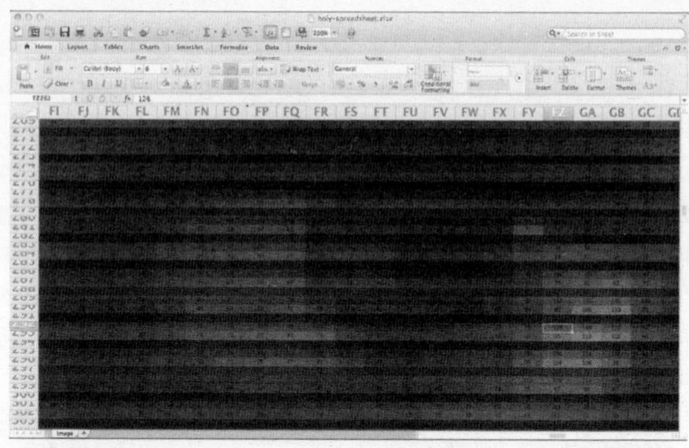

put a screen under a microscope. Staring back at you will be a spreadsheet of red, green and blue cells.

Whenever you take a photo, you are actually just taking a spreadsheet. While the numbers behind a digital photo are generally kept out of sight, however, there are a few clues that they are there. Because each pixel is three RGB values of eight binary digits each – a total of twenty-four 1s and 0s – if you read through the manual on a digital camera or TV, you might see the phrase '24-bit colour' to describe this, where 'bit' refers to a single 1 or 0. Likewise, in photo-editing software, if you go into the colour-editing options, there will be three sliders, one each for red, green and blue, which go from 0 to 255. As we saw before, it's sometimes easier to convert binary (base-2)

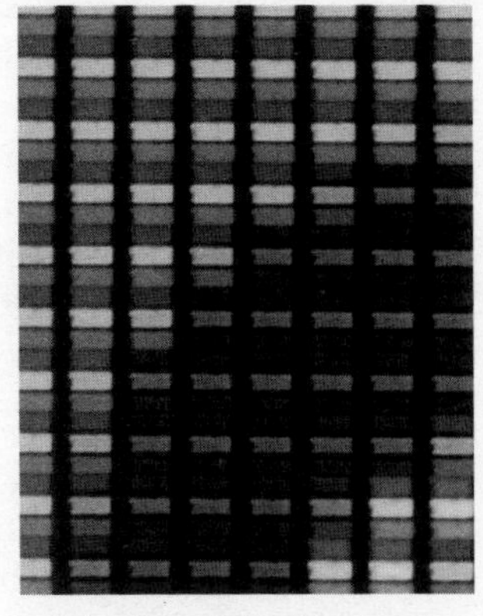

This is an iPhone5 screen at about a 400 × zoom.

numbers to hexadecimal base-16 for humans to read, so RGB colours are frequently displayed as three hexadecimal numbers. And that gives us the whole spectrum of colours, from roses (FF, 00, 00) to violets (00, 00, FF). Romance isn't dead, just digitized.

There's another hint to the maths going on in the background in file names. The format used to store an image as 1s and 0s is called a bitmap, because it's a way of mapping the bits – all the 1s and 0s – of a data file into colours in an image. Whenever you see a file with the extension .bmp, that's a reference to the fact that it is a function which takes in 1s and 0s and converts them into a picture.

Bits and Bytes

When dealing with binary numbers, each digit is typically referred to as a bit. However, keeping track of so many binary bits can get tiresome, and most of the time they are grouped in batches of eight digits anyway. So, to make life easier, each section of eight bits is called one byte (one byte can be represented by two hexadecimal digits). So seven bytes of data is really fifty-six 1s and 0s.

To add further confusion, the prefixes of kilo-, mega- and giga-, etc., are different to what we normally use. For our regular base-10 numbers, each of these is a thousand times bigger, because it's based on 10^3. Data sizes, though, keep to the powers-of-2 theme and are multiples of $2^{10} = 1,024$. A megabyte is not a million bytes, but rather $2^{20} = 1,048,576$ bytes. So seven megabytes of data is actually 58,720,256 1s and 0s. Glad I could clear that up.

And this is exactly how Tupper's Self-referential Formula works, the equation which, when plotted, makes a picture of its own equation. It is a bitmap function for converting a number into an image. The images produced by Tupper's formula are black and white pictures 106 pixels wide by 17 pixels high. If you take a 106 × 17 grid and place a 1 in the squares you want to be black and a 0 in the squares you want to be white, rotating the image and reading the digits off left

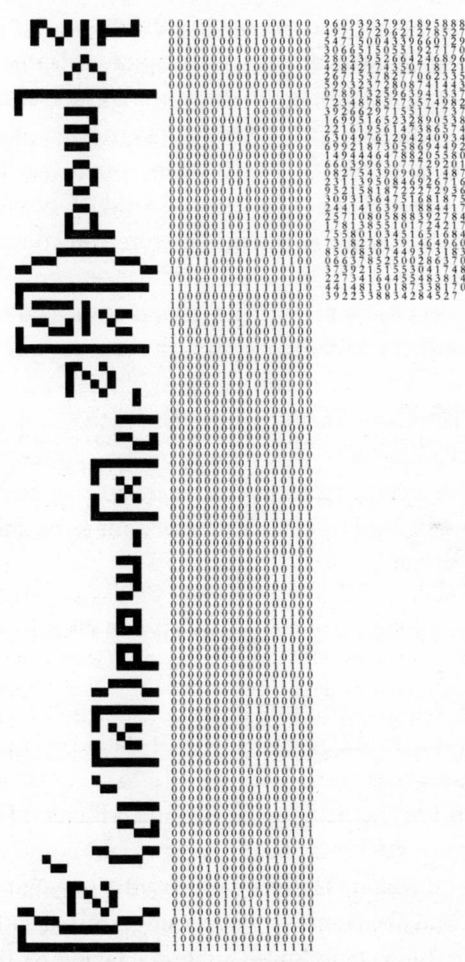

Tupper's Self-referential Formula as a binary number and multiplied by 17 in base-10.

to right, working down the image, will give you a 1,802-digit binary number. If you convert that number into base-10, then multiply it by 17, you get the value k, which was on the vertical axis when I plotted the formula earlier. If you go to any value of k on the vertical axis, the plot will be of the image corresponding to the binary version of $k/17$.

We can now turn messages and pictures into numbers and, of course, it doesn't stop there: things such as music can be digitized as well (effectively converting a plot of the soundwave into coordinates). Once something has been turned into numbers, a whole new world of possibilities is opened up. If you had a physical picture and you wanted to show someone else, you would need to somehow send that physical object to them. With a digital picture, you can transmit the numbers to them nearly instantly, anywhere in the world. You can also make as many copies as you want, instantly. Turning something into data makes it much more versatile. And because it is now made of numbers, we can add error-detecting patterns. Or even go one step further and find a way not just to detect errors but to be able to correct them as well.

How to Solve a Problem Like a Computer

Almost as soon as computers were first built, there was a need for a way to fix mistakes in their data. Bell Labs in the United States had one of the world's first computers. One weekend in 1947, a mathematician named Richard Hamming set it running on Friday night so it would complete some important calculations before Monday. But almost as soon as he left it alone, the machine made a single mistake, rendering everything else that followed useless. Having to break the bad news to his colleagues on Monday morning, Hamming started thinking about how to avoid this problem in the future. By 1950, he had finished his seminal paper 'Error-detecting and Error-correcting Codes'. He had found a way that computers could detect and fix their own mistakes.

Much like Turing, Hamming first worked on computers in secrecy during the Second World War. He was based at Los Alamos, working on the Manhattan Project (alongside Ulam), using rudimentary computers to complete calculations for the physicists. One of his jobs was to double-check the physicists' calculations so that when they tested the first ever nuclear detonation the energy released would not be sufficient to set the entire atmosphere of the Earth on fire. (Setting the Earth's atmosphere on fire is widely regarded as a 'very bad thing'.) The numbers checked out, but he was concerned about the physicists' assumptions regarding oxygen–nitrogen interactions. It seemed to him that it might still be a possibility, but his concerns were paid no heed. He was not much consoled by a friend, who saw him worrying and said, 'Never mind, Hamming, no one will ever blame you.'

After the war, he began work on some of the first computers ever built, which were rather unreliable. He estimated that the circuits he was working on had one logic-gate failure every 2 to 3 million uses, enough to make long calculations impossible. He also had the foresight to realize that one day we would need a way to correct computer error when 'signalling in the presence of noise where it is either impossible or uneconomical to reduce the effect of the noise on the signal'. (Translated into normal language, that's roughly 'trying to talk in a loud room where you can't just tell everyone else to shut up'.) Written before computers were communicating with each other across a room, let alone around the world and across the solar system, this was amazingly prescient of

Hamming.* And this is exactly what we use his codes for today. Without a cheap and robust form of data transfer, even in the face of noise and disruption, our modern technology would not be possible.

In his 1950 paper, Hamming introduced the simplest check digit possible, a parity check: a single extra 1 or 0 on the end of a binary number to make sure it has an even number of 1s in it. This means that if a binary number has an odd number of 1s, you know it has not been transmitted correctly. But this is still only error detection: we want to be able to correct errors as well. This can be done by putting the binary numbers in a grid and making sure that not only do all the rows have an even number of 1s but all the columns do too. If one digit is changed, it is then possible to find exactly which one it was and switch it back.

0 1 0 0 0 0 1 1 1
0 1 1 0 1 1 1 1 0
0 1 1 0 0 1 0 0 1
0 1 1 0 0 1 0 1 0
0 1 1 0 0 1 1 1 0
0 1 1 0 1 1 1 1 0
0 1 1 1 0 0 1 0 0
0 1 1 0 0 1 0 0 1
0 0 1 0 0 0 1 1 1

One of these digits is wrong. Can you find it?

But the real breakthrough concept in Hamming's paper is something that is now known as the Hamming distance. This is getting right down to how data is transmitted. Let's say I have a long list of 1s and 0s that I need to read out to you over a low-quality phone line. I'm a bit worried that, because the signal isn't great, you might mishear a 0 or 1. So my plan is to say each digit four times. So, to send the simple 010, I would say '0000 1111 0000'. Then, if you hear me say something like '1011', you can be pretty sure that it was meant to be '1111' but one of the digits was misheard.

* There were some hints, though. In 1940, Bell Labs had demonstrated an electromechanical calculator which could be operated from 300 miles away. Hamming was at the right company at the right time.

In terms of terminology, it's convenient to talk about a single piece of information in a code as a codeword. So, in the system where you send each bit four times, 0000 is a valid codeword, as is 1111. Any other four-bit 'word' which is not 0000 or 1111 must by default have a mistake in it (it can take a while to get used to seeing 'word' refer to a group of digits, but it makes computer scientists really happy). In my example, 1011 is not a valid codeword, so we correct it to be 1111. Sadly, if we received the word 0110, even though we know there has been an error, we can't fix it. The reason is that 0110 is the same 'distance' from both 0000 and 1111: in both cases it would take two mistakes to get to 0110.

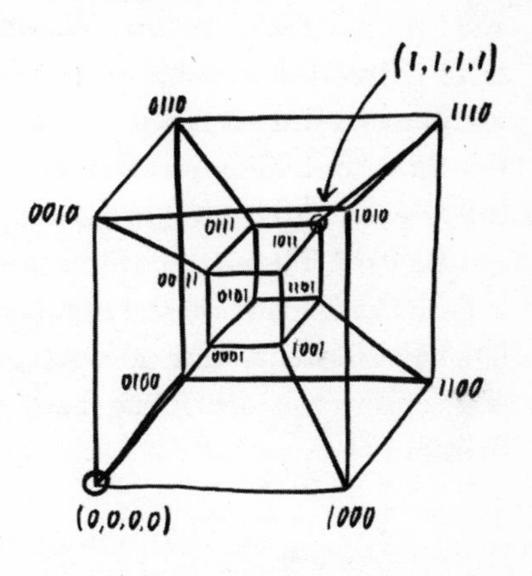

The distance a word can travel away from a valid codeword and still be corrected is the essence of the Hamming distance. This can have a literal interpretation as a distance as well: 0000 and 1111 are the coordinates for opposite corners of a 4D cube. The word 1011 is a different corner, closer to 1111 than to 0000. The word 0110 is the same distance away from both valid corners on the hypercube. If we repeated each digit five times, we'd be working on a 5D hypercube, and the extra distance between the 00000 and 11111 corners means we would be able to correct up to two errors.

This method of repeating digits is a blunt-force method of error correction, but it does work. The problem is that it is hugely inefficient: you have to send four or five times as much data as the actual message you want to transmit. It is possible to do much, much better by adding far fewer check digits to provide the same amount of error correction. The problem is effectively how to position the valid codewords on a high-dimension cube so that each one has as much space around it as possible. You can imagine a hypersphere around each codeword which encloses all the mistake codewords that can be corrected. In Hamming's words, 'The problem is that of packing the maximum number of points in a unit n-dimensional cube.'

If that all sounds a bit esoteric, there is actually a very accessible method of highly efficient error correction which you will have seen before. You may even have done some error correction yourself in your leisure time. That error correction is solving a sudoku. A sudoku is a grid of numbers that has three different overlapping mathematical patterns: one for the rows, one for the columns and one for sections of the grid. Because of these patterns, if you are given a sudoku with loads of the numbers missing, you can calculate what those missing numbers are. This is a form of error correction: you are filling back in data which has gone missing.

		4	5		2	6	8	
	8	9	7		1		2	
	7		4	8	3			
8			6	4	7			
		3	2	1				
		7	8	3				
	3	6		2	4			1
	5	8			6	2		4
2	4					9	7	6

This is almost exactly the same form of error correction as that used by mobile-phone text messages. Before your phone sends a text message, it turns your message into numbers, puts those numbers in a grid, then adds some extra check digits to include three different overlapping patterns into those digits. People can send a text message on a mobile phone with no idea this is happening in the background; they safely assume that it will arrive at the recipient's phone, anywhere else on the planet, without any words getting lost. In reality, given the number of phone-masts and systems it has to go through, mistakes will happen. But because of the mathematical patterns, the mistakes can be fixed. It's an illusion that text messages work so well. There is just a lot of maths fixing the errors before the user even notices.

I once met a collection of mathematicians who work for Hewlett Packard in the UK, and I got chatting to one of them, Miranda Mowbray. It turns out she was involved in developing one of the early error-correction methods used to send computer data over telephone cables. It was originally thought that twisted wires in unshielded telephone cables would be too cheap and low-quality to network computers reliably, but Miranda helped to find a mathematical way to make the error correction work and the world has had affordable networks ever since. Her code took binary data, split it into codewords of five digits and then converted each codeword into a six-digit ternary number. With this method, amazing speeds of 100 megabits of data every second could be sent over telephone lines. It was an honour to be able to thank her in person!

Which brings me back finally to my scarf. When I finished decoding it, the whole message was a quote my mother found in an interview I'd done. It reads: 'Maths is fun, keep doing maths.' (Certainly sounds like me.) Or rather, because she only used capital letters, it reads 'MATHSISFUNKEEP-DOINGMATHS'. Except it wasn't quite that. When my mother was laboriously knitting the scarf bit by bit, she made a mistake. Part way down, she accidentally swapped around a 1 and a 0. This bumped a U into a V, and so the message actually claims 'MATHS IS FVN.'

When I pointed this out, it really did ruin her fvn. My mother is quite the pedant and so she was disappointed to learn she had misknit. She has unravelled bigger things to fix smaller mistakes before. I was able to stop her from trying to fix my scarf by pointing out that she didn't need to. You see, to make the scarf long enough, the same message repeats itself four times over. And she had only made a mistake in one of those four versions. Which means that, if I calculate the average value across all four versions, I will get the original, accurate message without the mistake. My mum had by chance knitted me the world's first error-correcting scarf.

The Finite Nature of Data

There is an interesting side-effect from taking the infinite range of nuanced colours available in nature and converting them all to one of only 16,777,216 possible RGB values ($2^8 \times 2^8 \times 2^8 = 16,777,216$). Encoding something in numbers takes the infinite possibilities of the world around us and traps them in a finite range of values.

For example, the iPhone5 has a screen with a resolution of 1,136 pixels by 640 pixels, a total of 727,040 different pixels. Each of those 727,040 pixels must be one of the 16,777,216 possible RGB colours, which means that the screen on an iPhone can only show $16,777,216^{727,040}$ different images. Don't get me wrong, that is a massive number. Written out in base-10, it has 5,252,661 digits. But it is finite. Every image you have ever seen on an iPhone is just one of those pre-determined values.

That is also part of how Tupper's Self-referential Formula works. As you work your way up the vertical axis past every possible k value, it prints out every possible arrangement of $106 \times 17 = 1,802$ black and white pixels. All 2.9×10^{542} of them. So the fact that it produces its own formula along the way is not that amazing: it also produces every possible formula that can be drawn in a 106×17 grid.

The compact disc may now be outdated technology, but a lot of music albums are still released on CD. A standard 700-megabyte CD is actually 703.125 megabytes (a rare case of the music industry giving something extra away for free), which is a total of 5,898,240,000 1s and 0s. By my calculations, the number of possible different CDs in base-10 would have 1,775,547,162 digits. Which is also the number of corners a hypercube would have in 5,898,240,000 dimensions. So whenever a musician claims they have written a new album, all they have really done is choose a corner on a very high-dimension hypercube!

Seventeen

RIDICULOUS NUMBERS

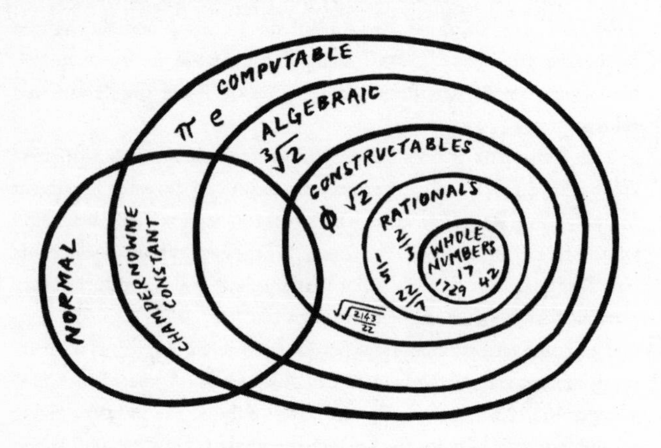

$\theta = 1.306377883863080690468614492602605$
$7129167845851567136443680537599664340$
$05376682659882150140370119739570729\ldots$

Try cubing this number. Now raise it to the power of 9. Then 27. Keep raising it to powers which are themselves powers of 3 and see what you spot in the answers.

This number's name is Mills' constant, and its symbol is

traditionally the Greek letter theta (θ). It was proved to exist in 1947 by the Princeton mathematician William Mills in a maths paper only one page long, but he had no idea what its value was. There are actually several different Mills' constants, but this is the smallest possible one and was calculated to the first 6,850 digits in 2005. This number is interesting because it can produce endless prime numbers. You will have noticed that raising Mills' constant to a power which is a power of 3 and rounding the result down to the nearest whole number will always give you a prime result. Yes, against all expectations, it is a number which can singlehandedly produce an infinite string of primes.

$\lfloor \theta^{3^n} \rfloor$ = A PRIME NUMBER

$\theta^3 = 2.229494772 49\ldots$

$\theta^9 = 11.0820313699\ldots$

$\theta^{27} = 1361.00000108\ldots$

$\theta^{81} = 2521008887.00000000000000000004195850241$

$\theta^{243} = 16022236204009818131831320183.0000000\ldots$

When I first saw Mills' constant, my response was the same as most people's: disbelief and wonder at how this could even be possible. It turns out that it cheats. Mills' constant has been deliberately built to have this property. To calculate its values, you start with the primes then work backwards to find what value of θ will produce them. (Like throwing a dart at a wall and then painting the dartboard around it.) Which makes it useless at finding prime numbers, as you need to know the primes first to then calculate Mills' constant.

This is also not guaranteed to actually *be* Mills' constant, it's just the smallest number we know of which *could* be Mills' constant. We know that such a number exists, and this is the best candidate we have at the moment, but it may be something else, a larger number. The uncertainty is because this value for Mills' constant works only if there is definitely a prime number between every pair of cube numbers, and we don't know for sure that primes are dense enough for that to be guaranteed. Amazingly, the Riemann Hypothesis implies they will, because, if it is true, the primes will be dense enough for there to be one between each pair of cube numbers. So this value of Mills' constant assumes that the Riemann Hypothesis is true. It is one of the many things in maths which will be proven to be correct . . . the moment someone can prove the Riemann Hypothesis.

Mills' constant is a great reminder that, if we want a number with a certain property, we can simply construct it ourselves – which is not to say it isn't a real number: it's there on the number line, as real as any other number such as π, $\sqrt{5}$ or 7. It's easy to forget just how many numbers there are, because we use so few of them in everyday life. But the number line contains not only every number humans have ever used but, also, any string of digits you care to write out corresponds to a number somewhere on that line.

To try to understand the menagerie of numbers, mathematicians like to put them into categories, which sounds like a useful thing to do. We've already used whole numbers as well as fractions, which are what I call the 'well-behaved' numbers. They have the decency to have only a few digits (such as $\frac{1}{4} = 0.25$), or they have a set string of digits which neatly repeats for ever ($\frac{1}{7} = 0.142857\ 142857\ 142857 \ldots$). But terminating and repeating digits are really the same thing; it just depends on what base system you're using. The fraction

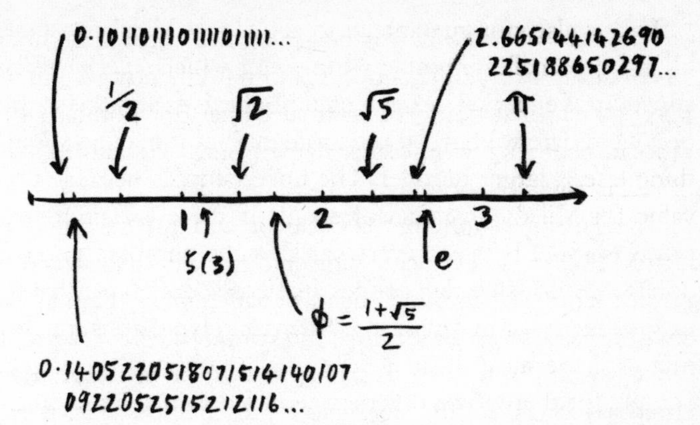

Some important and some unimportant numbers on the number line.

½ may be a nice neat 0.5 in base-10, but if you move it to base-3 you get the repeating digits 0.111111. . . Likewise, ⅔ is 0.6666. . . in base-10 but 0.2 in base-3. The point is, whether repeating or terminating, their digits are predictable: they are well-behaved.

The irrational numbers, on the other hand, don't play by the rules. Numbers such as $\sqrt{2}$, π and the golden ratio (φ) cannot be written as fractions and refuse to have easily predictable digits. This is why $\sqrt{2}$ was an interesting number for NASA to calculate: until you calculate it, you don't know what digit will be next. Likewise, π makes an interesting memorization challenge: the digits are unpredictable. The current world record of memorizing and reciting the first 67,890 digits of π is therefore more impressive than my ability to memorize and recite the first 67,890 digits of ⅔.

But not all irrational numbers were created equal: some are more irrational than others. At least numbers such as $\sqrt{2}$ and the golden ratio φ are the solutions to neat equations. If you solve $x^2 = 2$, you can get an answer of the square root of 2; likewise, the golden ratio is a solution to the equation $x = 1 + 1/x$. If it is possible to calculate a number algebraically from an equation featuring only powers multiplied by rational numbers (formal name: polynomial with rational coefficients; informal name: a neat equation), it is called an algebraic number.

The left-over numbers such as π and e (the number, $e = 2.71828\ldots$) defy being written down algebraically. You cannot write a neat, finite equation which gives them as an answer. Aside from allocating it the symbol π, you will never see π written in a form that does not go on for ever. You could say it transcends algebraic representation. Yes, there are ways to calculate π, but they involve things like an infinite series of numbers. The same goes for e. These numbers are called transcendental numbers, and they are mysterious and slippery entities. They also vastly outnumber other numbers. The relative handful of rational and algebraic numbers on the number line are swimming in a vast sea of transcendental numbers. And it's a sea we know very little about.

Despite their ubiquity on the number line, transcendentals are surprisingly hard to pin down. It took until 1873 to prove that e was transcendental, making it the first number we knew for definite was. The poster-child of maths, π, didn't join the transcendental fold until 1882. Even today, we know that at least one of $e + \pi$ and $e \times \pi$ is transcendental, but we have no idea which. On David Hilbert's 1900 list of important maths problems to solve, one of them involved checking if e^{π} is transcendental, and since 1934 we have known that it is. However, e^{e}, π^{π} and π^{e} are still open problems. Transcendentals are really hard to find in the wild.

Finally Proving What Cannot be Drawn

Dust your pencil, compass and ruler off and draw yourself a right-angled triangle with short sides that are 1 and 2 units long. Thanks to Pythagoras, this means the long side will be $\sqrt{5}$ long. If you extend and divide this, you can draw a line which is the same length as the golden ratio: $\varphi = (1 + \sqrt{5})/2$. With a compass and straight edge, it's possible to draw any shape which involves adding and subtracting, multiplying and dividing, squaring and square roots – which means that you can draw a line the same length as any rational number, some of the algebraic numbers and none of the transcendental numbers.

The proof in 1882 that π is a transcendental number put a 2,000-year-old problem to rest: for any given circle, can you draw a square of the same surface area using a compass and a straight edge? Since 1882, we know that no, you can't. To draw a square the area of a circle you need to be able to draw a line π units long, and you cannot draw transcendental numbers.

Algebraic numbers are a bit more subtle; it depends if they require anything more complicated than a square root. The algebraic numbers which use only square roots are called the constructible numbers, and they can be drawn with a compass and straight edge. Any algebraic number with a cube root or more cannot be drawn. Both trisecting an angle and doubling a cube involve taking a cube root. The subtleties of algebraic and transcendental numbers finally proved that these problems, which had been bothering mathematicians since the times of the ancient Greeks, were definitely impossible.

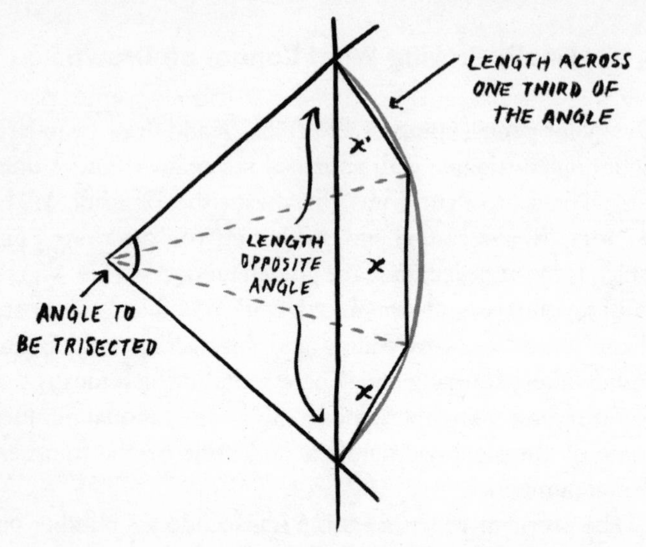

Trisecting the angle means starting with the length opposite the angle and finding the length across a third of the angle, x. This is the same as solving the equation $4x^3 - 3x = n$.

Because you can draw a length of $(1 + \sqrt{5})/2$, it means you can draw an exact pentagon. Set your compass length to 10cm, and a pentagon will fit in the middle of an A3 piece of paper (or, with a bit of cunning, on a landscape piece of A4). If you measure the golden-ratio line it should come in at slightly below 16.2cm (matching the ideal length of 16.1803... cm). We now know that we can draw any regular polygon exactly, if it can be reduced to drawing a line length which is a constructible number.

It was Gauss (the effortless summer of numbers 1 to 100) who first realized that drawing an exact regular polygon could be done for shapes which have a prime number of sides of the form $2^{2^n} + 1$. Often mistaken for Mersenne primes (which are the superficially similar $2^n - 1$), these are

Draw your own perfect pentagram

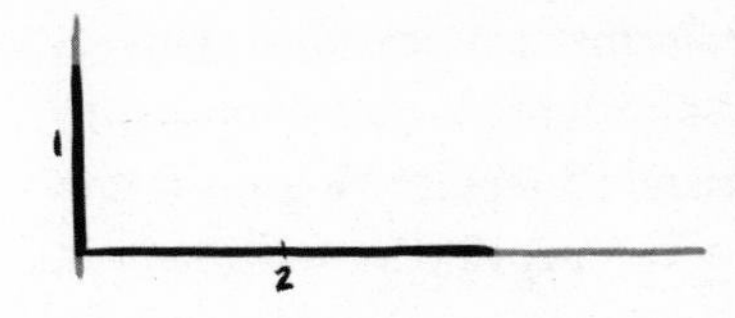

1. Draw a line 1 unit long at right angles to the 2-unit line.

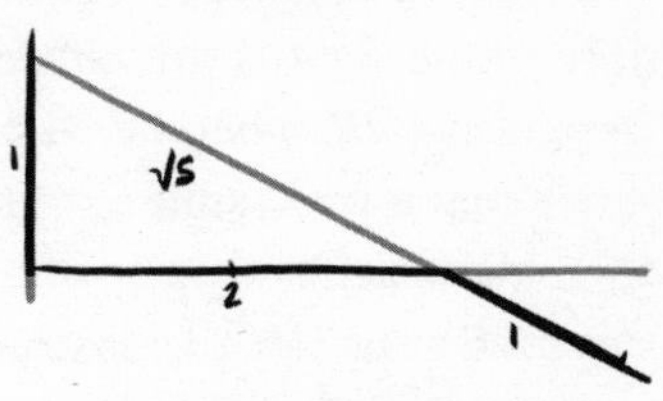

2. Making it a triangle gives you a line $\sqrt{5}$ long, and you can extend this by 1 unit.

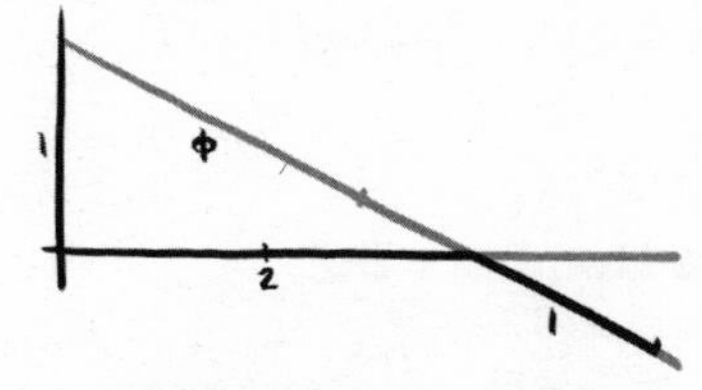

3. Divide this in half to get the golden ratio $\varphi = (1 + \sqrt{5})/2$.

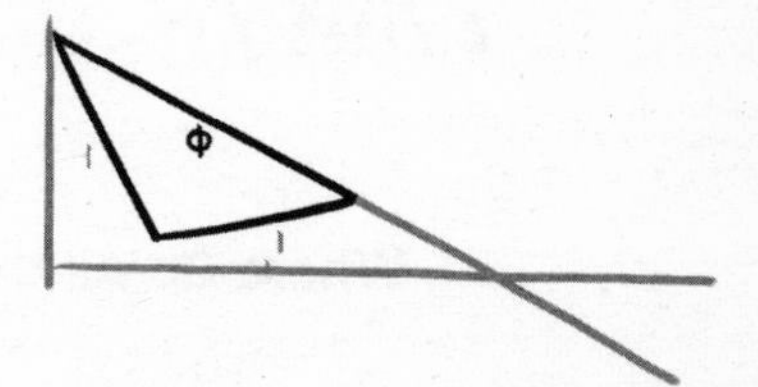

4. Add two 1-unit lines to make the start of the pentagon.

5. A 1-unit line and a φ-unit line give you the next corner of the pentagon.

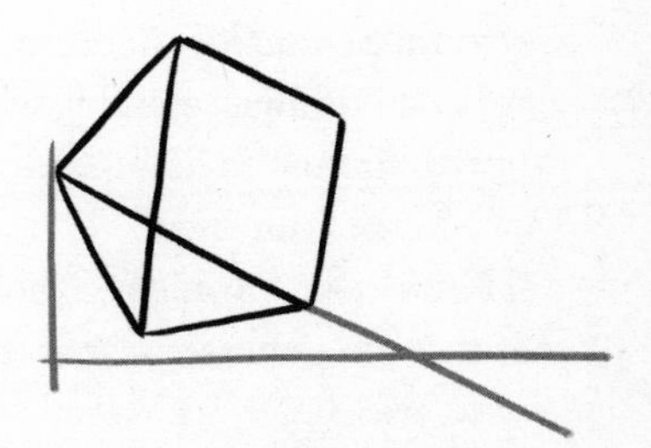

6. Finish it off with two final 1-unit lines.

called Fermat primes. What Gauss also spotted, but did not prove (that came later, in 1837), was that any regular shape you wish to construct must have a number of sides whose only prime factors are either Fermat primes or 2. This all started in 1796 when he was only eighteen and managed to prove that it is possible to construct a regular 17-gon. To get an angle which is exactly one seventeenth of a full rotation, you need to be able to construct the monstrosity below which, importantly, needs only square roots.

$$\sqrt{34-2\sqrt{17}}+\sqrt{17}-1+$$

$$2\sqrt{17+3\sqrt{17}-\sqrt{34-2\sqrt{17}}-2\sqrt{34+2\sqrt{17}}}$$

Where Do All the Numbers Go?

It's time to make your own transcendental number! In 1934, two different mathematicians simultaneously stumbled on a way to crank out the previously elusive transcendental numbers. Both the Russian Alexander Gelfond and the German Theodor Schneider discovered the amicably named Gelfond–Schneider theorem. Stated simply, if you take two algebraic numbers *m* and *n*, where m is not 0 or 1 and *n* is not rational, then m^n is definitely transcendental.* So here's mine: $13^{2\sqrt{2}}$. Make your own!

Here's the problem, though: the Gelfond–Schneider method can generate transcendental numbers, but it cannot

* This is how we discovered e^π is transcendental. Finish this chapter first, then check out how it was done in 'The Answers at the Back of the Book'.

check if any given number is transcendental. So let's start with some semi-arbitrary numbers. Watch, I'll whip one up right now. I'll call it 'the number Matt just made up'. Here it is: 0.10110111011110111110111110. . . and it goes on like that for ever. It's a base-10 number made out of groups of 1s that get progressively longer. It's probably transcendental. I think. I don't think you can write it down as a fraction and I doubt it's the solution to a neat equation. If you want to make your own, come up with a system for writing down endless digits in such a way that they never fall into a repeating pattern. However, unless it fits into the Gelfond–Schneider family (or one of a few other transcendental families), proving it is transcendental is super-hard.

There is a number similar to 'the number Matt just made up', and it's called the Champernowne constant, after British mathematician and economist David Champernowne, a contemporary of Turing at Cambridge University. In fact, Champernowne was only sixteen days younger than Turing and the two were good friends. Champernowne made his number in base-10 by listing all the numbers in order to form its decimal digits. Therefore, the Champernowne constant = 0.1234567891011121314151617181920212223. . . These digits go on for ever and they never start repeating.

What makes the Champernowne constant special is that we *know* it's transcendental. It is one of those rare cases where we've pinned a number down as definitely being transcendental. This is a recurring theme when it comes to categorizing numbers: there's a vast gap between how many numbers are transcendental and how many we have proved to be transcendental. The problem is that, even though mathematicians can come up with categories for numbers, it's really hard to know which category a given number goes in.

There is another category of numbers – normal

numbers – which we know are incredibly prevalent yet are super-hard to find. A normal number is an irrational number whose digits contain every possible string of digits with equal likelihood. So if $\sqrt{2}$ were a normal number, it's as likely to contain the ten digits 1405220518 somewhere in it as it is 0715141401 or any other list of ten digits. But we don't know if $\sqrt{2}$ is normal. The total number of numbers which have been shown to be normal is currently: 0.

But we can construct what I call 'artificial' normal numbers. These are numbers which are normal, such as the Champernowne constant, but they were deliberately made to have that property. Another artificial normal number is the Copeland–Erdős constant, which is the same idea, except it lists only the base-10 primes, giving a value of 0.23571113171923. . . Mathematicians are yet to prove that any number we already knew about is normal. Since 1909, mathematicians have known that almost all numbers are normal, but proving it for any one non-artificial example is so difficult it hasn't been done yet.

The classic example of a number which everyone desperately wants to be normal is π. We've checked the first nearly 30 million digits, and they seem to be normal, with every string of digits being equally likely. But we don't know for sure. If it is, then any number at all will be in there somewhere. My name, Matt, turned into numbers as 13012020 (m = 13, a = 01, etc.) appears in π, starting at the 291,496,384th digit. If π is normal, your name, your favourite song lyric, the complete description of what you will have for lunch tomorrow, is in there somewhere. As it will be in any normal number. 'Matt' also starts at the 301,480,410th digit of $\sqrt{2}$, the 312,366,242nd digit of e and the 137,673,084th digit of the golden ratio. Any string of digits you choose, however long, will appear in almost all numbers. More numbers contain the complete works of William Shakespeare than don't.

Turing then went on to prove that there are numbers which cannot be written down by a machine; there are numbers which cannot be computed. So not only are the whole numbers and rationals we use in everyday life an insignificant part of the total sea of numbers but, even when they are combined with algebraic and computable transcendental numbers, we are still only on the surface. Surrounding us is an unimaginable volume of uncomputable numbers: the dark numbers which we know are there but which we cannot grasp.

The Maths Smack-down

Here are a couple of quick questions:

1. Find me a number such that, when it is cubed and said number is added to this cube, the result is 5.
2. A man sells a sapphire for 500 ducats, making a profit of the cube root of his capital. How much is the profit?

It's strange to realize that all the numbers we take for granted were actually discovered at some point in the past and, before that, mathematicians had to make do without them. That must have made life difficult. It wasn't until the 1600s* that European mathematicians were, on the whole, prepared to accept that negative numbers existed, so, if you were trying to solve problems like the questions above, you had to do it using only positive numbers.

Not only that, but you couldn't use algebra. The vast

* If mathematical knowledge is a tidy stack of building blocks, history is like trying to herd porridge. Which is why I'm not a historian; all history here should be taken as my best understanding or interpretation of events.

majority of what we consider standard mathematical notation didn't exist in the early 1500s. The square-root symbol √ had been used a few times, but was unknown to most mathematicians; the multiplication sign × would not appear until 1618; and the π symbol for pi would not be around until 1706. Even the now-ubiquitous plus sign + was only introduced in 1360, by the always-ahead-of-his-time Nicole Oresme (which must have been useful for writing out his divergent harmonic series proof). As for algebra itself, the first time letters were used to represent numbers was 1591. Before that, all equations had to be written out 'the long way'.

The Greeks were great at geometry and diagrams, but had written all their maths in long, descriptive sentences. This was the Rhetorical Period in mathematics, and it lasted right up until the new maths of the Renaissance demanded more powerful forms of expression. There was a cross-over period where mathematicians were trying to solve advanced equations, but they were trying to do it without mathematical notation. They had to use long-winded, elaborate methods. This had the interesting side-effect that, for a brief time in the early 1500s, solving equations became a spectator sport – and led to one of the greatest maths smack-downs in history.

In maths, there have been some amazing rivalries: Newton vs Leibniz, Bernoulli vs Bernoulli, Fermat vs Everyone Else. But none of them compares to the one that occurred in Italy in February 1535. In one corner of the 'rumble in the algebra' ring was Italian mathematician Niccolò Fontana, better known by his nickname, Tartaglia, which translates roughly as 'The Stammerer'. Honestly, he even had a wrestling-sounding stage name. (I'm not making this stuff up, by the way.) He grew up near Venice and when his home town Brescia was ransacked by French troops, twelve-year-old Niccolò was hit by a sword to the face, which sliced through his jaw and palate, and he was left for dead. However, he survived, grew up and sprouted a manly beard to hide his hideous facial scars. The injuries, though, left him with a permanent stammer. Tartaglia taught himself maths from a young age, moved to Venice and became a renowned mathematical powerhouse.

In the other corner is Fior, student of the famous mathematician Scipione del Ferro. The parents of del Ferro had been in the paper-making business; a boom high-tech industry, given the recent invention of the printing press (it was the fifteenth-century information super-highway). He became a lecturer in arithmetic and geometry at the University of Bologna and made some incredible mathematical discoveries – only he didn't tell anyone else about them. He passed on his secrets to a few of his students, including Fior, before dying in 1526. Fior now felt the responsibility of defending his master's maths techniques against any newcomers (they had a kind of Obi-Wan Kenobi/Luke Skywalker thing going on). The stage was set.

Cut to the 1530s, and Tartaglia is strutting around town claiming that he's the best mathematician at solving equations since and including the ancient Greeks. He claims he can solve equations no one else has ever solved before, including a new type of equation called the cubic. Fior steps forward and says that his master, del Ferro, was solving cubic equations long before Tartaglia ever did, and he is now the master of equation solving. Tartaglia calls bullcrap. Fior gets all up in his face. There is only one way to solve this: an equation-off. Tartaglia and Fior will each write down thirty equations for the other to solve. These equations will be given to a middleman, who will pass them on, as well as make them publicly available. The crowds will then wait to see who can solve their rival's challenges. The atmosphere was tense, as you can imagine.

Fior and Tartaglia could not have got into this situation at any other time. They (and all other mathematicians) were trying to solve advanced maths equations using outdated numbers and symbols which were no longer fit for purpose. Any earlier, and it would have been impossible to solve equations like these; any later, and solving them would be too easy. It was the mental gymnastics needed to avoid directly using negative numbers that made solving these equations such a challenge, and a spectator sport.

We take negative numbers and the number 0 for granted these days, but they too had to be discovered before they could be used. The first recorded attempt to deal with 0 and negative numbers was in the early 600s, by the Indian mathematician Brahmagupta. But Europeans didn't take to them. When Fibonacci was extolling the virtues of using a base-10 number system, he called the digits 1, 2, 3, 4, 5, 6, 7, 8, 9 numbers, whereas the digit 0 was merely a sign. The Italian mathematicians were thus going to great lengths to solve these equations without using 0 as a real number. It wasn't until the 1600s that it gradually became accepted widely by mathematicians.

Negative numbers took even longer to gain acceptance. Around 1600, Thomas Harriot (of the cannonball-stacking problem) was using negative numbers to solve equations but, as late as the 1800s, mathematicians such as Francis Maseres described negative numbers as 'mere nonsense or unintelligible jargon'. Maseres refused to believe that −5 was a valid solution to $\sqrt{25}$. He argued that saying $-5 \times -5 = 25$ was the same as saying $5 \times 5 = 25$ and the negative signs had 'no meaning or significance'. Having grown up with negative numbers, this sounds bizarre to us, but as recently as two centuries ago some mathematicians were still in denial about negative numbers.

What I call the Negative Revolution started at the time of Fior and Tartaglia in 1535 and lasted until Thomas Harriot's death in 1621. It then took the following two centuries for mathematics to fully embrace 0 and negative numbers as being as real as any positive number. Fior and Tartaglia's maths face-off was a product of how close mathematics was to the Negative Revolution. They were also in the dying days of the Rhetorical Period started by the Greeks and right on the cusp of the Symbolic Era (again, a term I've made up) which was about to sweep Europe. In its wake, mathematicians would be writing mathematics using symbols and equations instead of long, descriptive sentences.

The cubic equations which Fior and Tartaglia were fighting over are ones which have a cubed term in them, so a challenge like the first question set above – 'Find me a number such that, when it is cubed, and the said number is added to this cube, the result is 5' – is really asking you to solve the equation $x^3 + x = 5$. These sorts of problems aren't easy to solve, and somewhere during the working out you're going to use loads of negative numbers. This was avoided in the 1500s by splitting cubic equations into a plethora of different subcategories.

Today we would consider things such as $x^3 + mx = n$ and $x^3 = mx + n$ to be the same cubic equation, because putting the mx on the other side is the same as having a negative value for m. For Fior and Tartaglia, though, they were completely different problems. In fact, Fior could only solve the $x^3 + mx = n$ type, so he set Tartaglia thirty of those. Tartaglia, however, had mastered all sorts of different methods, so he gave Fior a range of cubics to solve. Despite what is for us a trivial distinction, Fior couldn't solve the other types. Tartaglia raced through his equations in two hours, proving he could solve both the equations Fior had set him, as well as all the other types he had challenged Fior with. A decisive win for Tartaglia then.

The Reals are Really Real

Since the Negative Revolution, 0 and negative numbers have become so omnipresent in maths we call all the number line – positive, 0 and negative – the real numbers, or reals for short. We no longer question that something as simple as a square root can give you two different real answers, one positive and one negative. Because $3 \times 3 = 9$ and $-3 \times -3 = 9$, we accept that $\sqrt{9} = 3$ or -3. Or, in shorthand, $\sqrt{9} = \pm 3$. Likewise, $\sqrt{2} = \pm 1.41421...$ Both are equally valid answers. They are both real.

But, even now, calling them part of the reals still disturbs some people. Surely negative numbers are not as real as positive numbers? They must, somehow, be inferior to positive numbers when it comes to the real world. In the real world, you can have four ducks or four cups of tea, but you cannot have negative-four ducks or four negative cups of tea. If you could have negative-four ducks and you introduced them to five positive ducks, they would cancel out with a bang, leaving one solitary very confused duck.

However, physicists ignore negative numbers at their own peril. It turns out that a negative solution to an equation is not only mathematically lovely but can be as physically valid as a positive one. In the early 1900s, the subtlety of negative solutions to square roots caught physicists completely off guard.

In 1928, the British physicist Paul Dirac was trying to find a mathematical way to calculate the energy of an electron at any speed up to (but not including) the speed of light, and he succeeded with what is now known as the Dirac equation. But it involved squaring the charge of the electron – which means it not only has a positive solution, but a negative one as well. In the paper presenting his results, Dirac adds a passing comment addressing this, pointing out that 'The wave equation thus refers equally well to an electron with charge e as to one with charge –e' (where e is used by physicists to represent the charge of an electron, not the mathematical constant *e* we all know and love). Suddenly a pesky negative has appeared. But surely it was a quirk of maths and had no real physical significance . . .

$$\left[\left(ih\frac{\partial}{c\,\partial t}+\frac{e}{c}A_0\right)^2+\Sigma_r\left(-ih\frac{\partial}{\partial x_r}+\frac{e}{c}A_r\right)^2+m^2c^2\right]\psi=0$$

Wave equation from 'The Quantum Theory of the Electron'

Not only had an electron with opposite charge never been seen, its existence was preposterous. That aside, solutions for it were appearing in maths next to the normal solutions for the garden-variety electron. These solutions could not simply be ignored, though, as without them, the rest of the equation broke. Over the next few years, Dirac occasionally discussed possible ramifications of these other solutions to his equation, and in 1931 published a paper that dealt with the problem head on. Here he took the physical implications of the mathematical solutions seriously and analysed a particle he called the anti-electron. This was the first discussion of what eventually became known as antimatter.

And, it turns out, the mathematics was absolutely correct. Dirac's paper was published on 1 September 1931 and, less

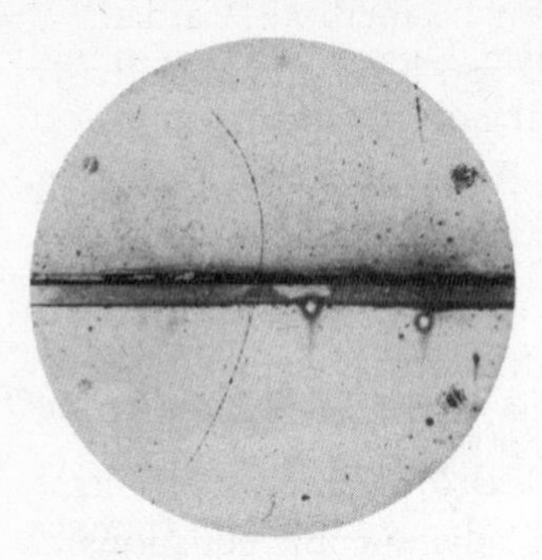

The first ever photo of an antielectron. The solid bar in the middle of the circle is a sheet of lead, and the thin curved line is an antielectron passing through it.

than a year later, on 2 August 1932, an anti-electron was spotted in the wild. It was discovered by American physicist Carl Anderson; he called it a positron, and was able to prove it existed by nothing less than taking a photo of it. Whereas modern physicists can build their own particle accelerators, Anderson had to build his particle detector and then wait patiently for it to be accidentally hit by one of the hugely energetic cosmic rays that bombard the Earth from space. Across the early 1930s, he took 1,300 photographic plates of cosmic rays hitting his detector, each producing a range of particles as they collided with the atoms therein. In fifteen of those photos the distinctive trail of a positive electron can be seen. A negative solution in a mathematical equation had accurately predicted the existence of a fundamental particle in the real world. The negative numbers are real and they are here to stay.

Thinking outside the Line

There was still one problem. Negative numbers were not the only new numbers that were appearing as sixteenth-century mathematics tried to solve more and more complicated equations. The mathematician Jerome Cardan (officially, Girolamo Cardano – he was Italian, after all) had a keen interest in Tartaglia's win over Fior, and invited him to come and visit so he could learn some of his methods. Cardan was of good maths stock (his father had been a maths advisor to Leonardo da Vinci), but he used his maths ability to make money from gambling, as well as from working as an unlicensed doctor. He managed to extract these new methods from a reluctant-to-tell Tartaglia and set about using them himself. Along the way, he noticed that something very strange was happening.

Negative numbers were the least of Cardan's problems. What he noticed was that while solving cubic equations he had to take the square root of a negative number. Namely, he came across a step in his working out which required the value $5 + \sqrt{-15}$. But this made absolutely no sense. There is no possible real number which can equal $\sqrt{-15}$, because all reals, positive or negative, always give a positive answer when squared. But then, later on, $5 + \sqrt{-15}$ was multiplied by $5 - \sqrt{-15}$ and so gave the answer 40.

This deeply disturbed mathematicians: these sorts of numbers should not exist, yet they were a vital step along the way to reaching answers which did exist. The square root of a negative number had the decency to go away again before the final solution to an equation, but what did it mean while it was there? Cardan described numbers he didn't understand as 'mental tortures'. What he didn't realize was that he was the first person to use a whole new type of number. Mathematicians have since embraced them, and we know them today as imaginary numbers. An imaginary number is written as i and has the unusual property that $i \times i = -1$. So now, $\sqrt{-1} = i$ and $\sqrt{-15} = i\sqrt{15}$.

Imaginary numbers had been taken on board by mathematicians pretty much by the early 1700s and, much like negative numbers, have proved themselves to be extremely useful, both in solving abstract mathematical equations and in practical applications. If you look back at Dirac's wave equation, you will see the number *i* hiding out in there. Paired up with normal real numbers, they give us what are called complex numbers, such as 4 + 2*i*. Modern fields such as quantum mechanics in physics and electrical engineering make extensive use of complex numbers, even though we do not know what an imaginary number actually is.

Once mathematicians realized how useful imaginary numbers could be, they needed a place to put them. The number line was already full with all the real numbers, and so they put them off to the side on a new number line at right angles to the original one. Because complex numbers have two parts, they can be thought of as two-dimensional numbers, and they each correspond to a point on a flat surface. What was once a 1D number line became a 2D complex plane. Every single complex number had its own home somewhere on that plane.

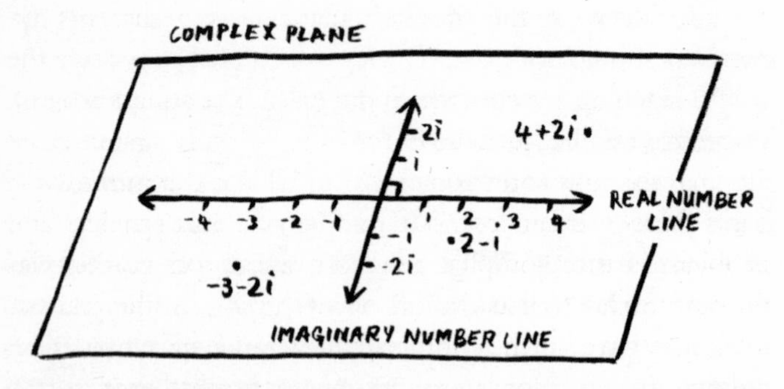

Expanding the 1D number line to become a 2D surface was a gift for mathematicians. Now they could go back and reinvestigate all the old maths, but replacing the old real numbers with modern complex numbers. Having long used the factorial function on regular numbers (such as

$5! = 5 \times 4 \times 3 \times 2 \times 1$), the race was now on to find an equivalent to factorials which worked for complex numbers. This was given the new name 'the gamma function' and had the upper-case gamma letter (Γ) as its symbol. In 1729, Euler did some initial work on the gamma function (as, possibly, did Daniel Bernoulli, son of Johann and nephew of Jacob) and Gauss tidied it up in the early 1800s. The resulting function still had the same recursive relationship that, for any complex number z, $\Gamma(z) = (z - 1) \times \Gamma(z - 1)$, only it could now start with any input you wanted.

Embedded within the gamma function is the factorial function, so it still gives the same results for whole numbers, only with the unfortunate offset that $\Gamma(n + 1) = n!$. Extending the factorial function now not only covered the complex plane but also filled in the gaps between the whole numbers. My favourite value is that $\Gamma(\frac{1}{2}) = \sqrt{\pi}$ (as always, π makes a surprise guest appearance). The values for $\Gamma(\frac{1}{3})$, $\Gamma(\frac{1}{4})$ and $\Gamma(\frac{1}{6})$ are also some of the previous few proven transcendental numbers.

When you plot a complex function on the complex plane, you end up with a surface covering it. The very first plots of this gamma function surface had to be done laboriously by hand. Now we can computer-generate them quickly. The problem is that complex functions can produce complex outputs. So, as well as the two values going in (the real and imaginary parts of the complex number), we have two values coming out. To plot it properly would require four axes: it would be a 4D plot. What I've shown you here is, once again, a 3D approximation of what should be a 4D display. We can safely assume that Hyperthetical would be much better at complex functions, as they could intuitively look at their fully detailed plots.

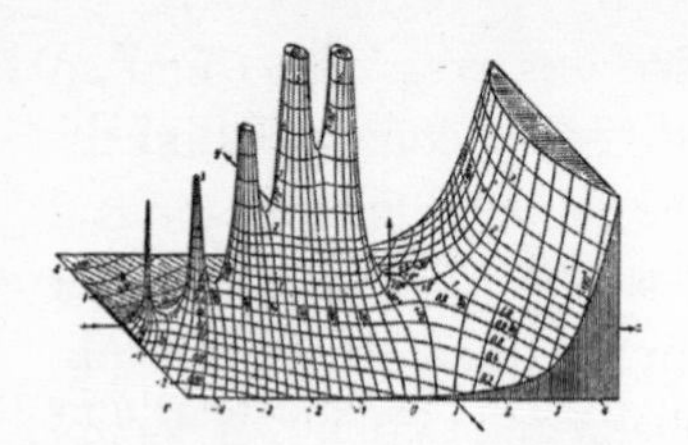
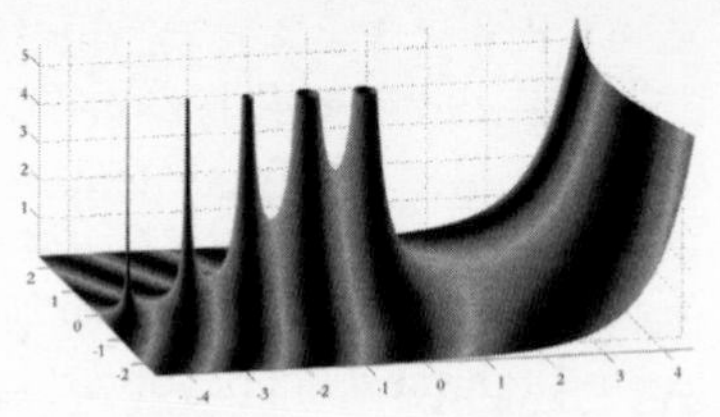

A hand-drawn sketch from 1909 of the gamma function, and a modern 3D plot.

If this looks familiar, it's because the Riemann zeta function is a complex function as well. Exactly as the gamma function is the complex expansion on the factorial function, the Riemann zeta function is the complex version of the Basel function we saw before (to revisit the Basel function, turn to page 298; to stay and fight the Riemann zeta function, turn to page 393). When I said that it takes two numbers as its inputs and returns two numbers as its outputs, I meant that it was a complex function. The Riemann zeta function is the Basel problem function generalized to include the complex numbers as well as missing parts of the real-number line. This is how it is able to sneak around to the negative numbers and extract values from them: it goes via the complex plane. If you come at $\zeta(-1)$ from the side, you can get Ramanujan's value of $1 + 2 + 3 + \ldots = {}^{-1}\!/_{12}$. Without complex numbers, all the breakthroughs that the Riemann Hypothesis has made possible would not be within our reach.

Given the power of complex numbers, it's an obvious question to ask if there are even stranger numbers further out. We can't continue the pattern by taking the square root of *i*, because all the roots of complex numbers are themselves complex numbers (the square root of *i* is $\frac{1}{\sqrt{2}} + \frac{i}{\sqrt{2}}$). It looks like, once we have expanded the reals to include the complex numbers, we have found all the numbers there are. But then the Irish mathematician William Hamilton (of graph-path fame) stepped up and showed the way forward.

Hamilton had already worked on understanding complex numbers as a kind of 2D number, and he had spent ages trying to find numbers beyond the complex. He expected them to be a sort of 3D equivalent, numbers made of three parts instead of two, but he met with absolutely zero success. Then, as he was out for a walk one day in 1843, inspiration hit. In his words, 'An electric circuit seemed to close, and a spark flashed forth.' Hamilton was so excited that he stopped mid-walk and wrote down this equation on the closest thing to hand – which happened to be a bridge. This may be the earliest ever occurrence of maths graffiti street art.

Hamilton realized that the next set of numbers would be 4D and consist of the real numbers, along with three other types of imaginary numbers: i, j and k. The imaginary number i is still exactly the same imaginary number as in complex numbers, only it's now joined by j and k. Both j and k have the same property (i.e. squaring them gives −1), but with the added property that all three of them multiplied together also equals −1. This means that if you multiply any two of them together, you get the third as the result: $i \times j = k$, and so on. This is the subtlety which makes them work and is why there is no 3D equivalent to complex numbers; having two different types of imaginary number merely replicates the same $i^2 = -1$ property, whereas having three gives the new $i \times j \times k = -1$ property.

$$i^2 = j^2 = k^2 = i \times j \times k = -1$$

×	1	i	j	k
1	1	i	j	k
i	i	-1	k	-j
j	j	-k	-1	i
k	k	j	-i	-1

It was this relationship between i, j and k which Hamilton wrote on Broome Bridge, just outside

Dublin. Once a year, mathematicians still make a pilgrimage to the bridge (a 'pilgrim-bridge') and write the equation back on with chalk. Should you be there on any other day, there is still a plaque on the bridge commemorating Hamilton's discovery of what he called quaternions. These quaternions have since settled happily into maths next to the complex numbers, and have even found some practical applications of their own: they are used in modern computer graphics calculations.

That said, the quaternions were not immediately embraced by mathematicians. Lord Kelvin admitted they were ingenious but described them as 'an unmixed evil to those who have touched them in any way'. They definitely upset Oxford mathematician Charles Dodgson. As he contributed nothing of note to mathematics, we would not be discussing Dodgson today if he hadn't also written fiction under the pen-name Lewis Carroll. Few people realize that Lewis Carroll's day-job was as a mathematician, including people at the time, among them Queen Victoria. Legend says she enjoyed *Alice's Adventures in Wonderland* so much she allowed him to dedicate the next book to her, and was rather miffed when it arrived. It was a maths text book entitled 'An Elementary Treatise on Determinants'.

As a mathematician, Carroll was extremely conservative, and he disliked modern mathematical concepts such as non-Euclidean geometry and complex numbers. It has since been suggested that he ridiculed these absurd concepts via the silliness in *Alice's Adventures in Wonderland.* Victorian literature and maths expert Melanie Bayley (also of Oxford University) has suggested that the three participants in the Mad Hatter's tea party are meant to directly satirize the i, j and k components of Hamilton's quaternions. With the Mad Hatter making non-commutative statements such as 'Why, you might just as well say that "I see what I eat" is the same thing as "I eat what I see"!' you can certainly see why. But, true or not, it certainly represented the mood at the time. Many mathematicians saw this increasing abstraction of maths beyond physical reality as a Cheshire Cat disappearing from the real world, leaving behind only a grin.

x	1	i	j	k	m	n	p	q
1	1	i	j	k	m	n	p	q
i	i	-1	m	q	-j	p	-n	-k
j	j	-m	-1	n	i	-k	q	-p
k	k	-q	-n	-1	p	j	-m	i
m	m	j	-i	-p	-1	q	k	-n
n	n	-p	k	-j	-q	-1	i	m
p	p	n	-q	m	-k	-i	-1	j
q	q	k	p	-i	n	-m	-j	-i

Like quaternions, order of multiplication of octonions changes the sign of the answer! $j \times m = i$, *whereas* $m \times j = -i$

It may now seem that we do run a very real risk of disappearing down the rabbit hole of more and more imaginary numbers, but it turns out they come to an abrupt halt. After the four-dimensional quaternions come the eight-dimensional octonions, and then that's it. No more. The only groups of numbers which work this way are the reals, the complex numbers, the quaternions and the octonions. All the imaginary types of numbers have now been discovered, even if we still find them strange and confusing.

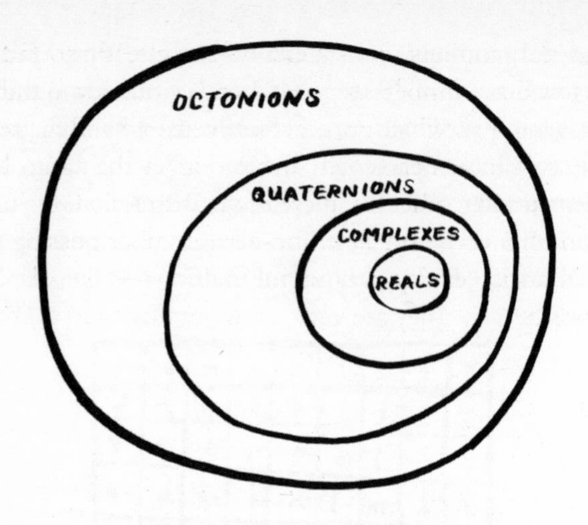

A Surreal Addendum

I didn't want to give the impression that we now know all there is to know about types of numbers. As always, maths is never complete and, as recently as 1969, the mathematician John Conway gave us a whole new type of number, the surreal number. The surreal numbers have the unusual distinction of being possibly the only piece of new mathematics which was first published in a work of fiction. Conway did not write an academic maths paper on surreal numbers; rather, they first appeared in a mini-novel written by computer scientist Donald Knuth entitled *Surreal Numbers: How Two Ex-Students Turned on to Pure Mathematics and Found Total Happiness.*

As opposed to the real numbers, which we discovered by starting with counting whole numbers then looking between them to find the rationals and the irrationals, the surreal numbers start by making sets of numbers and looking at which numbers appear as the middle point between two sets of other surreal numbers. Not only does this then give you all the real numbers with which we are already so familiar, but a few new numbers appear. My favourite new number is a surreal number which appears between zero and the rest of the reals. Which means it is slightly bigger than zero but is smaller than any other number. Named after the Greek letter epsilon (ε) it is the smallest non-zero number possible. The mathematician Erdős was in the habit of calling children epsilons because they are very small but non-zero humans.

Eighteen

TO INFINITY AND BEYOND

Throughout this book, the concept of infinity has been mentioned in hushed tones. In almost every other chapter, there has been a mention of something happening 'infinitely' or the answer to some sum being 'infinitely large', but no further attention has been paid to it. This is because infinity is like a loose thread on the coat of mathematics. If you start pulling on it, not only is it longer than expected, but everything else starts to unravel and suddenly you're standing there, mathematically naked, wishing you could push that thread back in and continue happily ignoring it.

But not us. We're going to grab that thread. It is time to meet the shadow that has haunted this book: infinity. We're finally ready to go to infinity and, indeed, beyond.

Making and doing things infinitely can get a bit tricky, but we'll give it a go. Find yourself a box and a lot of balls and number the balls, starting 1, 2, 3, 4 . . . You get the picture. The game is to put the balls in the box one at a time, starting with ball number 1, but, whenever you put in a square-number ball, you take out its square root and put it away in a drawer or somewhere safe. This means that the first move is a bit odd, because when you put in ball 1, 1 is its own square root, so you take it straight back out. Then you can put in balls 2 and 3. When ball 4 goes in, you take ball 2 back out and put it in the drawer. Then balls 5 through to 8 go in, before adding ball 9 and taking out ball 3. The question is: if you simply keep doing this, which balls will end up in the box? How many are safe in the drawer?

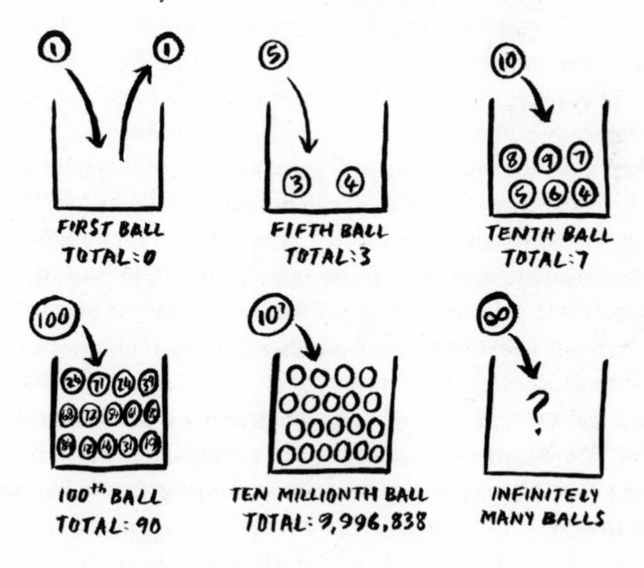

The surprise is that the box ends up completely empty. Now, this doesn't make sense, because, as you're going through the process, the number of balls in the box always goes up. At each move, you either add a ball, or you add one and take one away at the same time: the number of balls either increases or stays the same. But all the balls end up in the drawer, not the box. So when are all the balls dumped back out of the box?

We know all the balls end up in the drawer because every number can be squared. For each ball you put into the box, there is another ball out there which is the square of its number, and that square-number ball has to be added sooner or later. So, every ball you put in will eventually be taken out, in theory leaving the box empty. There is no number so big that it cannot be squared. But yet, at every step, there are always more balls in the box than in the drawer.

This paradox – that there are always more balls in the box, but yet they end up in the drawer – is caused by our expectation that infinity behaves like a really, really big number. But infinity is not a number. Infinity is nowhere to be found on the number line. People seem to have the idea that if you keep counting up along the number line past bigger and bigger numbers, in the end, the numbers just give up. They can't be bothered to go on, and there'll be an infinity sign (∞) there to mark the end of the number line. This is not the case. There is *always* a bigger number. Infinity is not safely contained at the end of the number line. Infinity is not a big number.

Infinity is actually a measure of how many numbers there are. The number line never stops, so we say it is infinitely long. Just like we said before that the number 'five' is the size of any set of five things, infinity is the size of a never-ending set of things. Imagine the set of whole numbers; that's how big infinity is. That's when the balls get dumped out of the box: when you go from moving up the never-ending number line to looking at the whole number line at once, when you go from thinking about a very big number to thinking about all the numbers.

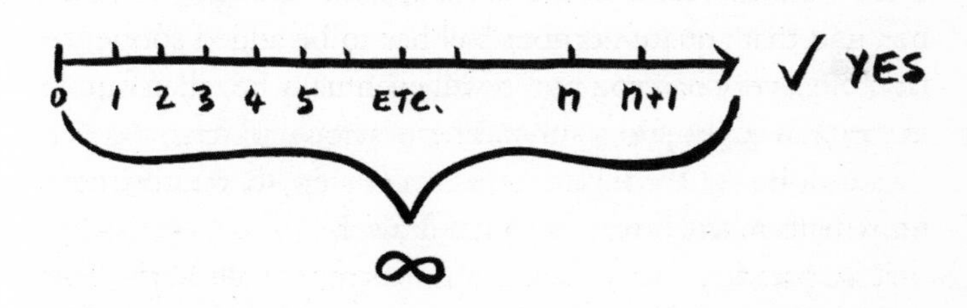

There are infinitely many whole numbers, and every single one of them is the square root of a whole number. Instead of imagining putting the balls into the box and the drawer, you could imagine doing it all in one step. All the balls that are square roots go straight into the drawer, and all the balls that are not go in the box. Not a single ball would be put in the box.

Our big problem is that infinity doesn't make any sense as far as our intuition is concerned, and there are no footholds. Maybe our brain doesn't *like* dealing with higher dimensions, but at least we're OK dealing with lower dimensions. The human brain is happy working with nice finite numbers. As soon as you start working with infinity, things start happening that are completely counter-intuitive. Finite numbers are not a preparation for infinity in the same way that 3D shapes readied us for 4D shapes. If anything, our affinity with the finite gives us a false sense of security when we encounter the infinite. We have to let go of all intuition; otherwise, it will lead us astray. Mathematical logic can be our only guide in a world that no longer has any link to our physical reality.

Doing maths without intuition is like taking a submarine below the surface of an ocean. Back in the maths jungle, our intuition allowed us to look around and make sense of our surroundings. Back then, enough of the maths directly linked to the world around us. Descending into the deep, however, means we can no longer see where we are going. We have left our physical universe and plunged into a world of pure abstract thought. On a submarine, you have to rely solely on the readouts of the instruments, and, likewise, we now have only mathematical results to guide us. But if we are fastidious and check every step, and trust the conclusions we deduce from what our mathematical instruments are telling us, all will be well.

Here we go.

Infinity Upsets People

> I caught the sound of a man airing the preposterous notion that the sum of all primes approaches infinity.
>
> – complaint to BBC Radio 4

Infinity can make people angry. It seems that many people would like to leave the infinite thread where it is and not make a fuss. They want 'infinity' to stay as a nebulous concept meaning 'something very, very big'; a vague term to represent the biggest thing possible. When I went on BBC

Radio 4's programme *More or Less* and discussed infinity, the BBC received letters of complaint. And this anger is nothing new.

Infinity earned its rightful place as a legitimate mathematical concept towards the end of the nineteenth century. The two big players behind this were German mathematicians Georg Cantor and his champion David Hilbert. No longer just accepting infinity as a general notion but actually investigating it rigorously did not go down well. Contemporary mathematicians described Cantor as a 'corrupter of youth'. This had a detrimental effect on Cantor, who already suffered from depression. Thankfully, Hilbert saw the power of what Cantor had done. He described Cantor's work as 'the finest product of mathematical genius and one of the supreme achievements of purely intellectual human activity', and famously said, 'no one shall expel us from the Paradise that Cantor has created'.

Cantor moved from Russia to Germany with his family when he was eleven, and later studied mathematics in Zurich and Berlin before – after a quick stint as a schoolteacher – settling down as a mathematician at a university in Halle in 1869. During the 1870s and 1880s, he wrote a series of papers which laid the foundations of our current understanding of infinity. Hilbert was, slightly, Cantor's junior. He joined the University of Göttingen in 1895, having been given the job by none other than Felix Klein (of 4D bottle fame). He stayed there until his retirement in 1930, which was just long enough for him to have possibly passed a young Herbert Seifert in the corridors (no doubt lost in thoughts of knot surfaces).

Flying in the face of what most mathematicians believed at the time, Cantor showed that it was possible to measure infinity through the use of sets. He proved that there's actually more than one infinity, and that some infinities are bigger than others. Amidst the outrage from other mathematicians, Hilbert saw the power of what Cantor had done. To this day, the best way to understand the advances Cantor made in studying infinity is something called the Hilbert Hotel. It is an infinite hotel, and it contains some unexpected surprises.

Is There Always Room at the Inn-finite?

Ideally imagined as an infinitely long corridor, the Hilbert Hotel starts with Room 1, then carries on through Rooms 2, 3, 4 . . . and so on, through all the infinite whole numbers. The hotel is completely empty, then an infinite coach arrives full of infinitely many people. These are wearing badges labelled 'Guest 1', Guest 2', and infinitely so on; there is one guest for each of the infinitely many whole numbers. The concierge comes out to meet them and direct them to their rooms. His job is easy: he holds up a sign saying 'Go to the same number room as your guest number.' The infinite hotel is now full.

Our challenge is to try to outwit the concierge and find a group of guests which cannot be accommodated. If we can find a set of people bigger than the Hilbert Hotel, then they must represent a bigger infinity. Matching items in two sets like this is the basis of all counting. If you have a collection of coins on the table in front of you and you want to know how many there are, you can use your fingers as a pre-known set to work out how big the set of coins is. If you start putting your fingers on the coins so that you have one finger on each coin, and one coin under each finger, you know there are ten coins (or as many coins as however many fingers you may have). If you have fingers left over, then you know your set of fingers is bigger than the set of coins. What we are doing with the Hilbert Hotel is finding a set of guests which cannot be matched up with the infinite set of rooms. If all the rooms are full and some guests are left over, we know the guests must form a bigger infinite set.

We'll start easy and just have one extra guest show up late to Hilbert's Hotel. This is no problem for the concierge. As long as the infinitely many residents don't mind a bit of moving around, then the person who is in Room 1 could pack up their stuff and move down to Room 2. Its current occupant vacates Room 2 and moves into Room 3, displacing that person into Room 4. In one move, all the infinitely many guests can walk out of their rooms and into the next one up, leaving Room 1 empty for the new guest. This can be repeated for any other guests who arrive after the hotel is already full, without it ever becoming 'more full'. An infinite set plus any finite set still gives the same infinity.

Now a second infinite bus pulls up outside the already-full hotel and an infinite group of people get off the bus looking for a place to stay. The concierge is briefly flustered. He can't send them to the same room numbers as their guest numbers: those rooms are already full. Neither can he repeat the previous method for adding a finite number of new guests, as there are infinitely many new guests, so the process will never end. He needs one quick and effective way to get them all into the hotel at once.

Thankfully, the concierge is up to the task. He suggests that each one of the always-happy-to-move guests in the hotel could leave their current rooms and relocate to the room with a number twice as big. So the person in Room 1 walks to Room 2 at exactly the same time the occupant from Room 2 is heading to Room 4; the guest in Room 3 is aiming for Room 6, waving to the occupant from Room 4, who is heading to Room 8. All the infinitely many guests leave their rooms, and enter only the even-numbered rooms, leaving all the infinitely many odd-numbered rooms for the new guests. Again, we can repeat this as many times as we want: the hotel never becomes more full. Finitely many copies of an infinite set gives the same infinity.

Now the bus we have been waiting for arrives. It's a massive infinite bus containing all the rationals, which we met in the last chapter. Every guest has a badge with a different fraction on it, and they all want to stay in the Hilbert Hotel. We'll give the concierge a fighting chance and start with a completely empty hotel. It feels like this should be an impossible task: there are more rational numbers than whole numbers. In a very real sense, the rationals are more 'dense' than the whole numbers. Between any two rational numbers, you can always find another rational. You can take any two fractions, average them to get a new fraction between them, and repeat this endlessly. Whole numbers, on the other hand, are sparse; there are no more whole numbers between 6 and 7. As you zoom in on the number line, the dense rationals go on for ever; the sparse whole numbers thin out.

It turns out, however, there is a way of solving this problem. The concierge wins again. As the guests get off the bus they see that the concierge has made an infinite table showing each of them where their room is. He did this by making an infinite grid with the whole numbers counting along the horizontal and up the vertical directions. Each entry in the table is then the rational number made by using the vertical coordinate as the numerator (the number on top of the fraction) and the horizontal coordinate as the denominator (the number on the bottom). Every single rational number must be on there, as every single guest can find their numerator and denominator and look themselves up in the table. The concierge has started in the bottom-left corner of this grid and numbered the entries systematically, zigzagging back and forth. Every single entry will be given its own unique whole number and this will be the room number for the corresponding rational guest.

Once again, all the infinitely many guests disembark the bus simultaneously, look at the grid together to spot their room number, and then head into the hotel to get comfortable. Using this method, it's possible to pair up the infinite set of rationals (all possible fractions) perfectly with the infinite set of integers (all possible whole numbers). They

17	17/1	17/2	17/3	17/4	17/5	17/6	17/7	17/8	17/9	17/10	17/11	17/12	17/13	17/14	17/15	17/16	17/17
16	16/1	16/2	16/3	16/4	16/5	16/6	16/7	16/8	16/9	16/10	16/11	16/12	16/13	16/14	16/15	16/16	16/17
15	15/1	15/2	15/3	15/4	15/5	15/6	15/7	15/8	15/9	15/10	15/11	15/12	15/13	15/14	15/15	15/16	15/17
14	14/1	14/2	14/3	14/4	14/5	14/6	14/7	14/8	14/9	14/10	14/11	14/12	14/13	14/14	14/15	14/16	14/17
13	13/1	13/2	13/3	13/4	13/5	13/6	13/7	13/8	13/9	13/10	13/11	13/12	13/13	13/14	13/15	13/16	13/17
12	12/1	12/2	12/3	12/4	12/5	12/6	12/7	12/8	12/9	12/10	12/11	12/12	12/13	12/14	12/15	12/16	12/17
11	11/1	11/2	11/3	11/4	11/5	11/6	11/7	11/8	11/9	11/10	11/11	11/12	11/13	11/14	11/15	11/16	11/17
10	10/1	10/2	10/3	10/4	10/5	10/6	10/7	10/8	10/9	10/10	10/11	10/12	10/13	10/14	10/15	10/16	10/17
9	9/1	9/2	9/3	9/4	9/5	9/6	9/7	9/8	9/9	9/10	9/11	9/12	9/13	9/14	9/15	9/16	9/17
8	8/1	8/2	8/3	8/4	8/5	8/6	8/7	8/8	8/9	8/10	8/11	8/12	8/13	8/14	8/15	8/16	8/17
7	7/1	7/2	7/3	7/4	7/5	7/6	7/7	7/8	7/9	7/10	7/11	7/12	7/13	7/14	7/15	7/16	7/17
6	6/1	6/2	6/3	6/4	6/5	6/6	6/7	6/8	6/9	6/10	6/11	6/12	6/13	6/14	6/15	6/16	6/17
5	5/1	5/2	5/3	5/4	5/5	5/6	5/7	5/8	5/9	5/10	5/11	5/12	5/13	5/14	5/15	5/16	5/17
4	4/1	4/2	4/3	4/4	4/5	4/6	4/7	4/8	4/9	4/10	4/11	4/12	4/13	4/14	4/15	4/16	4/17
3	3/1	3/2	3/3	3/4	3/5	3/6	3/7	3/8	3/9	3/10	3/11	3/12	3/13	3/14	3/15	3/16	3/17
2	2/1	2/2	2/3	2/4	2/5	2/6	2/7	2/8	2/9	2/10	2/11	2/12	2/13	2/14	2/15	2/16	2/17
1	1/1	1/2	1/3	1/4	1/5	1/6	1/7	1/8	1/9	1/10	1/11	1/12	1/13	1/14	1/15	1/16	1/17
	1	2	3	4	5	6	7	8	9	10	11	12	13	14	15	16	17

Continuing this grid will list every possible rational number.

They can then be systematically ordered, and easily numbered.

are both the same-size infinity, regardless of what our intuition told us 'must' be true. This was what Cantor first proved in 1873.

This method is a bit messy, though. The problem is that this grid with all of the rational numbers at its intersections has a lot of repetition to it. The person whose badge says 'Guest ½' will read their number off from the intersection 1 up and 2 across, but the room allocated to 2 up and 4 across will go empty because $2/4$ is the same as ½ and that guest already has a room. While this does fit all the rationals into the hotel, it leaves a lot of rooms empty. It would be nice to have a way to systematically match the rational numbers perfectly to the integers. For that, we're going to need an all-new number sequence.

Try writing out the Fibonacci sequence again, but this time with a slight change. Instead of adding pairs of numbers to provide the next number on the list, alternate with also copying just one number to the end. That is, instead of always adding two numbers in the sequence and placing the result at the end of the line, every other time, simply move a single figure. This sequence will start with 1, 1, like the Fibonacci sequence, but then it becomes wildly different.

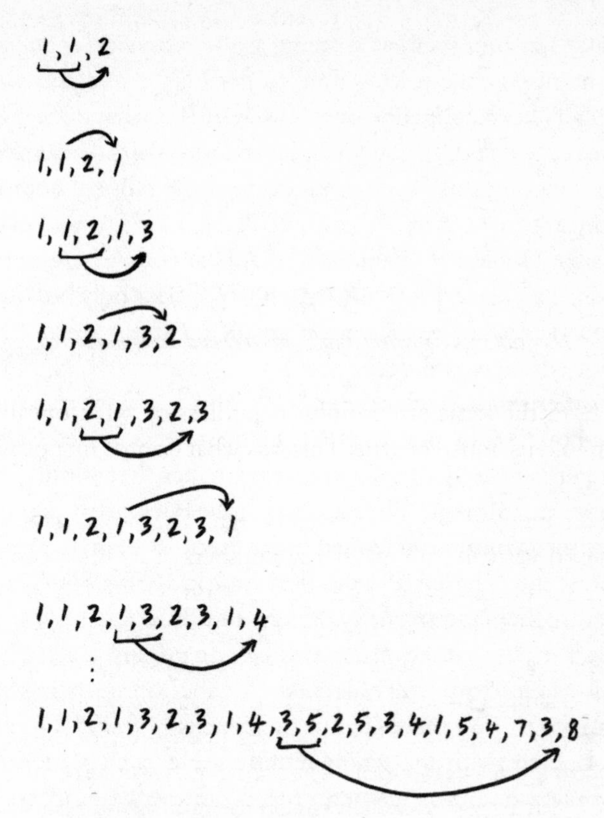

Alternating between copying one number or summing two.

The sequence moves forward with a kind of syncopated beat, and grows faster than your working-out moves along. As you are working your way along the sequence, the end of the sequence is racing off away from you!

1, 1, 2, 1, 3, 2, 3, 1, 4, 3, 5, 2, 5, 3, 4, 1, 5,
4, 7, 3, 8, 5, 7, 2, 7, 5, 8, 3, 7, 4, 5, 9, 4, 11, 7 . . .

The Stern–Brocot sequence.

This may not seem as amazing as the Fibonacci sequence: the numbers take a long time to get bigger and the same numbers keep appearing over and over. But what does make it amazing is that, as you go along and turn pairs of numbers into fractions, this list contains every single rational number. Those fractions start $1/1$, $1/2$, $2/1$, $1/3$, $3/2$, $2/3$, $3/1$, $1/4$ . . . and every possible fraction is guaranteed to appear somewhere in this infinite sequence. Not only that, it will appear only once and it's guaranteed to be in its most simple form (e.g. once $1/2$ has appeared, there will never be a fraction such as $2/4$, $3/6$, $10/20$ or any other fraction which equals $1/2$). This sequence was published in 2000 by the American mathematicians Neil Calkin and Herbert Wilf in their four-page paper 'Recounting the Rationals'. (More in 'The Answers at the Back of the Book'.) They magnanimously named it the Stern–Brocot sequence, because it was based on work first done in 1858–1860 by two other mathematicians named Stern and Brocot.

So, the rationals are also the same-size infinity. Even when a bus arrives with every algebraic number on it, there's still room for those guests. Cantor the concierge found a way to match them with the whole numbers. He knew that every algebraic number is a solution to a finite equation,* so he found a way to list every possible finite equation by its height (he developed a function to combine the powers and coefficients of any such equation and produce a whole number, which he called the height). While one equation may have more than one solution, like children in a family they can be ranked from oldest to youngest. This means, as every algebraic-numbered guest gets off the infinite bus, they can find their parent equation by its height, and their position

* I should specify 'polynomial equation with rational coefficients' before some pedant points out a finite equation like $x - \pi = 0$.

among their sibling solutions gives them their own whole-numbered room.

It looks like every infinite set we can think of still fits in the hotel; they are all the same-size infinity. More or less, this was the working relationship mathematicians had with infinity up until the 1870s. They were aware that some things were infinite, and accepted that infinity kind of existed as a thing in its own right, but all infinite things were equal. Anything which went on for ever was the same-size infinity. But then Cantor found a busful of guests that was too big to fit in the Hilbert Hotel. He was the first person to find something bigger than infinity.

We're Going to Need a Bigger Bus

For me, the discovery that there are different-sized infinities is one of the most fabulous pieces of mathematics. Our brains cannot even really conceive of what infinity is, let alone differentiate it from the concept of a bigger infinity. Yet Cantor managed to prove that such things exist. Instead of lumping all things which continue for ever as the same infinity, it turns out that some of them go on for ever at different rates.

On the horizon, a bus comes rumbling towards the Hilbert Hotel: a bus with too many people in it each to have a room. It pulls up, the doors open and, inside, are the real numbers. And not even all the reals, only the real numbers from 0 to 1. Right at the front of the bus is Guest 0, minding their own business, and then on the back seat is Guest 1. Between them is every possible decimal number. Along with rational and algebraic guests, somewhere in this bus is a guest with any possible string of decimal places you can imagine.

Cantor managed to prove that any possible way you assign these guests to the whole-numbered rooms in the hotel will leave at least one guest without a room. And if there is no way to pair up two sets without having anything left over, then one set of things must be bigger than the other. Amazingly, Cantor didn't show this only for various clever pairings that he could think of, he proved it for every single pairing possible. It's not that humans are merely not clever enough to find a way to get the reals into the hotel, there's simply no such possible solution.

After this bus has arrived, a member of the reception staff comes running out of the Hilbert Hotel. Horrified at the thought of losing business, they claim to have a way to systematically fit the reals into the hotel: that is, they claim they can produce a list of all of the real numbers and match each one up with its own whole number. But Cantor the concierge knows he can take any such list imaginable and always find at least one real number which has been missed off the list. No matter what list the receptionist claims to have, there will always be some guests missed off it.

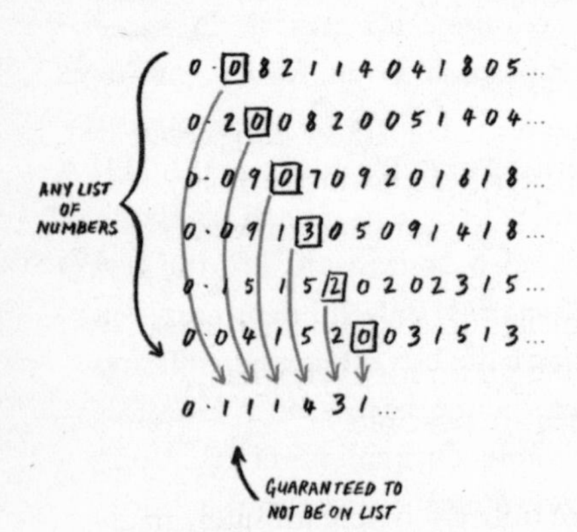

To be systematic, we can increase each digit by 1, with a 9 becoming a 0

What Cantor does is look at whichever guest is in Room 1 at the top of the list, and give his 'missing guest' a digit at the first decimal place that is different to the guest in Room 1. Then he makes the second digit of the missing guest different to the second digit of the guest in Room 2, and so on. When he's finished, for any list of guests he will now have a missing guest who is not on the list. For any room in the Hilbert Hotel, room *n*, the missing guest will have a different digit at the *n*th decimal place compared to the guest who was assigned that room. The infinite hotel is infinitely full, yet there is definitely a missing guest with no room. The set of the reals is a bigger infinity than the set of the whole numbers.

With more than one infinity on the scene, we now need different names. The old infinity we know and love, the infinity of the whole numbers, is known as countably infinite, because you can count off all the items in a countably infinite set using the whole numbers. The infinity of the real numbers is uncountably infinite. When something is described as being 'countless', then, technically, it's being described as the bigger infinity of the reals. Because this uncountable infinity applies to the continuous stretch of reals between 0 and 1 (whereas whole numbers, rationals and algebraic numbers all have gaps between them, the reals do not), it is thought of as the continuous infinity.

Also, with more than one infinity available, it means that whenever we call something 'infinite', we need to ask: but how infinite is it? Here is one new infinite situation and two old favourites:

- There are infinitely many ways to cut a pizza into pieces so they do not all touch the centre, but which infinity?
- What infinity of different shapes of constant width are there?
- A game of chess is normally forced to end if it goes on too long. But if it could go on for ever, what infinity of different games are possible?

We can show that there are only countably infinitely many ways to divide a pizza up into pieces which do not all touch the centre.* As we saw, cutting a pizza starts with an odd-sided regular shape of constant width, there is a countably infinite supply of them (one for every odd whole number) and each can be split only into a whole number of pieces. Every step may have infinitely many options, but they are whole-numbered, discrete options. So the result is countably infinite.

However, there are uncountably many possible shapes of constant width. There is only the one regular equilateral triangle, but there are infinitely many triangles if they do not have to be regular. You can change a triangle by changing the length of one of its sides, and for this there is a continuous range of possibilities. A side of a triangle could be any real number between, say, 1 and 2 units long, which is uncountably infinitely many. If something can be linked to a continuous set of options, we know there are uncountably infinitely many of them.

The number of possible chess games makes things even more interesting, because for each move there are only a finite number of pieces you can move, and they can only go to a set number of new positions on the board. It feels like the range of possible games is countably infinite. It seems like there are no continuous options to link a chess game to; all the positions are discrete. But we can link the moves directly to the real numbers. Imagine a game in which one player has a single queen, which means that at each move they can choose between putting it on a black or a white square. This is the same as choosing between a 1 and a 0. So the player can choose any real number, convert it into binary and move their piece according to those digits. Because this now means

* Update: Since the first edition of this book I have been shown a new method by mathematician Stephen Worsley, in which a 'wedge' attached to each piece can be varied continuously in size between 0 and 1. This gives uncountably many solutions!

that every real number will give you a different possible chess game, there are uncountably infinitely many of them. No wonder they have a limit on the number of moves allowed.*

The Fallout from Infinity

Part of what disturbed other mathematicians was that Cantor did not discover that there are two different infinities, but rather that there are many more. This whole new world of different-sized infinities was named using the Hebrew letter aleph (ℵ). Aleph-null, also known as aleph-zero, is the smallest of the infinities and is the size of the set of whole numbers. It is written as the aleph symbol with a subscript zero: $\aleph_0$. The next-biggest infinity is $\aleph_1$, followed by $\aleph_2$, and so on for bigger infinities. This goes on for infinitely many different-sized infinities. No wonder mathematicians had difficulty swallowing them all.

There was also one big unknown: what was $\aleph_1$? Cantor had shown that the uncountable infinity of a continuous line was bigger than the smallest infinity of the countable numbers $\aleph_0$, but he couldn't prove that it was definitely the next one up. The uncountable infinity of the reals may be $\aleph_1$, or there could be a different infinity, one no one had found yet, which was smaller than the continuous infinity but bigger than the countable infinity $\aleph_0$. Cantor's gut feeling was that the real numbers did form the next infinity, $\aleph_1$, and this became known as the Continuum Hypothesis. Try as they

* The World Chess Federation (FIDE) regulations 9.6b forces a draw on a game if their are 'any consecutive series of 75 moves that have been completed by each player without the movement of any pawn and without any capture.' The resulting movement forward of pawns, and capture of pieces, will spell a finite end to any game.

might, mathematicians could not prove or disprove the Continuum Hypothesis.

When David Hilbert gave his list of maths questions that needed to be solved in 1900, while the Riemann Hypothesis was down at number eight, the number-one problem was the Continuum Hypothesis. It was top of his list. But, unlike the Riemann Hypothesis, the Continuum Hypothesis was resolved during the twentieth century, but not in a way that mathematicians liked. It wasn't so much solved as it was destroyed. And it all started with the second problem on Hilbert's list.

Hilbert was a big fan of axioms (the statements mathematicians assume to be true). Ever since Euclid wrote *The Elements* and began by listing his starting axiomatic assumptions and proving everything else from there, mathematicians have been obsessed with axioms. Hilbert's big contribution to mathematics was to take the assumptions of Euclid, as well as the constraints of compass and straight-edge constructions, and find the complete set of twenty axioms from which all geometry follows. It was the first time anyone had been that explicit about how axioms are used in mathematics, and Hilbert is often said to have had the greatest influence in geometry after Euclid. The number-two problem on the list was to see if anyone could take what he had done for geometry and expand it to all maths. He wanted to know if it was possible to find a complete set of axioms upon which all mathematics could safely rest.

The first two questions on Hilbert's list would soon be answered by a mathematician named Kurt Gödel, but not in a way other mathematicians appreciated. Gödel, born in 1906 (so just a few years after Hilbert had issued his list), was an Austrian mathematician who, like Cantor, suffered from depression. He met his wife in a nightclub and they fled to the US in 1940, travelling via Russia and Japan, after Hitler had subsumed Austria into Germany in 1938. Having visited Princeton previously, upon arriving in the US Gödel proceeded to live and work there (becoming a close friend of an obscure fellow refugee named Albert Einstein) until his death in 1978. However, it was his work at the University of Vienna in 1931 for which he is famous. Only two years after he completed his studies, and aged only twenty-five, he published his infamous incompleteness theorems. Mathematics was never the same again.

In a single paper, Gödel proved that mathematics would never be complete. His first incompleteness theorem proved that for any set of axioms you could devise which is useful (namely, includes basic arithmetic), there will always be something you cannot prove with them.* No matter how good your axioms are, there will be some theorem which cannot be proved true or false. This can be fixed by adding an extra axiom to clear the matter up, but then there will be yet another theorem which is slightly beyond the reach of mathematics. It turns out that mathematics can never prove everything.

This is exactly what the problem was with the Continuum Hypothesis. At the beginning of the twentieth century, mathematicians had settled on a collection of nine axioms for dealing with infinities, collectively called ZFC – short for Zermelo–Fraenkel set theory with Choice. (See 'The Answers at the Back of the Book'.) But the Continuum Hypothesis was one of those theorems which cannot be proved either way. It was undecidable within the axioms of ZFC. It's the same problem Euclid had with the parallel lines postulate: it isn't possible to prove that parallel lines either exist or don't

* Gödel's second incompleteness theorem was that any consistent set of axioms cannot prove itself to be complete without an extra axiom being added.

exist from the other axioms. You either assume parallel lines exist and get one type of geometry, or assume they don't and get a different geometry. Both options are equally valid.

Mathematics had a choice: it could either assume that the reals form $\aleph_1$ or it could assume that they don't. They had a choice as to what the next infinity after the countable numbers is, determined by what new axiom they would like to add to ZFC. The problem is that axioms are supposed to be things which are self-evidently true, and there is no obvious option here. The debate over a tenth axiom still rages to this day. If you like your Continuum Hypothesis false, then there are things called forcing axioms, which basically force it to be wrong. There is a working candidate called Martin's axiom, which was proposed in 1970. If you want the Continuum Hypothesis to be true, you need to find what is called an inner model axiom to make this happen. These are harder to come by, but it's conjectured that an axiom going by the name of Ultimate L will do the trick. The big question is: which one do we want? Either choice of axiom will produce fruitful, but different, results.

This is the 'Paradise' which Hilbert proclaimed Cantor had created: a paradise where there are an infinite number of different-sized infinities and the question of 'which is the second smallest?' is undecidable: you can have it either way. You can see why contemporary mathematicians considered it a nightmare. As a subject founded on proof and certainty, being cut adrift like this left mathematicians feeling seasick. But we have since acclimatized to the uncertain world of the transfinite and, excitingly, it continues to be an active area of mathematical research (and debate) to this very day. Like a new island solidifying out of lava, we are lucky still to be able to see a new area of maths seething and taking shape.

n + 1

THE SUBSEQUENT CHAPTER

We're here. At the end of the book. Starting with games based on whole numbers and simple shapes, we've explored new worlds, from knots to graphs, and ventured thousands of dimensions beyond our own reality. We've seen how some of this maths is used to make our modern technology possible, and even how some of these modern computers are proving mathematical theorems for us. We reached and surpassed infinity and eventually discovered that, no matter how much mathematics humans discover, there will always be a few proofs just out of reach.

I, for one, find Gödel's incompleteness theorems rather comforting. It means that mathematicians will never be complete. There will always be something else which is undecidable with the current axions. Should the human species survive another few million years and continue churning out mathematics at the rate we've done for the past few thousand years, we still won't have covered it all. There will always be work for all of the future mathematicians. As always, some of that work will go on to be incredibly useful for the rest of civilization, and much of it will remain the pointless but endlessly amusing plaything of academics.

There's an unexplained mystery behind all of this, which I've been delicately avoiding throughout the book. If maths is the consequence of games and puzzles, the result of pure intellectual thought, why does it end up being so practically useful? I keep promoting maths as a bit of fun, yet no one can ignore that mathematical techniques are the workhorse of modern technology. In reality, mathematics is a serious industrial endeavour. There's a tension between what I claim to be the origin of maths and where it ends up being used.

And the truth of the matter is: we don't know.

In 1980, Richard Hamming, of code fame, wrote a paper called 'The Unreasonable Effectiveness of Mathematics'. It was based on a 1960 paper of almost the same name by the Hungarian Nobel Prize-winning physicist and mathematician Eugene Wigner. They were both highly pragmatic mathematicians who did 'applied' maths deliberately, and their papers argued that there is no obvious reason why mathematics, as a subject born out of pure human thought, should be so useful in the real world. The phrase 'unreasonable effectiveness' has since come to represent this enigma.

And it's not only basic mathematics that seems to match the reality around us; even highly abstract concepts raise their head. Scientists have always required mathematics to describe their theories, and as those theories have become more complicated, more elaborate and more advanced mathematical ideas have been required. However, in the early twentieth century, something changed fundamentally. The maths required was becoming more abstract and bizarre at an amazing rate, far beyond what anyone had expected.

> Non-Euclidean geometry and non-commutative algebra, which were at one time considered to be purely fictions of the mind and pastimes for logical thinkers, have now been found to be very necessary for the description of general facts of the physical world.
>
> 'Quantised singularities in the Electric Field', P. Dirac 1931

Dirac's experience of abstract mathematics predicting the existence of the very real anti-electron led him to conclude that there was now a new way of doing physics. He suggested that theoretical physicists initially needn't worry about doing experiments, but should play around with mathematical theories and find ones for which the maths worked out nicely. To have beautiful maths behind an untested theory was a kind of 'tick of approval', because it seems the universe likes elegant maths, regardless of how abstract it may be. Such mathematically pleasant theories could then be put to the experimental test.

String theory is a perfect example of this. Unlike the 4D space-time of Einstein, some variations on string theory require 11D space-time, which would give us more spatial dimensions. However, there has never been any evidence whatsoever that string theory is correct. We have zero observations of anything beyond 4D space-time. It's just that the mathematics behind string theory is so incredibly neat, it all lines up so nicely and elegantly it seems almost wasteful for the universe not to use it.

Because mathematics is so useful at describing the universe around us, there's a temptation to reverse the direction and see how physics can inform mathematics, a temptation to insist that one version of mathematical truth must be the 'real' one because it matches the world around us. But, regardless of how many dimensions our universe really is, it doesn't change the way the maths works. Similarly, we could insist on knowing if our universe is Euclidean or not; if we zoomed out far enough, would parallel lines still be possible? Appealing to our universe could settle the Parallel Postulate, but this is putting the application before the horse. Mathematics exists separate from physical reality – it is not moulded by it.

Our Hyperthericals could live in a wildly different universe to us and we would all discover the same mathematics. And only a small part of that mathematics would apply to the reality around each of us. But that doesn't invalidate the rest of it. For all we know, one being's abstract maths could describe the other's physical reality. I can imagine 4D non-Euclidean Hyperthericals trying to get their minds round the absurd concept of 3D creatures who exist in a world of parallel lines and knots. Their thought experiment could be our reality, and vice versa. But the maths is the same. It is all equally true.

It's my hope that this book has whetted your appetite for recreational mathematics. Having started with counting and shapes, we have taken one very narrow and specific path to the limits of human knowledge. There are countless detours we could have taken, and I hope you do read some further books about the many, many topics I couldn't fit in this finite book. And I hope that, like me, you take comfort in the fact that the world of mathematics is without bound, and humans can continue exploring it and discovering new things for all time to come.

So now you've finished this book, be sure to keep doing maths. Hang paintings in new and dangerous ways, cut pizzas and cakes into impeccably fair slices and bet a friend that their glass is further around than it is tall. And, of course, you must show at least one person what happens when you cut a Möbius loop in half.

The Answers at the Back of the Book

Chapter 1: Can You Digit?

Why are the powers of 2 the only numbers that cannot be written as a sum of consecutive numbers?

To prove this, we start by showing that all other numbers can be written as a sum of consecutive numbers.

Odd numbers are an easy first target. If a number is odd, then it is twice some number plus 1, which we can write as $2n + 1$, for some value of n. This means it is the sum of the consecutive numbers n and $n + 1$. For example, the number 17 (which is $2 \times 8 + 1$) is $8 + 9$.

Right: now for the even numbers. If an even number is a multiple of 3, we know it is equal to some $3n$. So we can write it as the sum $(n - 1) + n + (n + 1)$. In fact, this works for any odd factor: if the number is divisible by 5, then it is equal to $(n - 2) + (n - 1) + n + (n + 1) + (n + 2)$. For example, 100 is 20×5, so it is equal to $18 + 19 + 20 + 21 + 22$. Note that this method will go slight awry for numbers with only one 'relatively big' odd factor, such as $22 = 2 \times 11$. There will still be a run of 11 consecutive numbers that add to 22, but because they are centered on 2, some will be negative. In this case: $-3 + -2 + -1 + 0 + 1 + 2 + 3 + 4 + 5 + 6 + 7$. But don't worry, the negative values always cancel with the starting positive values and leave something nice like $4 + 5 + 6 + 7$.

This means that any number which contains an odd factor can be written as a sum of consecutive numbers. The only numbers which don't have an odd factor are exactly those whose prime factorization has only even factors in it. And, since 2 is the only even prime, this is just numbers like 2; 2×2; $2 \times 2 \times 2$; and so on.

But this only gets us halfway.

For completeness, we need to prove that these numbers which are powers of two cannot be written as the sum of consecutive numbers. To do that, we'll prove that no sum of consecutive numbers can give a total which is a power of two.

Any sum which has an odd number of consecutive numbers, we don't have to worry about, as the total will always be divisible by an odd number.

Let's look at what happens if we add up an even number of consecutive numbers:

$n + (n + 1) + (n + 2) + (n + 3) = 4n + 6$
$n + (n + 1) + (n + 2) + (n + 3) + (n + 4) + (n + 5) = 6n + 15$
$n + (n + 1) + (n + 2) + (n + 3) + (n + 4) + (n + 5) + (n + 6)$
$+ (n + 7) = 8n + 28$

You can see a pattern forming: for a sum of k consecutive numbers, we get k lots of n, plus the sum of the integers 1 to $(k - 1)$. These are our old friends the triangle numbers! Since we know triangle numbers that are the sum of the first $(k - 1)$ numbers can be written as $k(k - 1)/2$, we know the whole number can be written $nk + {}^{k}\!/\!_{2}(k - 1)$. Then, since k is even, this can be rewritten as a product of two whole numbers ${}^{k}\!/\!_{2}(2n + k - 1)$, and the right-hand term will be an odd number (again, since k is even). This means the total can't be a power of two, as it definitely has an odd factor.

A list of small narcissistic numbers

It's possible to find these using simple computer programs or a spreadsheet. If you want to check your answers, I've put a list of them below. They can also be found in the *On-line Encyclopedia of Integer Sequences* (*OEIS*) as sequence A005188. The OEIS is an amazing collection of almost every number sequence known to humankind. Whenever you see a number with an 'A' like A002193, look it up at oeis.org.

2 digits: no values
3 digits: 153; 370; 371; 407
4 digits: 1,634; 8,208; 9,474
5 digits: 54,748; 92,727; 93,084

When the 37 trick works

I claimed the 'repeat the same digit n times and then dividing it by the sum of its digits' trick works when n is a factor of $b - 1$ for any base b. And it does always work if that is the case, but it also works in a few extra situations.

To be complete, it works whenever n is a factor of $1 + b + b^2 + b^3 + \ldots + b^{n-1}$.

Phew. It's a relief to have that off my chest.

Not a Munchhausen number

Some people will claim that 438,579,088 is a base-10 Munchhausen number because $0^0 = 0$. Those people are wrong.

Chapter 2: Making Shapes

The proof is in the pentagon

Tying a knot in a straight strip of paper gives a regular pentagon. To prove it is perfectly regular, we need to show that all five interior angles are exactly 108°, which we can do by proving they are all the same (the sum of the interior angles of any pentagon is 540°).

If we fold up the pentagon knot, crease the edges and then unfold it, you can see the fold lines across the strip. All we need to do is prove these angles are all the same.

By unfolding and refolding the pentagon, we can compare angles on different parts of the strip. In the images on the next page, the two angles marked A lie on top of each other when the pentagon is folded up. Then we can use the fact that the sides of the strip are parallel, and deduce that two more angles are also equal to A.

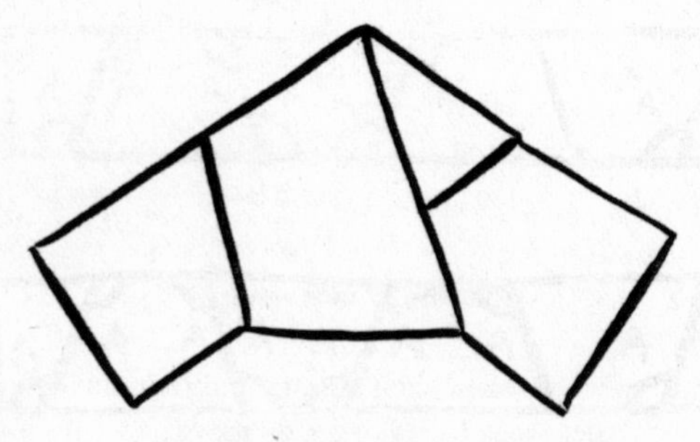

Paper strip with a flattened knot making a pentagon.

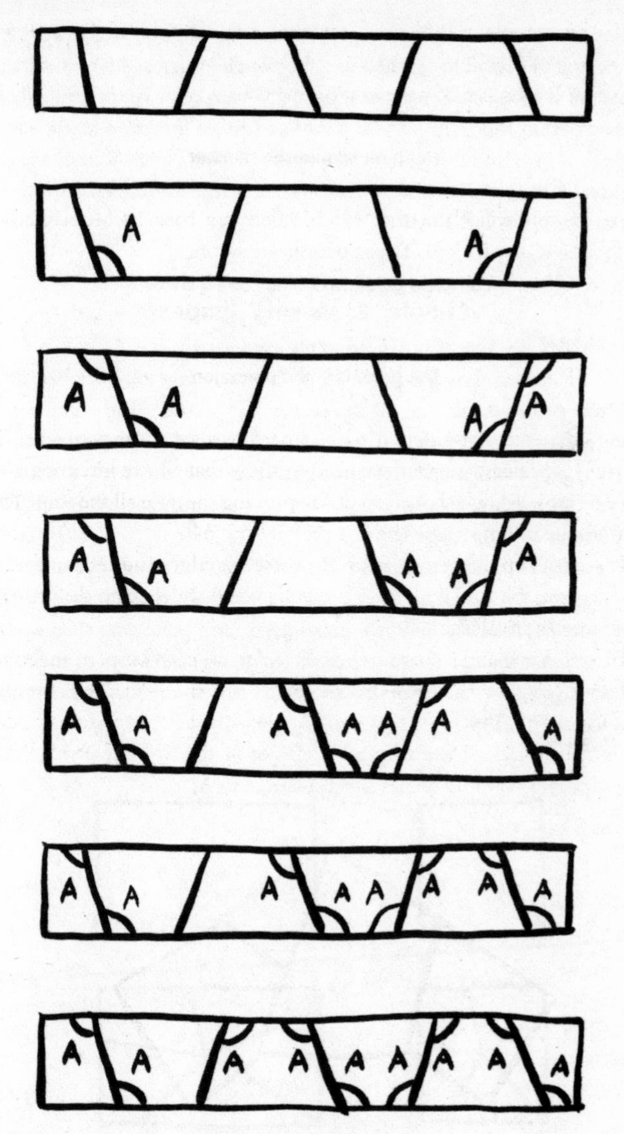

Steps to determining all the angles.

Repeating the folding-up-and-comparing-angles step, we find another angle equal to A; then use the parallel sides again to show the next one is also equal. We can repeat this step once more until all the lines crossing the strip can be shown to meet the edge at the same angle A. This also means that the lengths of these lines (and hence the lengths of the edges of the pentagon) will be the same, as the width of the strip is constant. Ta-da.

Complete guide to cutting cuboid cakes

OK, that's a lie. This is not a complete guide at all.

For my solution to the problem of how to cut a cube-shaped cake into five pieces (such that all the pieces are of the same volume and have the same surface area of icing) I used vertical cuts into the centre of the cake, from points equally spaced around the edge. I tried to extend this method to any cuboid-shaped cake and failed. Sort of. The cake becomes a massive mess.

For square-topped cakes, instead of splitting the whole perimeter of the cake into five equal parts we could have initially divided each of the four sides of the cake into five equal parts, and then each slice would have included four of those five parts (there are four sides to the cake, and $4 \times 5 = 20$ total segments, so each fifth of the cake's perimeter uses four of these).

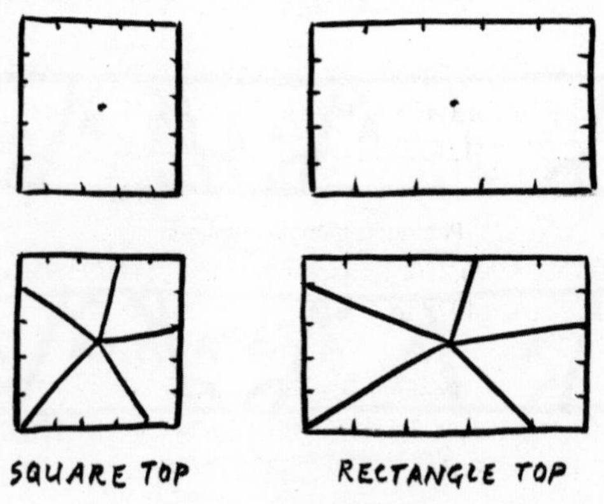

Cutting square and cuboid cakes.

For a general cuboid solution, we still split each of the four edges of the top rectangular face into five equal sections, but the sections on the longer faces of the cuboid will be longer than those on the other sides. For example, if your cuboid cake is 10cm × 15cm seen from above, you'd split the 10cm sides into five equal 2cm pieces, and the 15cm sides into five equal 3cm pieces. Then, as before, start from one corner and count four sections before marking your slice point, then count another four (even though some of these will be different lengths), and so on. The areas of all these sections are still exactly the same.

If your guests are not very observant, you might now get away with this. The area on the top of each piece is indeed identical, and the volumes of the pieces are all exactly the same. You will however come undone if so much as a single guest notices that the lengths around the perimeter are different, which means that the area of icing on the side is no longer fair!

Yes, the pyramid method would work on a cuboid cake, but it is still as much of a faff as ever. If anyone can find a better way to cut a cuboid cake fairly, let me know!

Chapter 3: Be There and Be Square

Multi-polygon numbers

Here is an incomplete (but still pretty comprehensive) list of multi-polygon numbers. Now with added zeroes! (which technically count.)

Square-triangle numbers:

0; 1; 36; 1,225; 41,616; 1,413,721; 48,024,900; 1,631,432,881; 55,420,693,056; 1,882,672,131,025l . . . (A001110 in *OEIS*)

Pentagon-triangle numbers:

0; 1; 210; 40,755; 7,906,276; 1,533,776,805; 297,544,793,910; 57,722,156,241,751 . . . (A014979 in *OEIS*)

Pentagon-square numbers:

0; 1; 9,801; 94,109,401; 903,638,458,801; 8,676,736,387,298,001; 83,314,021,887,196,947,001 . . . (A036353 in *OEIS*)

Hexagon-triangle (aka hexagon) numbers:

0; 1; 6; 15; 28; 45; 66; 91; 120; 153; 190; 231; 276; 325; 378; 435; 496; 561; 630; 703; 780 . . . (A000384 in *OEIS*)

Hexagon-square numbers:

1; 1,225; 1,413,721; 1,631,432,881; 1,882,672,131,025; 2,172,602,007, 770,041; 2,507,180,834,294,496,361 . . . (A046177 in *OEIS*)

Hexagon-pentagon numbers:

1; 40,755; 1,533,776,805; 57,722,156,241,751; 2,172,315,626,468,283, 465; 81,752,926,228,785,223,683,195 . . . (A046180 in *OEIS*)

Chapter 4: Shape Shifting

Proof that the Wobbler's centre of mass stays at the same height

Using these two positions of the Wobbler, it's possible to show that for the height of the centre of mass h to be the same in both cases, then the distance between the centres of the two discs d must be $r\sqrt{2}$. The algebra can get messy, but if you are super-careful you will always end up with the same answer. For the proof here, I have chosen each step to make the working out as simple as possible, which is almost cheating. So if at any step you wonder why I randomly multiply by something like $(\sqrt{2} + 1)$, it's because I know what's coming next, and I'm battening down the algebra in anticipation.

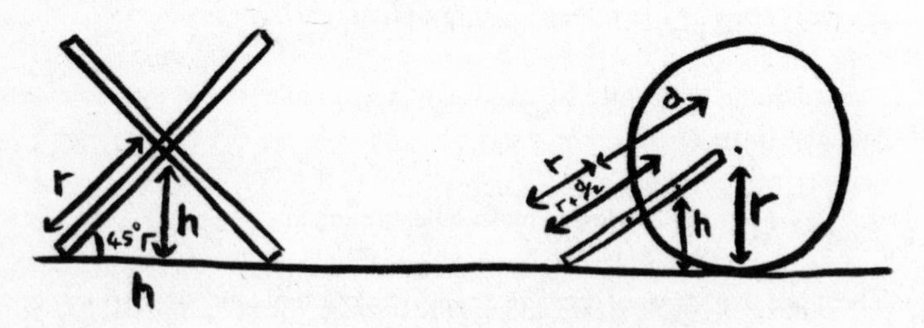

Centre of mass of the Wobbler, in two different positions.

Similar triangles give:

$$\frac{\mathrm{r}+\mathrm{d}}{\mathrm{r}}=\frac{\mathrm{r}+\frac{\mathrm{d}}{2}}{\mathrm{h}}$$

But $h = \frac{r}{\sqrt{2}}$, so

$$\frac{\mathrm{r}+\mathrm{d}}{r}=\frac{\mathrm{r}+\frac{\mathrm{d}}{2}}{\frac{\mathrm{r}}{\sqrt{2}}}$$

Multiply both sides by r:

$$r + d = \sqrt{2}(r + \frac{d}{2})$$

Then multiply both sides by $\sqrt{2}$:

$$\sqrt{2}r + \sqrt{2}d = 2r + d$$
$$(\sqrt{2} - 1)d = (2 - \sqrt{2})r$$

As a twist, multiply both sides by $(\sqrt{2} + 1)$:

$$(\sqrt{2} - 1)(\sqrt{2} + 1)d = (2 - \sqrt{2})(\sqrt{2} + 1)r$$
$$(2 - \sqrt{2} + \sqrt{2} - 1)d = (2\sqrt{2} - 2 + 2 - \sqrt{2})r$$
$$d = \sqrt{2}r$$

Disclaimer: This has actually only proved that if the centre of mass is the same height in those two specific positions, then the centres of the discs are $\sqrt{2}r$ apart. Proving that the centre of mass remains at the same height for all the positions between those is much, much harder. I also assume the centre of mass is in the middle of the intersection (because it is).

How to make a hexaflexagon

There are two types of hexaflexagon I mentioned, and one is simpler than the other. Making the second one is easier once you've made the first one, and I'll describe both here.

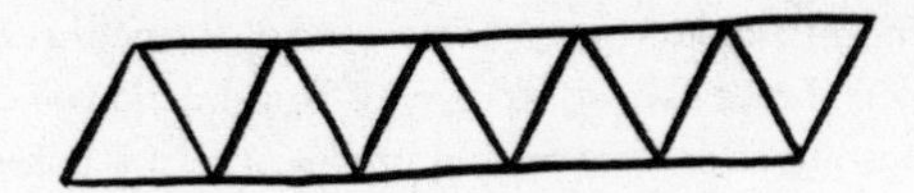

To make a trihexaflexagon, start with a strip of paper and fold it into equilateral triangles so you have ten triangles in a row, then cut off any excess. Place the strip of ten triangles horizontally on the table in front of you, so that the point of one triangle lies in the bottom-left corner, and another in the top right.

Take the seven triangles on the right and fold them up and over, on top of the three triangles on the left. Those seven triangles should now be pointing diagonally down and left. Next, take the four triangles from the end and fold them up and over as well. They will land on the left-most triangle, which you need to tuck them under. There should now be one triangle sticking out from behind the top-left corner of a hexagon. Fold this over and glue it to that original left-most triangle.

This is a trihexaflexagon, and to flex it you need to push together three alternating corners around the edge of the hexagon, making a point in the middle, and open it out from the other side. This should reveal the third face, and you can get back to the first two by flexing it again. There are three faces altogether.

To make a hexahexaflexagon, start with a longer strip of triangles (nineteen in total) and, before folding it into a hexagon as before, wrap the strip of triangles around into a spiral (you can wrap it around a ruler of the right width, or just fold every other edge between two triangles in the same direction). Once you've formed a spiral, you should have what looks like a strip of ten triangles, and you can follow the instructions above to fold it up and glue it into a hexagon. This hexaflexagon has six different faces, and you can flex it in the same way to find them all.

Chapter 6: Pack It Up, Pack It In

Thirty-one coins in a box

When we were talking about the most efficient way to pack coins, I mentioned that, although it has not yet been proved, for thirty-one two-pence coins the smallest square they have been fitted into has sides of 145.514mm. If you're trying to beat this record, that's what you're up against. The current world record arrangement is:

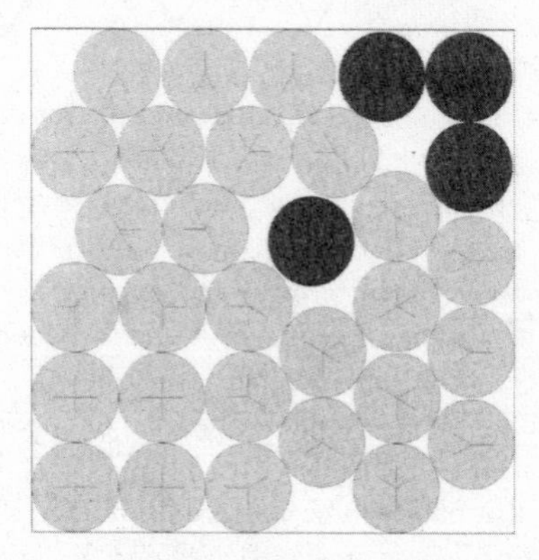

Fitting thirty-one coins into a square. Four of them still have room to jiggle about!

Chapter 7: Prime Time

Each prime squared is one more than a multiple of 24

I did this the long way when I first proved it. I started with the fact that all primes are one more or one less than a multiple of 6 (except 2 and 3), which means they are either in the form of $(6k + 1)$ or $(6k - 1)$. The number k can either be odd or even, so I swapped it for both $2m$ and $2m + 1$, which gives us the four options: $(12m + 1)$, $(12m + 7)$, $(12m - 1)$ and $(12m + 5)$. I squared each of them and demonstrated that they are each one more than a multiple of 24.

But there's an easier way.

A friend showed me that any odd number which is not a multiple of 3 will always square to be one more than a multiple of 24. Of course, all primes are odd, and all primes are not a multiple of 3 (except those meddling subprimes 2 and 3).

We'll start with our number n squared and subtract one: this can be written as $n^2 - 1$. To prove this is a multiple of 24, we can rewrite it as $(n + 1)(n - 1)$. (You can multiply this out to show that it equals $n^2 - 1$.)

$(n + 1)(n - 1)$ is the product of two even numbers (since n is an odd number, $n + 1$ and $n - 1$ must be even) and, in particular, one of those two even numbers must be a multiple of 4, as they're consecutive even numbers, and every other even number is divisible by 4. So, we know $(n + 1)(n - 1)$ is divisible by 8, as it's the product of something divisible by 4 and something divisible by 2.

We also know that of the three consecutive numbers $(n + 1)$, n and $(n - 1)$, exactly one of them must be divisible by 3. But that cannot be n, as it is a prime. It must be one of the other two. Hence, the product $(n + 1)(n - 1)$, as well as being divisible by 8, is also divisible by 3. Since 3 and 8 have no common factors, this means the whole thing is divisible by 24, as required.

This actually means that if you square any number which does not have 2 or 3 as a factor, it will be one more than a multiple of 24.

Chapter 8: Knot a Problem

Examples of Brunnian links

Borromean Rings (three-component Brunnian link).

A four-component Brunnian link.

A six-component Brunnian link.

Chapter 10: The Fourth Dimension

How many vertices and edges does a hypercube have?

A 1D line has a start and a stop, which we can give the coordinates (0) and (1). A square is slightly more exciting, as in 2D all coordinates have two numbers, so the four corners of a square are (0,0), (1,0), (0,1) and (1,1). Carrying this on, the coordinates of the eight vertices of a 3D cube are: (0,0,0), (1,0,0), (0,1,0), (0,0,1), (1,1,0), (1,0,1), (0,1,1) and (1,1,1). It makes sense for a cube to have eight vertices, because there are three coordinates and each of them has one of two possible values ($2 \times 2 \times 2 = 8$). For any number of dimensions n, there are n coordinates, so a hypercube will have 2^n vertices.

Now on to edges.

There are a few ways you can deduce that a 4-hypercube has thirty-two edges without making a model. My favourite is to realize that each vertex of a hypercube has four edges radiating out of it (just as all vertices on a cube feature three edges) and, since there are sixteen edges, $16 \times 4 = 64$. But each edge connects two different vertices, so we have counted them all twice. To undo this double-counting, we divide by 2 to get $64 \div 2 = 32$ different edges. The same logic works for the eight vertices of a 3D cube: $8 \times 3 \div 2 = 12$ edges. However, on the model, it's much quicker: you can just count them all.

A hypercube in n dimensions will have $(2^n \times n)/2 = n \times 2^{n-1}$ edges.

Chapter 11: The Algorithm Method

Alternative optimal stopping algorithm

There is an alternative version of the optimal stopping algorithm, which allows you to slowly lower your standards as you continue to screen potential candidates.

The optimal stopping algorithm works well on average, but that doesn't mean that it will work every time. For interviewing life partners, if you're running out of options, it's hard to let the statistics take their course until the very end without panicking and just settling for settling down. A compromise is to lower your standards systematically. The best method is to drop your expectations by one rank every additional $\sqrt{n}$ people you date. So you date and reject the first $\sqrt{n}$ people, accept anyone in the next $\sqrt{n}$ dates who comes in at first place in your ranking, then anyone in the top two places for the next $\sqrt{n}$, the top three for the next $\sqrt{n}$, and so on, until your standards hit rock bottom at the end of the queue.

Generalizing the Three Pile Trick

The Three Pile Trick involves asking someone to choose a card from a pack of twenty-seven cards, dealing the cards into three piles three times, asking the person who chose the card to indicate which pile their card is in each time, and then being able to determine which card the person chose. By carefully stacking the piles differently each time, you can manipulate their card so it ends up in any position of your choice in the deck.

This works because $27 = 3^3$, so three piles dealt three times will determine the answer uniquely. But this can also be achieved using piles of other sizes. For example, $49 = 7^2$, so we could take forty-nine cards and deal them into seven piles twice, and ask the person to indicate which of the seven piles their card is in each time, which would allow us to determine its location. Similarly, we can use other ways to split up the number – for a deck of fifty-two cards, we can deal them into thirteen piles of four, and then into four piles of thirteen, and that will be enough to find the card – and to position it anywhere we want in the deck, purely by the order in which we re-stack the piles.

Chapter 12: How to Build a Computer

Domino-computer-circuit layouts

As I mentioned, I have built a working domino computer, and here we present some diagrams of how the circuits work, so you can have a go

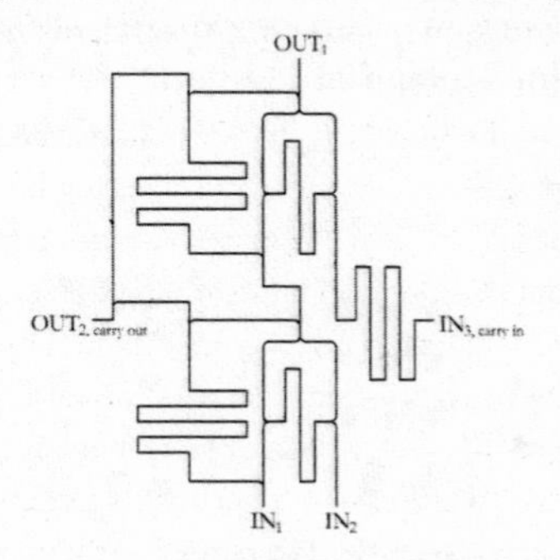

Two half adders make a full

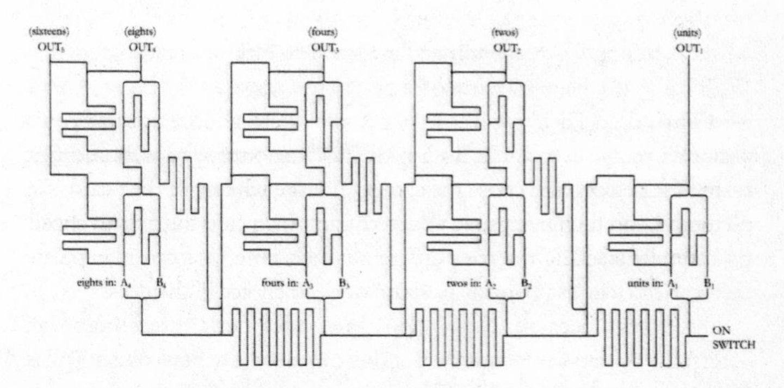

Four-digit binary adder (domino computer) to be made from dominoes.

at building them yourself. (Obviously, if you can work out how to do them yourself without reference to these diagrams, that's even better!)

Using the half-adder from p. 266 as a standard unit, you can combine it with 'blocks' to make a full adder. This is what we scaled up to make a four-digit adder.

Chapter 13: Number Mash-ups

Proof that the sum of the first n odd numbers equals n^2

This is a neat geometric proof demonstrating that, as you keep adding the odd numbers, the total is always a square. Each new odd number wraps perfectly around two of the sides.

However, we can develop another method by looking at sums of limited length: adding a sequence of numbers where each number is a multiple of the previous number. A sequence where you add a value to get from one term to the next (like when we summed consecutive

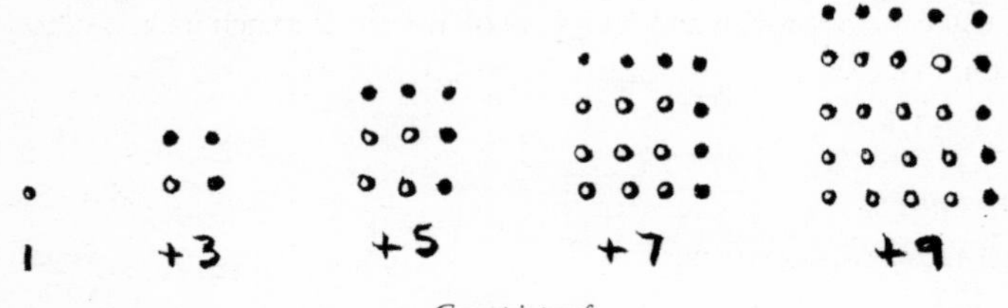

Geometric proof.

numbers to get triangle numbers) is called an arithmetic sequence, whereas multiplying each term to get the next is a geometric sequence. We'll try to calculate the sum of a geometric series in the most general of terms, so we'll just use r to represent the ratio between any two amounts in the sequence. We'll also start the sequence with the first term as a, to represent any amount. So the sequence goes a, ra, r^2a, r^3a, all the way up to r^na. Let's see what we can do with its sum, which we'll give the name T.

This is what we want to calculate:

$$a + ra + r^2a + r^3a + \ldots + r^{(n-1)}a + r^na = T$$

We can turn this into a new equation by multiplying both sides by *r*:

$$r \times (a + ra + r^2a + r^3a + \ldots + r^{(n-1)}a + r^na) = rT$$

Which simplifies down to:

$$ra + r^2a + r^3a + \ldots + r^{(n-1)}a + r^na + r^{(n+1)} = rT$$

In a cunning move, we can take this new form, and subtract the original equation from it:

$$(ra + r^2a + r^3a + \ldots + r^{(n-1)}a + r^na + r^{(n+1)}) -$$
$$(a + ra + r^2a + r^3a + \ldots + r^{(n-1)}a + r^na) = rT - T$$

You can go through and see which of the terms match up and cancel out to leave:

$$r^{(n+1)}a - a = rT - T$$

Both sides factorize nicely:

$$a(r^{(n+1)} - 1) = T(r - 1)$$

Giving us the neat formula:

$$T = \frac{a(r^{(n+1)} - 1)}{(r - 1)}$$

Like we did with the triangle numbers sum on page 292, you can see how the trick is to cancel the series of numbers out by subtracting it from itself, but you need to have made a change first so that there are still a few terms left afterwards. This is done by finding a way to bump all the terms along a bit, leaving a few spare at the ends.

Mersenne primes and perfect numbers

You now have all the tools required to prove why all Mersenne primes produce a perfect number. We've come across this relationship before, but we didn't prove it. For any Mersenne prime $2^n - 1$, there is a perfect number $(2^n - 1) \times 2^{(n-1)}$. Check this for a few of the smaller perfect numbers. Now, it is possible to prove that this formula works not only for the few cases we have already met, but that it will continue working for all other Mersenne primes in existence, with no exceptions. You know the only factors of $(2^n - 1)$ are itself and 1, because we are only using prime Mersenne numbers. The number $2^{(n-1)}$ is a power of two and so it has a lot of factors. The factors 1, 2, 4, 8, 16 . . . $2^{(n-2)}$, $2^{(n-1)}$, to be exact. All you need to do is combine these to list out every single factor of $(2^n - 1) \times 2^{(n-1)}$ except itself, and then prove that they add up to give you back the formula $(2^n - 1) \times 2^{(n-1)}$.

Good luck with that. If you get really stuck, I'll have a solution waiting for you on the website.

Reversing Fibonacci and Lucas numbers

If you did throw the Fibonacci and Lucas numbers into reverse, you should have produced lists of numbers going the other way, which alternate between positive and negative.

Fibonacci:
. . . 34, −21, 13, −8, 5, −3, 2, −1, 1, 0, 1, 1, 2, 3, 5, 8 . . .
Lucas:
. . . 47, −29, 18, −11, 7, −4, 3, −1, 2, 1, 3, 4, 7, 11, 18 . . .

In both cases, the ratios between consecutive new terms in the other direction approaches $^{-1}/_{\varphi}$.

Rearranging the Bernoulli-number equations

As mentioned in Chapter 13, I forced through $m = 3$ and $n = 4$ to calculate the sum of the first four cube numbers and, sure enough, the Bernoulli numbers gave me an answer of 100. Here's how it went:

The Bernoulli equation for finite sums is:

$$1 + 2^m + 3^m + \ldots + n^m = \frac{(B + n + 1)^{m+1} - B^{m+1}}{m + 1}$$

Setting $m = 3$ and $n = 4$, we get:

$$1 + 2^3 + 3^3 + 4^3 = \frac{(B + 5)^4 - B^4}{4}$$

$$= \frac{B^4 + 4B^3 \times 5 + 6B^2 \times 5^2 + 4B \times 5^3 + 5^4 - B^4}{4}$$

$$= \frac{B^4 + 20B^3 + 150B^2 + 500B \times 625 - B^4}{4}$$

$$= \frac{20B^3 + 150B^2 + 500B + 625}{4}$$

Substituting in the Bernoulli numbers gives:

$$= \frac{20 \times 0 + 150 \times (\tfrac{1}{6}) + 500 \times (-\tfrac{1}{2}) + 625}{4}$$

$$= \frac{25 - 250 + 625}{4}$$

$$= \frac{400}{4} = 100$$

Your bonus challenge is to set $m = 1$ and show that this equation rearranges to give the formula for the nth triangle number.

I also mentioned that it's possible to derive the value $\pi^2/6$ for the infinite sum with the value $m = 2$. Here's how that goes:

Bernoulli equation for infinite sums:

$$1 + \frac{1}{2^m} + \frac{1}{3^m} + \frac{1}{4^m} + \ldots = \frac{2^{m-1}|B^m|\pi^m}{m!}$$

For $m = 2$:

$$1 + \frac{1}{2^2} + \frac{1}{3^2} + \frac{1}{4^2} + \ldots = \frac{2^1|B^2|\pi^2}{2!}$$

$$= \frac{2 \times \frac{1}{6} \times \pi^2}{2}$$

Then the twos cancel, to leave $\pi^2/6$.

Chapter 15: Higher Dimensions

Spiky spheres

Most people would describe circles and spheres as being smooth objects, but we have calculated that in higher dimensions they bristle with points. When does this smoothness suddenly start to taper out into spikes? The simple answer is that these circles and spheres were actually a bit spiky all along – we just never noticed.

Even going from 2D to 3D, from a circle into a sphere, the shape becomes slightly more tapered, but in such a subtle way that we don't give it a second thought. If you slice off 25 per cent of the radius of a circle, you remove more of it than if you take the same 25 per cent off a sphere. A 3D sphere is already a bit spikier than a 2D circle.

Removing 25 per cent of a circle's radius from the end that touches the circumference takes away a segment containing 7.2147 per cent of the total surface area. If we do the same thing on a sphere and make a slice 25 per cent from the end of the radius and take off that cap, we remove

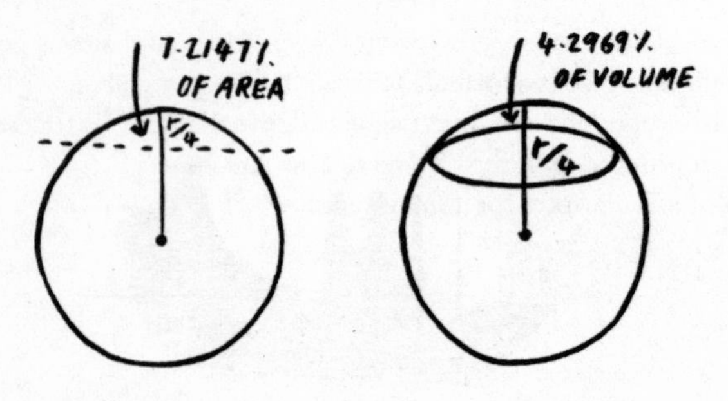

only 4.2969 per cent of the volume. Because the sphere is curving in more dimensions, it tapers more as you get towards the outside. This happens to a more and more exaggerated degree the higher up the dimensions you go, as higher-dimensional spheres have more dimensions to curve in.

Chapter 17: Ridiculous Numbers

Proving e^{π} is transcendental

It was the Gelfond–Schneider theorem which finally proved that e^{π} is transcendental, and here's how it was done: the theorem states that for two algebraic numbers m and n, m^n is definitely transcendental if m is not 0 or 1 and n is not rational. The values of $m = -1$ and $n = -i$ fulfil those criteria and so -1^{-i} must be transcendental. We can show that -1^{-i} is actually e^{π} in disguise by using the famous result that $e^{i\pi} = -1$ (which, sadly, there is not enough room to prove in this book) and a quick bit of algebraic massaging.

$$
\begin{aligned}
-1^{-i} &= (e^{i\pi})^{-i} \\
&= e^{(-i \times i)\pi} \\
&= e^{\pi}
\end{aligned}
$$

Imaginary and Real Shading

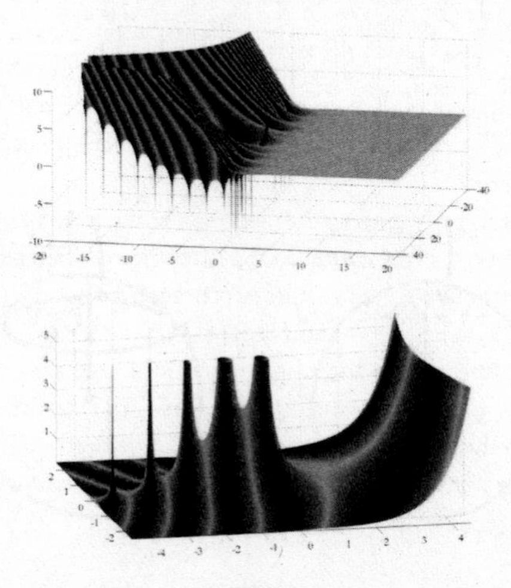

As mentioned, the plots of the complex functions are 3D approximations of what should be a 4D display. To take a dimension out of the equation, all of the complex outputs are reduced to a measure of how far they are from the middle of the complex plane, their 'magnitude'. This gives a sense of how large a complex number is, but not how imaginary or real it is.

To put this information back in, I've used colour as the fourth dimension. The darker the surface, the 'more imaginary' that output was. So the light grey parts are completely real, and the black sections are all imaginary, with shades in between. To get technical, the size of the angle that each complex number made with the real axis was shaded from grey to black.

Chapter 18: To Infinity and Beyond

Calkin–Wilf tree

In Chapter 18, I described a method for generating a list of all the rational numbers using a sequence similar to the Fibonacci sequence, where you start with (1,1) but then between each step in which you add two terms to get the next value, you also copy each single term by itself on to the end of the sequence. But that's not the only way to make a list of every rational number possible.

In 1999, the American mathematicians Neil Calkin and Herbert Wilf actually produced not one but two efficient ways to enumerate all the rationals in their four-page paper 'Recounting the Rationals'. While I encourage you to read their original work (it's easy to read and available free online), I'll summarize the two methods here, because they're so lovely.

The first is to grow a tree of fractions. At the base of the tree is the rational number 1, represented as the fraction $1/1$. The rule for this tree is that every fraction a/b splits into two branch fractions: $(a + b)/b$ and $a/(a + b)$. So $1/1$ splits into $2/1$ and $1/2$. If this is continued, every possible fraction will appear once and once only on the tree and, even better, it will always appear in its most simplified form. To convert this into a list which we can then number, the fractions can be read off in rows of the tree, or, as I prefer, the tree can be rearranged into a sprawling bush and the fractions read off in a spiral. Either way works.

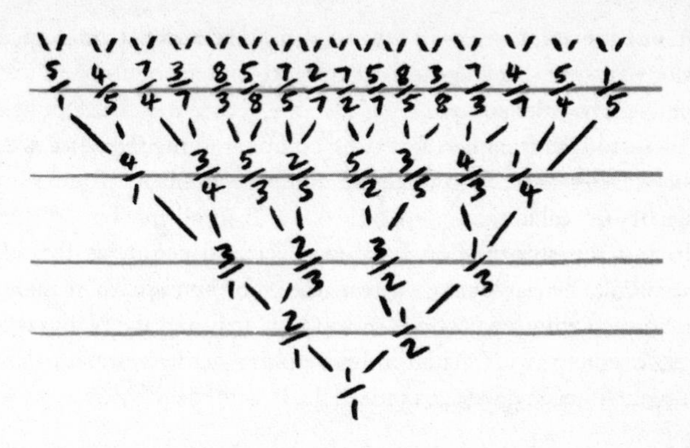

1/1 2/1 1/2 3/1 2/3 3/2 1/3 4/1 3/4 5/3 2/5 5/2 3/5 4/3 1/4 5/1 4/5 7/4 3/7 8/3 5/8

The Calkin–Wilf tree: rows (tree) method.

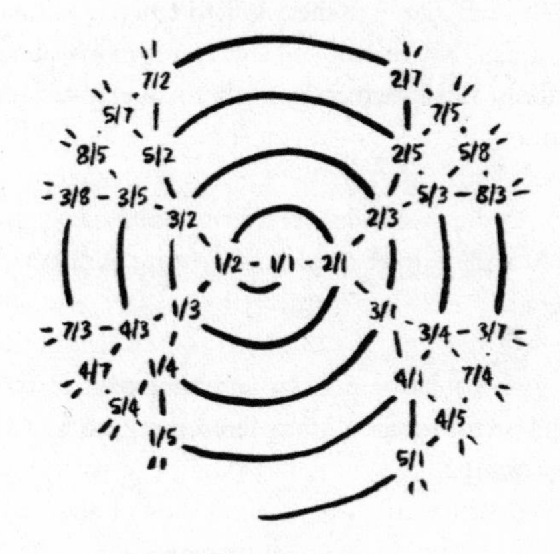

1/1 1/2 2/1 1/3 3/2 2/3 3/1 1/4 4/3 3/5 5/2 2/5 5/3 3/4 4/1 1/5 5/4 4/7 7/3 3/8 8/5 5/7

The Calkin–Wilf tree: spiral (bush) method.

The tree method alone would have made for an amazing piece of mathematics, but the authors went a step further. This fraction tree is actually the afterthought in the paper; the main result was the Stern–Brocot sequence we already met (A002487 in the *OEIS*). We have already seen the Fibonacci-esque method for calculating the numbers in the Stern–Brocot sequence, but there are other, more bizarre, options.

Stern–Brocot sequence: 1, 2, 1, 3, 2, 3, 1, 4, 3, 5, 2, 5, 3, 4, 1, 5, 4, 7 . . .

Calkin and Wilf explain that the nth number in the sequence is equal to the number of ways to write the number *n* in binary if you can use values of 0, 1 and 2 instead of only 0 and 1. In normal binary, there is one unique way to write each number, as is true for any base system, but if you allow values equal to or greater than the base, then there are more ways to write any given number. 'Five' in binary is 101, but we can now also write it as 021. There are two options, and you will see that the fifth value in the sequence, is indeed 2.

Axioms of ZFC, written in natural language

In case you were wondering what they were, here are the nine axioms of ZFC (Zermelo–Fraenkel set theory with Choice), along with a couple of bonus options. They're more usually stated in mathematical language, or using symbols, but I've translated them so you can understand what they're saying.

1. Axiom of extensionality

If two sets X and Y have exactly the same elements, then they are equivalent sets.

2. Axiom of the unordered pair

Any two objects can form an unordered pair. For any *a* and *b*, there is some set X = {a,b}.

3. Axiom of separation

For any property, you can split any set X into subsets of elements which do or don't have that property.

4. Axiom of union

For any two sets X and Y, there is a new union set Z made of all the elements from X and Y.

5. Axiom of the power set

For any set X, you can make a new set Y which is the set of all possible subsets of X.

6. Axiom of infinity

Infinite sets exist.

7. Axiom of replacement

You can apply a function to all the elements of any set X and generate a new set Y.

8. Axiom of foundation

If a set is not empty, then it must have a 'smallest' element.

9. Axiom of choice

Given a family of non-empty sets which have no elements in common, you can choose one item from each set to make a new set.

10. An undecided tenth axiom?

Either forcing axioms or the inner-model axiom.

Text and Image Credits

Credits and Permissions

Images

Grateful acknowledgment is given to the following for permission to reproduce copyrighted material.

Every effort has been made to contact copyright holders. The author and publisher would be glad to amend in future editions any errors or omissions brought to their attention. Unless otherwise indicated, photos are courtesy of the author.

p. 215–216: Made with Robert Webb's Stella software, http://www.software3d.com/Stella.php.
p. 256 © 2005 Antikythera Mechanism Research Project1
p. 259 © Getty Images, Science Museum
p. 269–270 and 273–274 © Jonathan Sanderson
p. 317 Produced with SeifertView, Jarke J. van Wijk, Eindhoven University of Technology
p. 433 © E. Specht
p. 455 Photo by Steve Ullathorne

Text

p. 48: From: http://apod.nasa.gov/htmltest/rjn_dig.html
p. 297: *Ramanujan: Letters and Commentary*, Srinivasa Ramanujan Aiyangar, contracted for clarity.
p. 340: Thomas Hales, *Cannonballs and Honeycombs.*
p. 388: P.A.M. Dirac, 1928, p. 2.
p. 389: 'The Positive Electron', Carl D. Anderson, 1933, p. 5.
p. 422: Paul Dirac, 'Quantised Singularities in the Electromagnetic Field', 1931.

Acknowledgements

This book would still be in the set of things-which-do-not-exist if it were not for the support of my wife Lucie Green. She also had the decency to be writing a book at the same time (but sufficiently out-of-phase), so we could take 'writing breaks' together and pretend we were on holiday. Fun fact: There is a photo of her hidden somewhere in this book (along with several other Easter eggs). Consider it a game of 'Where's Lucie?'

Thanks to my literary agent Will Francis for turning my nebula of ideas into a potential book and then my editor Helen Conford for shaping that book. Sarah Day was also invaluable for making sure my words could then be read by normal human beings. During these acknowledgments, I am going to attempt an exhaustive list of the people who showed me the maths in this book, and I apologise to the complimentary set of people I will forget.

Katie Steckles is my long-time maths support and helped immensely with research and corrections as the book came together. Charlie Turner then subjected the final book to the wrath of her pedantry and fact-checked the crap out of it. All remaining untruths are not her fault. It was Katie Steckles who named the pentagon-knot the 'emergency pentagon' as well as being the ring-leader when we tied five people up so removing any one person would cause everyone else to come undone (at the MathsJam Conference 2013). She was the first ever person to challenge me to a game of noughts-and-crosses on a klein bottle. The friends I was calculating square-triangle numbers with in the pub were Charlie Turner and Florencia Tettamanti. Charlie was also the hilarious mathematician who calculated that the factor-linking graph is traceable up to node 91.

Since the first edition of this book was printed, additional errors have been spotted by Loren Parker, Steve Mould, Dave Hilton, Michael Jacobs, John Lewis, Peter Giblin, Stephen Molinari and pedant extraordinaire Adam Atkinson.

The pizza problem was first told to me by the mathmo Colin Wright. He also showed me how to tie my shoes the maths way, and how to trap spiky higher-dimension spheres.

The polydivisible number puzzle was given to me by Alison Kiddle when I went to the dentist. The challenge of cutting a cube cake was given to me by Chris Lintott when he should have been filming the transit of Venus with my wife. The wobbler discs were first shown to me by the engineer Hugh Hunt. It was David Acheson who first showed me the sausage catastrophe conjecture.

The further research into Grafting Numbers was done by Robert Tanniru of Oakland University in Michigan. I recommend reading his paper 'A Short Note Introducing Grafting Numbers and Their Connection to Catalan Numbers'.

Chris Sangwin is the master of all things mechanical maths; he has helped me with shapes of constant width, the wobbler discs and drilling a square hole. Julia Collins helped me with all things knots, as well as doing the knitting and crocheting in the book (with Madeleine Shepherd). Joel Haddley invented the word 'heptagrin', and now we're all stuck with it. Not only did Christian Perfect design the Perfect Herschel Polyhedron, but he also helped me with much of the mathematical typography in this book. Miranda Mowbray's work on error-correcting codes was done with her colleagues Alistair Coles and David Cunningham.

The biologist Stephen Currey gave me all sorts of viruses. Thanks a lot, Stephen.

The Borromean rings were 3D printed for me by Laura Taalman. She works at the National Museum of Mathematics in New York which is also where I rode the square-wheeled bike. The story about Alan Turing was told to me by his last-ever student, Bernard Richards. I was interviewing him for a BBC Radio 4 programme produced by Roland Pease.

The 3D solids of constant width, amicable 220&284 heart keyrings, utility problem mug and 'I Chart You' valentines card would all not exist if were not for my partners-in-Maths-Gear Steve Mould and James Grime. Did I mention all those products and more are available for sale at mathsgear.co.uk?

Actually, my fellow Festival of the Spoken Nerd stars Steve Mould and Helen Arney deserve an acknowledgment for working with me while I was permanently semi-distracted writing this book. James Grime is also a good maths friend who showed me the proof that the utilities graph is not planar and Mill's Constant, among many other things.

Thanks to Robin Ince for booking me to do maths with 3,500 people at the Hammersmith Apollo. He has shown an unnerving dedication to the ability of maths to entertain and my ability to deliver that. Thanks to Rob Eastaway and Simon Singh for their advice about writing a maths book. Dave McCormick was my train spotting consultant. All French translations were done by my brother-in-law Ben Dixon. Merci!

The Domino Computer was designed with help from Katie Steckles, Paul Taylor, Andrew Taylor and Siân Fryer. Building it also involved the hard work of Ben Curtis, Becky Smedley, Mike Bell, Blair Lavelle, Andrew Pontzen, Chris Roberts, Ben Ashforth, Gillian Kiernan and David Julyan. As documented by Jonathan Sanderson and Elin Roberts.

Thanks also to Bradley Haran of Numberphile fame who has filmed me for YouTube with dominoes, pies, mile-long pieces of paper and the number zero.

Oh yeah, and there is a competition hidden somewhere in this book. If anyone wins it, I'll think of a suitable prize. Beware of the traps.

My ramblings about group theory were checked-over by Paul Taylor, but he correctly takes no responsibility for my gross oversimplification of a rich and exciting area of mathematics. Andrew Taylor is the conjurer of code who turns my silly ideas into working web apps. It was also Andrew who wrote a programme to generate the two-word, three-dimensional Index for this book.

The story about the rounding in inFAMOUS Second Son was told to me by Bruce Oberg of Sucker Punch Productions.

I was on the BBC Radio 4 programme *More or Less* (hosted by the fantastic Tim Harford) which solicited the complaint about infinity. I was talking about Brun's constant, of all things. The bridge pilgrimage is called the Hamilton Walk and goes from Dunsink Observatory to Broombridge. The annual event is organised by Maths Week Ireland (co-ordinated by Eoin Gill and Sheila Donegan).

Andrew Pontzen checked over the parts of the book where I dared to talk about physics. He agreed I had it mostly right but then mumbled something about how we'll never know if our universe is perfectly flat because of quantum mechanical limits in measurement.

Thanks to the mathematicians who checked over my descriptions of their work and offered helpful comments. Including but not limited to: Melanie Bayley, Neil Bearden, Jerry Bonnell, Ken Brakke, Robert Matthews, Scott Morrison, Robert Nemiroff , Edouard Oudet, Terence Tao, Robin Thomas, Timothy Trudgian and Denis Weaire.

Thanks to the team at Penguin, featuring Casiana Ionita, Rebecca Lee, Claire Mason and Imogen Scott as well as the marketing and publicity duo of Sue Amaradivakara and Ingrid Matts. Credit also goes to my agent Jo Wander and my admin ninja Sarah Cooper.

Huge thanks to the Department of Mathematical Sciences that I call home at Queen Mary University of London, and my many colleagues across the college. Including the fine people in the Centre of Public Engagement and the relentlessly supportive Peter McOwan.

Thanks to everyone who has ever attended a monthly MathsJam or the annual conference. You are a constant supply of inspiration, ideas and slave labour.

Website design is by the world's greatest pixel pusher: Simon Wright. Ever since we were at school together in the late 90s he has been making my ridiculous projects look more professional than they have any right to be.

All of the photographs were taken by Al Richardson, who showed them an undue but hugely appreciated level of care. The illustrations and design were done by Richard Green, who somehow turned my sketches into intelligent pictures. Some plots and graphs were done with the help of Ben Sparks and the fantastic software package GeoGebra. Most of the renders of 4D shapes were done using the program Stella4D developed by Robert Webb. Some were based on 4D animations produced by Davide Cervone. David Fletcher sprinkled asteroids on my torus. The 3D plots of the Gamma and Zeta functions were made by Edric Ellis using MATLAB.

My mother really did knit me the binary scarf and the Klein-bottle hat. She is amazing like that. My father is an accountant who brainwashed me into liking numbers from a young and vulnerable age. Effectively, this book is all their fault.

This is not a photograph of the author Matt Parker, this is a spreadsheet of him. It consists entirely of conditionally formatted cells in a spreadsheet with numbers from 0 to 255. The details are explained on pp. 356–7. Make your own spreadsheet of a photo at makeanddo4D.com. (Photo by Steve Ullathorne)

A B C

Index

1

Writing an index for a book is traditionally a long and labour-intensive task. So I used a computer programme to process mine automatically. Unsurprisingly, the most common word is 'the'. But if you exclude the 100 most common words in the English language, the word I use the most is 'maths' (444 times) followed by 'shape' (321 times). The most frequent non-maths word is 'cannot' (a discouraging 81 times).

2

Realizing that the frequencies of single words was not that useful, I generated a list of every two-word pair in the book and then selected a subset of those to be the index. For each pair of words, the index lists every location in the book. But the list is not exhaustive; I selected the word pairs I liked. Some will be useful, some insightful and some I just thought were particularly funny as a disembodied fragment.

3

Of course, a book is a 3D object, and so I thought it appropriate to have a 3D index. Why most books limit themselves to a one-dimensional index listing page numbers only is beyond me. For each listing in the index, you will find the page number for the distance into the book, as well as the horizontal and vertical coordinates on the page. Or, for word pairs which appear a few times on consecutive pages, the start and end page numbers of the line they carve out.

4

You can see an example grid on this page. For all other pages, you will have to estimate the distances. Even I drew the line at drawing grid lines on every page of the book.

5

Harriot, Thomas: p56, A2; p385–386